'This is a must-read for every Vice-Chancellor, President and University Trustee of the world's 17,000 Higher Education Institutions (HEIs). While less than two percent of the world's population will graduate from an HEI, these graduates will form the vast majority of the future decision makers in the private and public sectors and civil society at large. As such, our HEIs are crucial entry points in the global attempt to create a more sustainable future. Our current global HEI leaders and their respective professional organizations have a major new responsibility – addressing sustainability. Here is their handbook to "the University of the Future".'

*Charles Hopkins, UNESCO Chair in Education for Sustainable Development, York University, Canada*

'If we choose to survive, and come to live in tune with the biosphere's processes and limits, then universities will have made an indispensible contribution. But, to do this, they will first need to change and evolve. This is a welcome book about such processes, and one that all involved in higher education should read.'

*William Scott, Professor Emeritus, University of Bath, UK*

'This book is of great importance as it clearly illustrates which path universities still need to follow towards sustainability. However, the book also indicates how much progress has already been made by some universities, and this is encouraging!'

*Gerd Michelsen, UNESCO Chair in Higher Education for Sustainable Development, Leuphana University of Lüneburg, Germany*

'Universities have two responsibilities in relation to sustainability. The first is to educate their students, whatever their disciplines, so that they can be informed future citizens and promote sustainability in their personal and professional roles. The second is to deliver on their role as beacons of social responsibility and lead through example in their pursuit of sustainability as organisations. This book contributes substantially to the understanding of university leaders and policy makers in ensuring their organisations fulfil both these roles.'

*Nick Foskett, Professor and Vice Chancellor of Keele University, UK*

'*The Sustainable University* is an important contribution to the ongoing conversation about how higher education in all its aspects can transform to help create a sustainable future.'

*Geoff Chase, Dean, Undergraduate Studies and Director of the Center for Regional Sustainability, San Diego State University, USA*

'The contributors to this book are the people I believe in. They are the honest, courageous, humble, smart and tenacious agents of change toiling at the front lines in humankind's epic struggle to save itself. The quality of mind and heart woven into these pages is beautiful in its authenticity and ambition. The effect is a chorus of shared vision, shared struggle and shared insight worthy of any reader's time.'

*Leith Sharp, Founding Director of Harvard University's Office for Sustainability and Founding Chair of the Sustainable Futures Academy, USA*

'This volume contains a wealth of experiences and ideas that will help higher education institutions make a necessary and greater contribution to creating a more sustainable society. Universities should not only act as agents of change, but need to change themselves if they are to be sufficiently effective in this regard. This book shows us how.'

*Pamela S. Chasek, Professor, Government Department, and Director, International Studies Program, Manhattan College, USA*

'Like few other social institutions, universities have the freedom and the power to choose to be models of sustainable practices and relationships, and to make it part of their core business. *The Sustainable University* provides the rationale, strategies and exemplars for acting upon a positive choice.'

*John Fien, Professor of Sustainability, RMIT University, Melbourne, Australia*

# The Sustainable University

The direction of higher education is at a crossroads against a background of mounting sustainability-related issues and uncertainties. This book seeks to inspire positive change in higher education by exploring the rich notion of the 'sustainable university' and illustrating pathways through which its potential can be realised. Based on the experience of leading higher education institutions in the UK, the book outlines progress in the realisation of the concept of the sustainable university appropriate to the socio-economic and ecological conditions facing society and graduates.

Written by leading exponents of sustainability and sustainability education, this book brings together examples, insight, reflection and strategies from the experience of ten universities, widely recognised as leaders in developing sustainability in higher education. The book thus draws on a wealth of experience to provide reflective critical analysis of barriers, achievements, strategies and potential. It critically reviews the theory and practice involved in developing the sustainable university in a systemic and whole institutional manner, including the role of organisational learning.

While remaining mindful of the challenges of the current climate, *The Sustainable University* maps out new directions and lines of research as well as offering practical advice for researchers, students and professionals in the fields of management, leadership, organisational change, strategy and curriculum development who wish to take this work further.

**Stephen Sterling** is Professor of Sustainability Education and Head of Sustainability Education at Plymouth University, and former Senior Advisor on ESD to the Higher Education Academy (HEA). He is a member of the UNESCO Expert Reference Group for the Decade of Education for Sustainable Development (DESD) and was a founder of the Education for Sustainability Programme at London South Bank University.

**Larch Maxey** is a Research Fellow with the Centre for Sustainable Futures at Plymouth University and the founding Chair of Plymouth Incredible Edibles, a student, staff and community food initiative. He has co-run several sustainable education projects, was founding Chair of Swansea University's Sustainability Forum and is currently Director of Research with the Ecological Land Cooperative. He is the co-founder of the Transition Research Network.

**Heather Luna** is a freelance education for sustainability consultant, having worked with the HEA as a project coordinator from 2005 to 2011. She is a Tutor on the Ellen MacArthur Foundation and University of Bradford Postgraduate Certificate in the Circular Economy. She founded Sustainable Thornbury, a Transition Initiative, and is active in the collapsonomics and Dark Mountain dialogue and explorations.

**Routledge Studies in Sustainable Development**

**Institutional and Social Innovation for Sustainable Urban Development**
*Edited by Harald A. Mieg and Klaus Töpfer*

**The Sustainable University**
Progress and prospects
*Edited by Stephen Sterling, Larch Maxey and Heather Luna*

# The Sustainable University

Progress and prospects

**Edited by Stephen Sterling,
Larch Maxey and Heather Luna**

Foreword by Sara Parkin OBE

Routledge
Taylor & Francis Group

LONDON AND NEW YORK

earthscan
from Routledge

First published 2013
by Routledge
2 Park Square, Milton Park, Abingdon, Oxfordshire OX14 4RN

Simultaneously published in the USA and Canada
by Routledge
711 Third Avenue, New York, NY 10017

First issued in paperback 2014

*Routledge is an imprint of the Taylor & Francis Group, an informa business*

*British Library Cataloguing in Publication Data*
A catalogue record for this book is available from the British Library

*Library of Congress Cataloging-in-Publication Data*
The sustainable university : progress and prospects / [edited by] Stephen Sterling, Larch Maxey, Heather Luna.
pages cm. -- (Routledge studies in sustainable development ; 2)
1. Universities and colleges--Environmental aspects--Great Britain. 2. Environmental education--Great Britain. 3. Sustainable development--Great Britain. I. Sterling, Stephen R., editor of compilation. II. Maxey, Larch, editor of compilation. III. Luna, Heather, editor of compilation.
LB3241.4.G7S87 2013
378.1'9610941--dc23
2012033318

ISBN 978-0-415-62774-0 (hbk)
ISBN 978-1-138-80151-6 (pbk)
ISBN 978-0-203-10178-0 (ebk)

Typeset in Times New Roman by Saxon Graphics Ltd, Derby

# Contents

# Figures

# Tables

# Boxes

# Contributors

**John Blewitt** is Co-Director of the MSc in Social Responsibility and Sustainability of Aston Business School, Aston University (Birmingham). His research interests encompass issues relating to urban sustainability, media communications and public education. He is currently exploring the relationship between real places and virtual spaces in the context of the reimagining of the public library service as a public space for the realisation of the citizen's right to the city. He is also a Schumacher Institute Fellow and the author of *Understanding Sustainable Development* (Earthscan, 2008) and *Media Ecology and Conservation* (Green Books, 2010).

**Mary Bownes** is Senior Vice-Principal, External Engagement and Professor of Developmental Biology at the University of Edinburgh with strategic responsibility for postgraduate students and early career researchers, research training, widening participation, admissions and recruitment of undergraduate students, community relations, social responsibility and sustainability, and development and alumni.

**Judi Farren Bradley** is a Chartered Architect. She worked initially in commercial practice, with particular reference to work in conservation and heritage. She initially joined Kingston University as a principal lecturer within the School of Architecture and Landscape, where she runs the Postgraduate Diploma in Professional Practice (RIBA/ARB Part 3), but now shares this responsibility with that of Director of Research within C-SCAIPE, which has enabled her to further her research into professionalism and learning in practice.

**Debby Cotton** is Professor of Higher Education Pedagogy at Plymouth University. She has published widely on various aspects of pedagogic research including sustainability education, e-learning and fieldwork in higher education. Recent publications include an edited SEDA Special Publication on ESD in Educational Development (number 31).

**Marie K Harder** is Professor of Sustainable Waste Management at the University of Brighton as well as its university-wide Sustainable Development Facilitator. She recently coordinated an international EU:FP7 project on Values-based

Indicators for SD, and has considerable experience in applied research across discipline boundaries, working with businesses, government agencies, civil society and NGOs. Her work spans recycling and behaviour change, values and indicators. She has achieved over £2 million in funding, nearly 50 refereed publications and ten successful PhD supervisions. From September 2011 she will work at Fudan University, China most of the year, over five years.

**Peter Higgins** is Professor of Outdoor and Environmental Education at the University of Edinburgh, and teaches academic and practical elements of these fields. He is a member of a number of national and international panels and advisory groups on outdoor and sustainability education and an advisor to the Scottish ministers. He is a Scottish representative on the UNESCO programme Reorienting Teacher Education to Address Sustainable Development. He is author of several books and has published over 100 articles on the theory, philosophy and practice of outdoor education, environmental and sustainability education; most recently (jointly) *Learning Outside the Classroom*.

**Peter Hopkinson** is Professor of Innovation and Environmental Strategy at the University of Bradford. He is the Director of ESD for the Ecoversity programme and co-ordinates Ecoversity as part of his new role as Director of a new £6 million sustainable enterprise centre – a cross-institutional research, knowledge exchange and education initiative. He is also Academic Director of the university's academic partnership with the Ellen MacArthur Foundation, designed to promote the circular economy. Together with Peter James he established in 2001 the Higher Education Environmental Performance Improvement (HEEPI) project.

**Peter James** is Professor of Environmental Management at the University of Bradford, and co-director of the HEFCE-funded Higher Education Environmental Performance Improvement project (see www.goodcampus. org). This supports change for sustainability in UK universities, with a recent focus on IT and laboratories through its Suste-IT and S-Lab initiatives. His publications include *Driving Eco-Innovation, Sustainable Measures* and *The Green Bottom Line*. He has also been a member of the European Commission's Expert Groups on Environmental Technology, and IT and Energy Efficiency.

**Heather Luna** is a freelance Education for Sustainability Consultant, having worked with the Higher Education Academy (HEA) as ESD Project Coordinator from 2005 to 2011. Her publications include co-authoring the HEA Policy Think Tank piece *Universities and the Green Economy: Graduates for the Future* (2012). She is a Tutor on the Ellen MacArthur Foundation and University of Bradford Postgraduate Certificate in the Circular Economy, and was an economics lecturer in New York and Thornbury, England. She founded Sustainable Thornbury, a Transition Initiative, and is active in Re-evaluation Counselling and the Dark Mountain dialogue and explorations.

**Larch Maxey** is a Research Fellow with the Centre for Sustainable Futures at Plymouth University. He is founding Chair of Plymouth Incredible Edibles (a student, staff and community food initiative). He has been teaching, researching and practising sustainability since 1991. His 45 publications include *Low Impact Development: The Future in Our Hands* (2009) and *Small Is Successful* (2011). He has co-run several sustainable education projects, was founding Chair of Swansea University's Sustainability Forum and is currently Director of Research with the Ecological Land Cooperative. He is co-founder of the Transition Research Initiative.

**Robbie Nicol** is Senior Lecturer in Outdoor Environmental Education at the University of Edinburgh where he is Programme Director for the MSc in Outdoor Environmental and Sustainability Education. He is also a steering group member of the Universities Global Environment and Society Academy. His 50 publications are multidisciplinary in nature and include environmental philosophy, environmental education, outdoor education and sustainability education. He jointly authored *Learning Outside the Classroom* (2011).

**Sara Parkin** is Founder Director of Forum for the Future, where she designed the Leadership for Sustainable Development Masters course. She is Chair of the Richard Sandbrook Trust, sits on the board of the European Training Foundation and is a member of the Advisory Committee for Finance South-East Community Fund, as well as advising on Science in Society for the Living with Environmental Change research programme (2009–). She has written several books, the latest being *The Positive Deviant: Leadership for Sustainability in a Perverse World* (Earthscan, 2010). She is an honorary companion of the Institution of Civil Engineers and of the Institute of Energy, and became a Founding Fellow of the Engineering Council in 2009. In 2001 she was awarded an OBE for services to education and sustainability.

**Fiona Quinn** is a Research Assistant within the School of Surveying and Planning at Kingston University, having graduated from the School with both a degree in Property Planning and Development and a Masters in European Real Estate. She is working on a range of projects connected with real estate valuation and sustainability and she assists in teaching research methods; she also runs her own organic skin care business as part of her commitment to sustainability.

**James Ritson** is a graduate from Kingston University with both a degree and graduate diploma in Architecture. He specialises in sustainable and healthy buildings design and has published work in the areas of health and sustainability, sustainability and conservation and heritage of the existing built environment. He has previous practice experience mainly working on historic buildings. He is currently Lecturer and Researcher in the School of Surveying and Planning and he chairs and coordinates the ArchiLab Research Group.

**Alex Ryan** is Associate Director of Sustainability (Academic) at the University of Gloucestershire and teaches Religious Studies with the Open University. She has worked on projects for the UK Higher Education Academy exploring organisational approaches to sustainability, ESD and institution-wide curriculum change, interdisciplinary ESD and responses to ESD across the humanities and social sciences, as well as the *2008 Review of ESD in HE in Scotland.*

**Sarah Sayce** is Professor and Head of the School of Surveying and Planning at Kingston University. Having qualified as a chartered surveyor and worked in commercial practice both in consultancy and in the corporate sector, she became an academic quickly rising to her current position. She has been researching in the field of sustainability applied to real estate and to built environment education for many years and led the successful bid to establish the Centre for Sustainable Communities Achieved through Integrated Professional Education (C-SCAIPE) of which she is Chair.

**Chris Shiel** is Associate Professor at Bournemouth University and Director of the Centre for Global Perspectives. She champions global perspectives in higher education with an approach based on the development of global citizens who understand the need for sustainable development. She is currently leading a project to engage senior leaders and university Board Members in modelling leadership behaviour for sustainable development.

**David Somervell** trained as an architect and worked in community energy in Glasgow before joining the University of Edinburgh as Energy Manager in 1989. He now works as Sustainability Adviser seeking to integrate all aspects of the social responsibility and sustainability strategy implementation at the university. He has mentored the Transition Edinburgh University project since its inception in 2008. He co-edited *Educated Energy Management* (E & F N Spon, 1991), led the development of the university's three award-winning combined heat and power projects, and has published an account of this in *Clean Tech, Clean Profits* (Kogan Page, 2010).

**Stephen Sterling** is Professor of Sustainability Education and Head of Sustain-ability Education at Plymouth University, and former Senior Advisor on ESD to the Higher Education Academy. A member of the UNESCO Expert Reference Group for the Decade of ESD, and Distinguished Fellow of the Schumacher Institute, he has an extensive publications record including co-editor of *Sustainability Education: Perspectives and Practice Across Higher Education* (Earthscan, 2010). He was a founder of the Education for Sustainability Programme at London South Bank University, and recent roles include External Examiner for the Sustainable Development Programme at the University of St Andrews and member of the HEFCE Sustainable Development Steering Group.

**Ros Taylor** is the founding Director of Kingston University's Sustainability Hub. She established and chaired Kingston's Steering Group for Sustainability. An ecologist by background, she developed and led environmental programmes at Kingston, instigating a series of undergraduate and postgraduate environmental and sustainability degrees. She has spoken widely on sustainability, and was a regular contributor to BBC Radio 4's *Home Planet.*

**Daniella Tilbury** is University Director and Chair in Sustainability at the University of Gloucestershire. She is President of the Copernicus Alliance of Universities for Sustainability and facilitator of the Higher Education Treaty for Rio+20. She serves as Chair of the UNESCO Global Monitoring and Evaluation Group for the DESD and is a contributor to the GUNI *Higher Education in the World* 2011 report.

**Ros Wade** is Reader in Education for Sustainability (EfS) at London South Bank University (LSBU) and programme co-director of the international EfS Masters programme. She is also Chair of the London Regional Centre of Expertise (RCE) in ESD. Her publications include *Journeys around Education for Sustainability* (LSBU, 2008; co-edited with Jenneth Parker). Her recent research interests have been in the area of sustainability and political science and the role of learning communities of practice in EfS. She has been involved in a number of ESD capacity-building projects, most recently in South and East Africa.

**Rehema M White** is currently the Sustainable Development Masters Programme Director at the University of St Andrews. Her research on the theory and practice of sustainable development is informed by global experience and both natural and social sciences. She focuses on sustainable development governance, knowledge and sustainable development, and sustainability and art/craft.

# Foreword

*Sara Parkin OBE, Founder Director, Forum for the Future*

> The goal of education is to form the citizen. And the citizen is a person who, if need be, can re-found his civilisation.
>
> (Eugen Rosenstock-Huessy, 1888–1972)

At a time when thoughts about a higher purpose for universities seem out of place next to financial and league table worries, it may seem perverse to burden the universities with responsibility for the future of human civilisation. But I do – and for good reason.

The quote above comes from a 1933 refugee from Germany who became professor of history and social science at Dartmouth University, USA: a man who lived through and reflected upon breakdowns in civil relationships between human beings. Today, as the rising demands of the human population strain the capacity of nature to supply, similarly destructive competition for and conflict over resources and services seems inevitable. Yet, despite many decades of documenting the consequences of environmental degradation (most evidence provided by universities), our civilisation has not learnt either how to stop damaging activities or how to plan for peaceful resilience at times of rapid change.

In his epic *The Economics of Climate Change*, Nicholas Stern said climate change was a failure of the market, but he was wrong. The failure lies with leadership, most seriously the intellectual leadership of our politicians, businesspeople and education systems – and most heinously universities, where a great deal of leadership education goes on. This makes this book, *The Sustainable University,* as much an all-sector challenge as a practical guide to tackling the biggest threat to human civilisation ever – halting and reorienting the *un*sustainable habits that pervade all aspects of our lives.

This book proves we know what to do. It also provides many good ideas and examples of how universities can demonstrate sustainability-literate leadership in the way they are run and in how they prepare citizens to do likewise. Ultimately, the test of success has to be how a university (or any organisation for that matter) contributes to building natural, human and social capital, how ubiquitous that contribution is across the institution (i.e. not just in the energy manager's job description, or the odd module), and how seriously the university utilises its power to influence others.

The bottom line is that it is not nature that needs to change; it is us – a truth that needs to dawn more widely. As does implementing the core principles of human behaviour change: understanding the need for change; acquiring the necessary knowledge and skills that make different behaviours possible; and operating a system of recognition and reward that reinforces right behaviour. Arguably, progress towards sustainability-oriented behaviour is arrested somewhere between the first and second principle, though others are less generous than me in that analysis.

Will universities offer the intellectual leadership needed to shift our civilisation off its self-destructive course and on track for a sustainable future? Obviously they can, if they so choose. I can't see it coming from anywhere else. Moreover, becoming an exemplar of sustainability in teaching, research and institutional management just could be the best way to resolve those financial and league table worries.

## Note

Sara's latest book, *The Positive Deviant: Sustainability Leadership in a Perverse World*, was published by Earthscan in 2010.

## Reference

Stern, N (2007) *The Economics of Climate Change: The Stern Review*, Cambridge: Cambridge University Press

# Preface

This book, while based on UK experience, will be relevant to any university internationally that is seeking to respond to the rising agenda of sustainability, as many of the issues, barriers, opportunities and possibilities are common across higher education systems.

*The Sustainable University: Progress and Prospects* arises from a workshop organised by the Centre for Sustainable Futures (CSF)[1] at Plymouth University, which was held in Plymouth city in June 2011, and attended by representatives of eleven universities widely recognized as UK leaders in sustainability. The workshop and ensuing book were funded by Plymouth University's Pedagogic Research Institute and Observatory (PedRIO).

The book explores the possibility of a 'sustainable university', and all that this might mean and entail. The genesis workshop was held in the belief that:

- the sustainability agenda was highly important to the purposes, role and operation of universities in the current socio-economic and ecological conditions;
- attention to sustainability might weaken as the sector experienced a range of pressures and change;
- much of value had been achieved in the UK, experience gained and lessons learnt, which should be documented;
- the book would have strategic, academic, practical and – hopefully – inspirational value.

The original idea for the book emerged from the work, interest and commitment of colleagues across Plymouth University, who from the founding of the Centre for Sustainable Futures (CSF) in 2005, worked to place – as far as possible – sustainability at the heart of the university's identity and operation. CSF's mission was nothing if not ambitious: to develop the transformative potential of higher education at the university and beyond for building towards a sustainable future.

In 2010, CSF produced *Sustainability Education: Perspectives and Practice Across Higher Education* (Jones *et al.*, 2010), which sought to explore the implications of sustainability for the curriculum across a range of disciplines, drawing on Plymouth's experience and the work of the Higher Education Academy's subject centres. The current book is a follow-up to *Sustainability Education,* recognising that the challenge extends beyond curriculum to the identity and culture of the university as a whole, as reflected in CSF's original mission, and echoed in international best practice.

Seven years after CSF's beginning, Plymouth has a strong track record, affirmed by winning the Green Gown Award in November 2011 for 'Whole Institutional Change – Continuous Improvement', and coming a very close second in the 2012 People and Planet Green League, confirming Plymouth's place as overall top performer since the inception of the Green League in 2007. Sustainability is one of Plymouth University's key identifiers, and many policies, structures, research projects and courses to that end are now in place. Of course, Plymouth is not alone: in the period from 2005, when the Higher Education Funding Council for England (HEFCE) produced its first policy document on sustainable development and higher education (HEFCE, 2005), interest and activity around exploring the implications of sustainability for policy, practice and programmes in higher education has developed markedly, both nationally and internationally. The surge of activity internationally in higher education stimulated by the Rio+20 Summit held in June 2012, and the Global University Network for Innovation's compilation (GUNI, 2012) are healthy indicators of a resurgence of interest in the social, environmental and economic responsibility of higher education.

And yet, to be blunt, despite encouraging signs of change, sustainability remains a minority sport. Both nationally and internationally, only a small proportion of institutions are taking serious steps to push the boundaries of discovering what a fully engaged 'sustainable university' might mean. At Plymouth we are very pleased with what has been achieved, measured against most of the rest of the sector. But we are the first to recognise that this exploration and development only goes so far. We have not yet achieved the transformation that we envisaged some years ago; and while the university has made real strides, we wouldn't claim that it is yet a 'sustainable university' in its full sense, and as explored in this volume. Neither would the authors of chapters in this book make that claim for their own institutions, but their writing here is validated by the fact that all authors have been deeply engaged in significant efforts to embed and advance sustainability in those institutions. Each chapter, in different ways, presents a story of change.

The book is ambitious. The notion of a sustainable university is also ambitious. But such is the severity and urgency of the challenges that now face societies, and our graduates – as evidenced by a growing number of high-level reports – that a commensurate step-change is needed. What is curious is that such reports (see Introduction) rarely mention the role of education and learning, but commonly call for policy change without acknowledging the kinds of learning that realising such shifts might require. What is even more curious – and concerning – is that key policy papers on the future of higher education rarely mention the huge sustainability issues that these global reports continue to document. And so, between these two critical shortfalls, there lies a yawning gap between the commonly perceived and accepted purposes of higher education on the one hand, and on the other, the learning imperatives which need to be fulfilled if we have a chance of assuring a more sustainable future for ourselves and our children. Hence the need for this book.

Currently, and despite some positive evidence of change and response, the global higher education system as a whole remains maladapted to the conditions we face. Given that these conditions have been monitored and documented regularly over recent decades (not least by some university research centres), we might say, with a touch of irony, that higher education seems a slow learner. Yet its role is vital in both a remedial and innovative sense, if we are to realise a safer and more fulfilling future for all.

The paradox is that, ever since the 1972 UN Conference on the Human Environment, education has been held to be a key agent of change throughout a series of high-level agreements and declarations (see Tilbury, this volume), but it remains largely part of the problem of unsustainability. Positive reorientation and renewal is necessary and possible. We do not claim that this book has all the answers, but we do hope it will help raise the debate and assist the learning and change process among those already on the path – and, even more importantly, those still to venture forth.

*Stephen Sterling, Larch Maxey and Heather Luna*
Plymouth, June 2012

## Note

1   CSF began in 2005 as a Centre for Excellence in Teaching and Learning (CETL) funded by the Higher Education Funding Council for England.

## References

GUNI (Global Universities Network for Innovation) (2012) *Higher Education's Commitment to Sustainability: From Understanding to Action,* World in Higher Education Series No. 4, Barcelona, Spain: GUNI

HEFCE (2005) *Sustainable Development in Higher Education: Statement of Policy,* Higher Education Funding Council for England. Available at: www.hefce.ac.uk/pubs/hefce/2005/05_28/

Jones, P, Selby, D and Sterling, S (2010) *Sustainability Education: Perspectives and Practice Across Higher Education*, London: Earthscan

# Acknowledgements

We would like to thank all those who have contributed to making this book possible. In particular, we would like to thank all the chapter authors and other contributors for their time, patience and effort in researching and writing their contributions. In addition, we want to thank Sara Parkin for kindly agreeing to write a Foreword, the anonymous referees of the book proposal and also the reviewers who perused chapters: William Scott, Peter Corcoran, Janet Richardson, Leith Sharp, Justin Dillon, Joanna Blake and Marcus Grant.

The editors would also like to thank Jerome Satterthwaite for his editorial help, but more importantly, for holding the whole project together and patiently holding us to time. Lastly, thanks are due to Plymouth University's Pedagogic Research Institute and Observatory (PedRIO) for its support and for funding the book project.

# Selected abbreviations

| | |
|---|---|
| AASHE | Association for the Advancement of Sustainability in Higher Education |
| BiTC | Business in the Community |
| BREEAM | Building Research Establishment Environmental Assessment Method |
| CETL | Centre for Excellence in Teaching and Learning |
| CSF | Centre for Sustainable Futures (Plymouth University) |
| COPs | communities of practice |
| CSR | corporate social responsibility |
| DEFRA | Department for the Environment, Food and Rural Affairs |
| DIUS | Department for Innovation, Universities and Skills |
| EAUC | Environmental Association for Universities and Colleges |
| ESD | Education for Sustainable Development |
| EfS | Education for Sustainability |
| GA | Green Academy |
| GUNI | Global University Network for Innovation |
| HE | higher education |
| HEA | Higher Education Academy |
| HEEPI | Higher Education Environmental Performance Improvement project |
| HEFCE | Higher Education Funding Council for England |
| HEI | higher education institution |
| IAU | International Association of Universities |
| ICT | information and communications technology |
| LEED | Leadership in Energy and Environmental Design |
| LiFE | Learning in Future Environments |
| NUS | National Union of Students |
| OECD | Organisation for Economic Cooperation and Development |
| PedRIO | Plymouth University Pedagogic Research Institute and Observatory (PedRIO) |
| PVC | Pro Vice-Chancellor |
| RCE | Regional Centre of Expertise in Education for Sustainable Development |

| | |
|---|---|
| REF | Research Excellence Framework |
| SD | sustainable development |
| SMEs | small and medium enterprises |
| STEM | science, technology, engineering and mathematics |
| UCU | University and College Union |
| UNDESD | United Nations Decade of Education for Sustainable Development |
| UNEP | United Nations Environment Programme |
| UNICA | Network of Universities from the Capitals of Europe |
| UNU-IAS | United Nations University Institute of Advanced Studies |
| UUK | Universities UK |
| VC | Vice-Chancellor |
| WCED | World Commission on Environment and Development |

# Introduction

*Stephen Sterling and Larch Maxey*

> We are moving into a world that differs in fundamental ways from the one we
> have been familiar with during most of modern human history. This transition
> has profound consequences.
>
> (UNEP 3, 2012)

This book is about the immediate and longer-term future, and the ability – or
otherwise – of higher education (HE) to make a decisive contribution to its
social, economic and ecological security. It not intended as a specialist book: it
will have failed if it is viewed as such. It is for anyone interested in the purpose
and role of higher education in the contemporary world, and for those concerned
with our collective future. It was written in 2012, the year of Rio+20 Earth
Summit (the United Nations Conference on Sustainable Development –
UNCSD), the high-level conference held in Rio de Janeiro, when the world came
together to 'define pathways to a safer, more equitable, cleaner, greener and more
prosperous world for all' (UNCSD, 2012). In the same year of the first 'Earth
Summit' in Rio, 1992, the authors of the seminal *Limits to Growth* study of 1972
updated their research, defining the sustainable society as '[o]ne that can persist
over generations, one that is far-seeing enough, flexible enough, and wise
enough not to undermine either its physical or social systems of support'
(Meadows *et al.*, 1992: 209).

We are a very long way from that state. The same authors in a further update
(2005) warn of the dramatic consequences of humanity's impact continuing to
exceed the Earth's carrying capacity, while the United Nations Secretary-
General's High-level Panel (UNSGHP) on Global Sustainability, formed of 22
eminent members, notes that, 'We are reaching, and increasingly overstepping,
planetary boundaries' (2012: 90). The Panel further states: 'The signposts are
clear: we need to change dramatically, beginning with how we think about our
relationship to each other, to future generations and to the ecosystems that
support us' (UNSGHP, 2012: 1).

This book explores the challenge and potential of aligning universities'
purposes, policies and practices with the global economic, social and
ecological problematique that is rapidly defining the trajectories and

possibilities of our age. Its fundamental argument – shared by all the book's authors – is that there a serious mismatch between the purposive and operational norms of higher education as reflected and practised by most higher education institutions across the world, and the conditions of complexity, uncertainty and unsustainability that we as a global society face, and that our graduates will certainly encounter. This incongruence centres on the sustainability challenge. It is a far-reaching and urgent challenge concerning both present and future generations' quality of life in the context of unprecedented global conditions and systemic pressures.

Almost daily headlines around such issues as energy, water and food security, biodiversity and species loss, poverty and inequity, climate change and shifting and extreme weather patterns, employment issues, social justice and economic volatility fuel a renewed urgency and debate about sustainable development – how to live well, into the future, without eroding the Earth's ability to sustain present and future generations. Sustainability is about securing economic viability, social coherence and ecological integrity at local to global scales – where these system conditions are seen as deeply co-dependent rather than as separate dynamics. Sustainable development is the journey towards these conditions. Yet, according to the Royal Society, we now face a 'confluence of challenges' which, depending on our collective response over the next few decades, provides 'the opportunity to move towards a sustainable economy and a better world for the majority of humanity, or alternatively the risk of social, economic and environmental failures and catastrophes on a scale never imagined' (Royal Society, 2012: 105).

In addition to this research, a series of further authoritative reports emerged in 2012 designed to influence debate and decision-making, not least at the June Rio+20 summit. These include the United Nations Environment Programme's *Issues for the 21st Century* (UNEP, 2012); the World Economic Forum's *Global Risks* report (WEF, 2012); WWF's *Living Planet Report 2012* (Oerlemans *et al.,* 2012), and the Planet Under Pressure conference's *State of the Planet Declaration* (International Council for Science, 2012). All reflect a consensus on the scale, urgency and interrelatedness of the issues that define our age. Lester Brown, a long-time researcher and advocate of global sustainable pathways, summarises:

> we are addressing not just the future of humanity in an abstract sense, but the future of our families and our friends. No generation has faced a challenge with the complexity, scale, and urgency of the one that we face.
>
> (Brown, 2011: xi)

The UN Panel on Global Sustainability interprets this challenge in terms of necessary measures that will bolster resilience; strengthen global equity, including gender equity; transform how we value goods and services and measure growth; preserve valuable ecosystems; enhance collaboration, coherence and accountability across sectors and institutions; and create a common framework for global sustainability (UNSGHP, 2012: 1) – and stresses the importance of education in enabling this task.

This overarching context raises deep questions about the direction and adequacy of current policy and practice in HE both globally and in countries such as the UK; about the strange disconnect between the higher education agenda and the sustainability agenda, *despite* the pockets of excellent work conducted by some university research centres (for example, see Research Councils UK/UUK, 2011; University Alliance, 2012). Undoubtedly, the whole HE sector is already in a period of change, with questions being raised about its role, funding, structure and delivery mechanisms. Accordingly, the UK's universities association, Universities UK (UUK), had its Longer Term Strategy Network initiate a scenario development exercise, from October 2010 to July 2011, which sought to

> gain a better understanding of the drivers and forces that have the greatest potential to shape the present and the future, to consider how these might be anticipated or influenced to ensure that universities can continue to deliver highly valued outcomes in a future environment, and to gain greater insight and understanding into the present context.
>
> (UUK, 2011: 2)

However, among its perfectly legitimate concerns related to governance, funding, demand and quality, there is no apparent awareness of the broader global context, and no mention of the profound challenge of sustainability issues apart from a single mention of climate change, despite the subtitle of the report, 'Meeting the Challenges of the 21st Century'. This collective myopia is hard to understand, given that HE is expressly in the business of preparing people, often those with the brightest minds, for their individual and our collective futures.

The problem is not confined to higher education however. Change guru Otto Scharmer suggests that throughout society there is a 'massive institutional failure: we haven't learned to mold, bend, and transform our centuries-old collective patterns of thinking, conversing, and institutionalizing to fit the realities of today' (Scharmer, 2009: 3). We can invoke here the notion of 'systems failure'. According to Peters (1999: 124), failure can be considered to be of four types: objectives not met; undesirable side effects; designed failures; and inappropriate objectives. While current criticism of education, particularly in political debate, often centres on the first meaning, the sustainability challenges outlined above require HE policy and practice to be evaluated more broadly. Higher education policy and practice often demonstrates Peters' other three types of failure precisely because it commonly ignores sustainability. Some of the consequences are evidenced by UNEP's Foresight Process report, which points to the need for 'new models of governance', and to 'transform human capabilities for the 21st century' as the top two issues identified among 21 priorities for the current century (UNEP, 2012: v). Expanding on the latter point, the same report states that:

a huge effort is needed on all fronts before society is adequately equipped to deal with the sustainability challenge of the 21st century. 'Capabilities', in this sense, means the necessary job skills, modes of learning, management approaches and research efforts.

(UNEP, 2012: 7)

As a broad range of research and international reports indicate, society is headed inevitably towards a period of transition governed by planetary 'resource and sink' limits. This transition begs questions about educational assumptions, norms and purposes – examined or otherwise – based on long-held socio-economic assumptions and desiderata of unending expansion, progress and growth. The sustainability challenge will only be met if there is sufficient *response-ability* across society as a whole, that is, we find or develop sufficient ability and resolve to address the issues that face us, and on a scale commensurate with those issues. This applies at all levels: *personal, professional, institutional* and *societal*. The status and influence of higher education on all four levels means that universities have a unique and critically important role to play in helping counter negative trends, and nurturing positive and innovative solutions and initiatives.

There has, of course, been some welcome recognition of this fact. In the UK, *inter alia,* the existence of the Green Gowns Award, the People and Planet Green League, the Learning in Future Environments (LiFE) sustainability performance index, the Universities and Colleges Climate Commitment for Scotland, the Environmental Association for Universities and Colleges (EAUC), the support of the Higher Education Academy, the commitment of the Welsh Assembly Government, and the policy support of the HE funding councils has been and remains significant. Meanwhile, among key stakeholders, the UK government's Department for Business Innovation and Skills (BIS) maintains that: 'The transition to a low carbon world will transform our whole economy. It will change our industrial landscape, the supply chains of our businesses and the way we all live and work' (BIS, 2012). A Business in the Community report suggests that over 90 per cent of businesses already recognise that leadership skills for the transition to a sustainable economy are a critical business issue, and over 80 per cent think there is an urgent need to put more programmes in place, including in HE (BITC, 2010).

A critically important voice in this debate is that of students. This is especially true in England, which provides a rich case study of how these issues will play out given radical changes in funding of the HE sector. A national survey of student attitudes on sustainability carried out by the National Union of Students for the Higher Education Academy was published in 2010, and a follow-up survey of more than 3,000 students published in 2012. The findings of the second survey reinforce the conclusions of the first. More than two-thirds of 2011 first- and second-year respondents (66.6 per cent and 70.3 per cent respectively), as in 2010 (70 per cent), 'believe that sustainability should be covered by their university'. Secondly, '[t]here is a continued preference among students for a reframing of curriculum content rather than additional content or courses' (Drayson *et al.,* 2012: 4). Recognising the need for significant engagement of the sector, the Higher Education Funding Council for England (HEFCE), in its influential 2005 sustainable development policy document, envisioned that:

> Within the next 10 years, the higher education sector in this country will be recognised as a major contributor to society's efforts to achieve sustainability – through the skills and knowledge that its graduates learn and put into practice, and through its own strategies and operations.

As the end point of those ten years approaches, the *full* realisation of this vision remains elusive and remote, yet it is more salient and important than ever. The 2012 report of the influential People & Planet Green League, which assesses universities' environmental performance, notes significant improvement by an increasing number of universities since the launch of the League in 2007, but calls for deeper commitments across the sector to achieve a 'future-proof higher education sector' (Hazan, 2012: 25).

Internationally, the UN Decade of Education for Sustainable Development (DESD) (2005–14) has mandated, helped support and guided efforts to align educational policy and practice with the sustainability agenda (Wals, 2009, 2012). In preparation for Rio+20, an international grouping of agencies, organisations and student groups associated with higher education developed a 'People's Sustainability Treaty on Higher Education' (see Tilbury, this volume) (The Higher Education Treaty Circle, 2012). In essence, this document calls for radical change in higher education purposes, policies and practices to 'better serve the needs of current and future generations' (2012: 2). Given that, according to the OECD (2010), approximately 135 million students study worldwide in more than 17,000 universities and other institutions of post-secondary education, and this is expected to double by 2025 (Maslen, 2012), this is no small matter. Yet we have been here before, as the treaty itself acknowledges, with this discussion and process first taking significant form with the Talloires Declaration back in 1990 (driven by the United States-based University Leaders for a Sustainable Future – ULSF). Many agreements, declarations and accords have emerged since (see Tilbury, this volume), yet the gap between rhetoric (in the form of non-binding agreements) and matching practice remains (Bekessy *et al.,* 2007; Wals and Blewitt, 2010). At the same time there has been genuine progress internationally during these years of the UN DESD (Wals, 2009; Witthaus *et al.,* 2010; Tilbury, 2011; GUNI, 2012; Wals, 2012). Sustainability has a much higher profile in higher education discourse, policy-making, research and curriculum development than it did at the beginning of the DESD in 2005, both paralleling, and influenced by, the evolving status of sustainability and sustainable development in wider society over the same period. Yet the picture is mixed, as the most recent UNESCO DESD report states: 'colleges and universities around the world are beginning to make more systemic changes towards sustainability amidst educational reforms towards efficiency, accountability, privatisation, management and control that often hamper their possibilities to do so' (Wals, 2012: 6).

Some HE institutions (HEIs) are at the forefront of exploring whole institutional change based on an ethos of sustainability. But, at the same time, there are large parts of the higher education system, nationally and globally, that remain untouched. For many universities – concerned with such key areas as recruitment and retention, quality, internationalisation, employability, financial security and profile – the overarching context of sustainability, that is, the megatrends that will shape the future positively or negatively, is simply not recognised. Others see it as no more than a matter of 'greening' their estates and operations, of saving energy, reducing carbon footprints, of developing green travel plans, local purchasing and so on. The quality and extent of sustainability-related search is increasing and this is to be welcomed, yet too often 'sustainable development is just another course or research project as expendable as anything else if it does not pay its way' (Wals and Blewitt, 2010: 70).

Greening HE estates is valuable and necessary, but not sufficient, and often only marginally influences curriculum, teaching and learning – long recognised as the most intractable part of the university's operation (AASHE, 2010; Thomas, 2009). In the United States, despite a healthy campus greening movement, some three million students graduate each year, 'the vast majority of which are not equipped with a systemic understanding of sustainability principles and are thus ill prepared to create a sustainable society through their actions in their personal and professional lives' (Second Nature, 2012). The rise and increasing recognition of education for sustainable development (ESD) and the development of new ESD-related programmes is very important and welcome, but is also not in itself sufficient, not least as ESD can be easily siloed into limited corners of an institution's provision as just another 'subject', without recognition of its broader significance and implications.

Hence, in this book, we try to clarify and exemplify the difference between an *accommodative*, a *reformative*, and a *transformative* response by HEIs to sustainability (see Sterling, this volume, for more discussion). The first is a 'bolt-on' response where for example, new modules or courses are added to provision, or some estates greening is carried out, without any change sought beyond these steps; the second and further response is 'build-in', where existing assumptions and practices begin to be questioned and changes made in policy, practice and curricula. While these are positive changes, there are increasing calls for deeper responses, given the enormity of the social, economic and ecological challenges outlined above. Hence, the People's Sustainability Treaty, and many others (for example, Sterling, 2004; Jegatesen and Koshy, 2008; Ferrer-Balas, 2008; Fadeeva, 2010; Second Nature, 2012; Wals, 2012), call for a transformation of the sector, for a shift of culture and a redesign of organisational purpose towards sustainability, involving whole institutional change.

There is a shared expectation and conviction in much of the sustainability discourse that universities should not only act as agents of change, but need to change themselves if they are to be sufficiently effective in the first regard. In other words, while sustainability requires the development of sufficient learning and capacity *through* the effect of educational systems, this is dependent on the achievement of adequate learning and capacity *within* educational systems. To clarify further, we make a distinction between two arenas of learning: *designed learning* and *institutional learning.* Designed learning is the concern of all educational programmes: it is planned, resourced and provided for all the different student groups that experience higher education. Institutional learning refers to the social and organisational learning that the policy-makers and providers may themselves undergo or experience: senior managers, academic staff, support staff, and policy-makers and stakeholders. In the movement to align HE towards sustainability over recent years, it has become clear that substantive progress in *designed learning* is dependent on sufficient depth and extent of *institutional learning* and capacity building. This key point has been proven repeatedly in recent experience, and is evidenced in the chapters of this book. As Richmond states, sustainability education 'is not only a key priority *in* education, but also *for* education' (2010: 19).

The book is about how – and how far – learning in *both* arenas can be advanced deeply, and with such effect that higher education can fully assume its role in helping bring society closer to a more sustainable and equitable future. In essence, the book argues that a university can only contribute fully to a more sustainable future if it becomes more sustainable itself, if it strives and learns to become a *sustainable university.* Such a university embodies, critically explores and lives sustainability, rather than seeking to deliver it in various discrete curricula or research programmes without reference to its own ethos, practices and operation. As Foster says, deep sustainability is radically a learning process that means 'permanent adaptive responsiveness to a permanently changing, ever-emergent set of circumstances' (2008: 116).

In developing this book, its editors and authors have been under no illusions about the scale of the challenge, but have also been aware of the opportunity and positive renewal that sustainability presents to higher education. In many ways, the 'response-ability' of higher education parallels the 'response-ability' of civil society. While this might be expected – as education may be seen as a subsystem of society echoing its norms, values and expectations – higher education is uniquely placed to either slow or accelerate our collective ability to transition towards a genuinely more sustainable, safe and livable future than the one in prospect.

To help address this critical paradox, this book explores the rich notion of the *sustainable university* and illustrates pathways through which its potential can be realised, based largely on the experience of some of the universities that have taken a lead in responding to the sustainability agenda in the UK over recent years. While the chapters are UK-based, the norms and *modus operandi* of universities in the Westernised and globalised world have much in common: many of the arguments, issues, strategies and case studies presented here will have broad relevance and resonance internationally.

## Outlining the book

The book presents a series of linked chapters, which, taken together, attempt to:

- outline progress in the realisation of the concept of the 'sustainable university';
- critically review theory and practice in realising key aspects of what it means to develop the sustainable university in a systemic and whole institutional manner, including the role of organisational learning;
- provide reflective critical analysis of barriers, achievements, strategies and potential;
- consider prospects and map out possible new directions.

When the book was conceived at the planning workshop in June 2011, the writing team was keen to enrich the book by involving others' voices, not least to help reflect multiple perspectives and experience in a complex field. To that end, most of the chapters that follow also feature short vignettes (see shaded boxes) written by other authors. In most cases these vignettes are freestanding, though relevant to the chapter in which they are embedded. Further, the chapters are interrelated thematically, yet each can be read independently. The book can thus be read as a whole, or dipped into.

The book consists of three parts: Context, Aspects, and Institutional change. *Part 1: Context,* consists of three chapters where macro trends, policies and processes within which the sustainable university must take shape are reviewed. Stephen Sterling's chapter (Chapter 1) builds on this Introduction, beginning by outlining the contextual challenge for higher education, including alternative future scenarios. It argues that HE needs to be responsive and closely attuned to these changing times in order to help secure a safer, more sustainable society. The concept of the 'sustainable university' is explored, including short contributions from eight leaders involved in developing sustainability in higher education. The chapter finishes with a short case study of how systemic change has been engendered at Plymouth University.

John Blewitt's chapter (Chapter 2) focuses on what he argues is the most significant factor in understanding the gap between HE's potential and the urgency and scale of the challenge: the dominance of neoliberal policies and practices within HE. The chapter argues that by placing HE in the service of economic growth, neoliberalism has moved us further from the sustainable university. The way forward involves a radical challenge to neoliberal commitments to market force, drawing on the immense range of wider social movements committed to and working for this level of change.

Daniella Tilbury's chapter (Chapter 3) also suggests that social movements have a key role to play in what she describes as the 'rebooting' of HE for sustainability. This chapter reinforces the argument that HE's contribution to sustainability should be intimately linked to wider social, political and economic engagements so that higher education avoids the risk of playing snakes and ladders with political policy-making, moving to the more powerful scenario where HE and policy-making work together for social transformation on a global scale.

*Part 2: Aspects* consists of seven chapters which look in detail at aspects of the sustainable university, and initiatives that offer insight into its creation and potential. Ros Wade's chapter (Chapter 4) considers the role and significance of Regional Centres of Expertise in Education for Sustainable Development (RCEs), which aim to develop a local/global network of transformative learning for sustainability. The chapter shows that universities have been pivotal in developing RCEs worldwide and now benefit from the community of practice they offer. RCEs enable links between formal, non-formal and informal learning, and address the challenge of interdisciplinarity. Their flexible nature, particularly the wide partnership base from which most of them work, allows them to engage with local communities, responding to needs in a variety of ways, including through research and development and the development of communities of practice.

The importance of leadership roles in HE is the focus of Chris Shiel's contribution (Chapter 5). Drawing on experiences at Bournemouth University and on a HEFCE Leadership, Government and Management project, the chapter shows how senior leaders can create enabling environments, facilitating a culture where every member of an institution is empowered to contribute to sustainability transitions.

Marie Harder and Rehema White's chapter (Chapter 6) uses detailed case studies from their two universities (Brighton and St Andrews) to ask what 'community' means within a university and its role in building the sustainable university. The chapter argues that universities need to create a community of values to deal with the challenges of sustainability. They show that through the promotion of a community of values universities can enhance sustainability action, which in turn facilitates the further development of community and action. The chapter shows how such a 'virtuous circle' is affected by governance structures and other formal mechanisms, arguing that HE leadership has an important role in creating the conditions for, and fostering participation in, communities of values for sustainability.

Alex Ryan and Debby Cotton's chapter (Chapter 7) begins an exploration of pedagogy and curriculum for future sustainability by considering another form of HE community, concerned with developing Education for Sustainability (EfS) in the formal curriculum. The chapter uses this discussion, and the limited progress made to date in embedding EfS within curricula, to highlight the importance of systemic approaches to EfS that embrace broader pedagogic developments, innovation and institutional change. The chapter explores the implications of formal, hidden and informal curricula for EfS and shows that holistic approaches to pedagogy and curriculum development can place transformational learning at the heart of HE. The pedagogies and learning that result may not always be *labelled* as EfS, but they will seize the opportunities presented by the fundamental changes currently underway in HE to play their part within the sustainable university's full range of activities, including research and community engagement.

Rehema White's chapter (Chapter 8) proposes that sustainability research is well positioned to embrace innovation, emerging models and trends, including interdisciplinary and participatory approaches, mobilisation, building capacity and awareness, contributing to theory while having local impact and global relevance, synthesising different forms of knowledge, encouraging reflection and linking research and learning. The chapter argues that sustainability research is a process capable of transforming researchers, research participants and understandings of what research is or might be. Having offered its survey of the theory that should inform sustainable research, the chapter concludes with ten practical recommendations for promoting sustainability research, several of which echo HE community-building strategies addressed in Harder and White's chapter.

Peter Higgins, Robbie Nicol, David Somervell and Mary Bownes' chapter (Chapter 9) addresses student experience, drawing on Edinburgh University's engagement with sustainability through its campus, curriculum, communities and student-led activities such as the creation of the first Transition University Initiative. The chapter argues that student experience is an important and emerging issue within HE and one capable of both informing and reinforcing moves towards sustainability. The chapter gives in-depth examples of one HEI's attempts to become a sustainable university over two decades, highlighting lessons learned and the size of the challenge of pulling diverse areas of activity together into a holistic framework.

Sarah Sayce, Judi Farren Bradley, James Ritson and Fiona Quinn's chapter (Chapter 10) argues that stakeholders' well-being is central to the sustainable university, although it has been largely overlooked in sustainability rankings and HE initiatives. There are clear links with Blewitt's and others' chapters as it draws on well-being literature to consider the impact on students and staff of current macro-economic and other HE constraints and pressures, specifically the stress that results from financial pressure, the audit culture and technological changes. This chapter follows Higgins *et al.* in considering the role of estates and the built environment as offering opportunities to develop sustainable approaches that improve well-being. The chapter's recommendations emphasise that such initiatives need to be approached holistically as the match between sustainability and well-being spans all aspects of the sustainable university.

*Part 3: Institutional change* begins with Peter Hopkinson and Peter James' reflection (Chapter 11) on Bradford University's Ecoversity project, an attempt to implement whole institutional and systemic change of the kind advocated throughout this book. Ecoversity's four chronological phases are outlined, from scant sustainability engagement, through a managerial, estates-driven phase, to emergent change and democratic collaboration, and finally to the current consolidating phase. The chapter argues that a middle-out, rather than top-down or bottom-up approach, contributed to Ecoversity's success. The chapter ends with ten pointers for those wishing to apply Ecoversity's lessons to their own context.

Ros Taylor's chapter (Chapter 12) addresses Kingston University's experience of whole institutional change. As with Ecoversity, this took place over a decade, but unlike Bradford, it was driven from the grassroots. 'Managing upwards' as part of Kingston's opportunistic approach made it particularly important to map progress towards a shared vision, to communicate well and to foster fun in the process. This allowed multiple contributions to be effectively drawn together, gaining momentum, top-level buy-in and long-term funding. While the chapter emphasises the necessary specificity of each HEI's sustainability journey, it asserts that Kingston's is a repeatable journey, and outlines the key barriers and steps which all can draw upon.

Heather Luna and Larch Maxey's chapter (Chapter 13) draws lessons from the Green Academy – an institutional change programme funded by the UK's Higher Education Academy involving eight participant HEIs. It reports on the results of the first cohort two years after the programme began in 2010. Based on participant observation and interviews, it shows how valuable the programme was for individual participants and their HEIs, reporting extensively their candid and thought-provoking views. The chapter ends with recommendations for responses to the challenges and opportunities.

Stephen Sterling and Larch Maxey's conclusion (Chapter 14) to the book argues that higher education needs to build a culture of critical commitment. The themes of *critique*, *vision* and *design* are elaborated as a framework for the rethinking and reorientation that is needed.

The chapters contain a wealth of experiences and ideas that help bridge the gap between the role HE could play in societal sustainability transitions and the role it is currently playing. Extraordinary times call for extraordinary responses, in education as in other sectors of society. In higher education, we believe this response needs to be the emergence and flourishing of the sustainable university, and the nurturing of graduates, staff and communities similarly capable of flourishing in conditions of uncertainty and rapid change, people who are both willing and able to commit to contributing to a safer, saner and more livable future.

The authors of this book share a belief in the need for whole systems change, and this is reflected in their contributions. The chapters that follow do not try to cover all aspects of the sustainable university; the book is an invitation to become engaged in this extraordinarily urgent, important, exciting and rewarding task, to help close the gap between HE's practice and potential.

## References

AASHE (2010) *Sustainability Curriculum in Higher Education: A Call to Action*, Denver: Association for the Advancement of Sustainability in Higher Education Available at: www.campusresponsables.com/sites/default/files/Recommandations DDdanscurriculum-AASHE.pdf (accessed 15 June 2012)

Bekessy, S, Clarkson, R and Sampson, K (2007) 'The failure of non-binding declarations to achieve university sustainability: a need for accountability', *International Journal of Sustainability in Higher Education*, 8 (3): 301–16

BIS (2012) *Green Economy*, Business in the Community. Available at: www.bis.gov.uk/policies/business-sectors/low-carbon-business-opportunities (accessed 12 June 2012)

BITC (Business in the Community) (2010) *Leadership Skills for a Sustainable Economy*, London: BITC. Available at: www.bitc.org.uk/resources/publications/leadership_skills.html (accessed 5 October 2011)

Brown, L (2011) *World on the Edge: How to Prevent Environmental and Economic Collapse*. Earth Policy Institute, London: Earthscan

Drayson, R, Bone, E and Agombar, J (2012) *Student Attitudes Towards and Skills for Sustainable Development*, A report for the Higher Education Academy, York: Higher Education Academy. Available at: www.heacademy.ac.uk/assets/documents/esd/Student_attitudes_towards_and_skills_for_sustainable_development.pdf (accessed 23 June 2012)

Fadeeva, Z (2010) 'Perspectives on higher education for sustainable development: transformation for sustainability', in Witthaus, M, McCandless, K and Lambert, R (eds) *Tomorrow Today*, Leicester: Tudor Rose on behalf of UNESCO

Ferrer-Balas, D, Adachi, S, Banas, C, Davidson, A, Hoshikoshi, A, Mishra, Y, Onga, M and Otswals, M (2008) 'An international comparative analysis of sustainability transformation across seven universities', *International Journal of Sustainability in Higher Education*, 9 (3): 295–316

Foster, J (2008) *The Sustainability Mirage: Illusion and Reality in the Coming War on Climate Change*, London: Earthscan

GUNI (Global Universities Network for Innovation) (2012) *Higher Education's Commitment to Sustainability: From Understanding to Action*, World in Higher Education Series No. 4, Barcelona, Spain: GUNI

Hazan, L (ed.) (2012) *People and Planet Green League Report 2012: Driving UK Universities' Transition to a Fair and Sustainable Future*, Oxford: People and Planet

HEFCE (2005) *Sustainable Development in Higher Education: Statement of Policy*, Higher Education Funding Council for England. Available at: www.hefce.ac.uk/pubs/hefce/2005/05_28/ (accessed 2 May 2012)

Higher Education Treaty Circle (2012) *The People's Treaty on Higher Education for Sustainable Development*. Available at: www.copernicus-alliance.org (accessed 25 June 2012)

International Council for Science (2012) *State of the Planet Declaration*, Planet Under Presssure: New Knowledge Towards Solutions, 26–29 March, London. Available at: www.planetunderpressure2012.net/pdf/state_of_planet_declaration.pdf (accessed 25 May 2012)

Jegatesen, G and Koshy, K (2008) *Systemic Transformation of Higher Education Institutions for a Sustainable Future: The Role of Indicators*, Centre for Global Sustainability Studies, Universiti Sains Malaysia, Penang. Available at: http://umconference.um.edu.my/upload/163-1/Paper%2069.doc (accessed 5 June 2012)

Maslen, G (2012) 'Worldwide student numbers forecast to double by 2025', *University World News*, 209. Available at: www.universityworldnews.com/article.php?story=20120216105739999 (accessed 4 July 2012)

Meadows, D, Meadows, D and Randers, J (1992) *Beyond the Limits: Global Collapse or a Sustainable Future*, London: Earthscan

Meadows, D, Meadows, D and Randers, J (2005) *Limits to Growth: The 30-year Update*, London, Earthscan

OECD (2010) 'OECD launches first global assessment of higher education learning outcomes'. Available at: http://globalhighered.wordpress.com/2010/01/28/oecd-launches-first-global-assessment/ (accessed 10 April 2012)

Oerlemans, N, McLellan, R and Grooten, M (eds) (2012) *WWF Living Planet Report 2012*. Gland: WWF. Available at: http://assets.wwf.org.uk/downloads/lpr_2012_summary_booklet_final_7may2012.pdf (accessed 15 June 2012)

Peters, G (1999) 'A systems failures view of the UK National Commission into Higher Education Report', in Ison, R (1999) (ed.) *Systems Research and Behavioral Science*, John Wiley, 16 (2): 123–31

Research Councils UK/UUK (2011) *Big Ideas for the Future: UK Research That Will Have a Profound Effect on our Future*. Available at: www.rcuk.ac.uk/documents/publications/BigIdeasfortheFuturereport.pdf (accessed 25 June 2012)

Richmond, R (2010) 'Envisioning, coordinating and implementing the UN Decade of Education for Sustainable Development', in Witthaus, M, McCandless, K and Lambert, R (eds) *Tomorrow Today*, Leicester: Tudor Rose on behalf of UNESCO

Scharmer, O (2009) *Theory U: Leading from the Future as it Emerges: The Social Technology of Presencing*, San Francisco: Berrett-Koehler Publishers

Sterling, S (2004) 'Higher education, sustainability and the role of systemic learning', in Corcoran, P B and Wals, A E J (eds) *Higher Education and the Challenge of Sustainability: Contestation, Critique, Practice, and Promise*, The Netherlands: Kluwer Academic

Royal Society (2012) *People and the Planet*, The Royal Society Science Policy Centre report 01/12, London: The Royal Society. Available at: http://royalsociety.org/policy/projects/people-planet/report/ (accessed 1 June 2012)

Second Nature (2012) *Transforming Higher Education for a Sustainable Society*. Available at: www.secondnature.org/documents/brochure_final.pdf (accessed 12 June 2012)

Tilbury, D (2011) *Education for Sustainable Development: An Expert Review on Processes and Learning*, Paris, UNESCO. Available at: http://unesdoc.unesco.org/images/0019/001914/191442e.pdf (accessed 15 May 2012)

Thomas, I (2009) 'Critical thinking, transformative learning, sustainable education, and problem-based learning in universities', *Journal of Transformative Education*, July 2009, 7 (3): 245–64

UNCSD (2012) *About the Rio+20 Conference*. Available at: www.uncsd2012.org/rio20/about.html (accessed 15 May 2012)

UNEP (2012) *21 Issues for the 21st Century: Result of the UNEP Foresight Process on Emerging Environmental Issues* (Alcamo, J and Leonard, S [eds]), Nairobi, Kenya: United Nations Environment Programme (UNEP). Available at: www.unep.org/publications/ebooks/foresightreport/ (accessed 10 May 2012)

United Nations Secretary-General's High-level Panel on Global Sustainability (2012) *Resilient People, Resilient Planet: A future Worth Choosing*, New York: United Nations. Available at: www.un.org/gsp/report/ (accessed 15 June 2012)

University Alliance (2012) *Problem Solved: University Research Answering Today's Challenges*. Available at: www.unialliance.ac.uk/campaigns/problemsolved/ (accessed 29 June 2012)

UUK (2011) *Futures for Higher Education: Analysing Trends. Higher Education: Meeting the Challenges of the 21st Century*. Available at: www.universitiesuk.ac.uk/Publications/Documents/FuturesForHigherEducation.pdf (accessed 16 May 2012)

Wals, A (ed.) (2009) *Review of Contexts and Structures for Education for Sustainable Development 2009 (DESD, 2005–2014),* Paris: UNESCO. Available at: http://unesdoc. unesco.org/images/0018/001849/184944e.pdf (accessed 25 June 2012)

Wals, A (2012) *Shaping the Education of Tomorrow: 2012 Full-length Report on the UN Decade of Education for Sustainable Development,* Paris: UNESCO, DESD Monitoring and Evaluation

Wals, A and Blewitt, J (2010) 'Third-wave sustainability in higher education', in Jones, P, Selby, D and Sterling, S (2010) *Sustainability Education: Perspectives and Practice Across Higher Education,* London: Earthscan

WEF (World Economic Forum) (2012) *Global Risks 2012,* Seventh Edition, An Initiative of the Risk Response Network, WEF. Available at: www3.weforum.org/docs/WEF_ GlobalRisks_Report_2012.pdf (accessed 25 May 2012)

Witthaus, M, McCandless, K and Lambert, R (eds) (2010) *Tomorrow Today,* Leicester: Tudor Rose on behalf of UNESCO

# Part I
# Context

# 1 The sustainable university

## Challenge and response

*Stephen Sterling*

The trouble with our times is that the future is not what it used to be.
(Paul Valéry, 1871–1945, quoted in Ben-Ze'ev, 2000)

## Introduction

The world that today's graduates are entering is already – and will be increasingly – very different from that inherited by previous generations, including by many of today's university staffs when they were graduating years ago. If Valéry's paradoxical line was true in his own time, it is even truer today if numerous global forecasts and reports are given their deserved credence (Royal Society, 2012; UNEP, 2012; World Economic Forum, 2012; Oerlemans *et al.,* 2012; International Council for Science, 2012; UNSGHP, 2012). Current headlines, and increasingly everyday experience, underline that we live in interrelated local and global conditions of uncertainty, complexity and unsustainability; summed up by some as the 'triple crunch' of climate change, the end of cheap energy, and financial instability (NEF, 2008). These are times of contingency and emergence, not prediction and control, and the future in prospect is not a linear extrapolation of what has gone before. Indeed, we may be facing the collapse of systems that we have taken for granted for decades (Meadows *et al.,* 2005; Diamond, 2005; Homer-Dixon, 2012). The mainstream of higher education may, therefore, largely be educating for a future that 'no longer exists'.

Most of the millions of professional staff and some 135 million students (OECD, 2010) involved in higher education across the globe, including 2.5 million UK students, would agree that education is essentially about preparing for the future, both at individual and social levels. And yet, strangely, 'the future' – the planetary future and key trends that will affect people's lives in this century – hardly registers in most mainstream policy-making and practice in higher education, despite high-level calls over many years for a sufficient and appropriate response from higher education (see Tilbury, this volume). There are certainly significant exceptions and clear evidence of growing interest and engagement, not least through the catalytic effect that the Rio+20 summit had on parts of the international higher education community through the development of 'The Rio+20 Directory of Committed Deans and Chancellors 2012' and the *People's Sustainability Treaty for Higher Education* (see Tilbury, this volume). In the UK, funding council policies have had some evident success and impact over recent years in putting sustainable development on the higher education (HE) agenda. However, the fact remains that, aside from some notable pockets and centres of excellence, *most* HE research, taught programmes and initiatives make no reference to this overarching context, and sustainability – where it is acknowledged – is often seen as a special interest, or the province of campus management only, or is only understood in environmental terms. Yet, at the same time, global society and states are faced with a whole series of dynamic, interconnected and unprecedented challenges: in adjusting to an environment of no or low economic growth, rising unemployment, high energy costs, resource depletion and increasing competition, the need to mitigate and adapt to the effects of climate change, an alarming loss of biodiversity, ocean acidification and collapsing fish stocks, rising poverty, inequity and malnutrition, and a global population of 7 billion and rising (UNFPA, 2011) to name some of the critical issues. However, and simultaneously, the possibilities and opportunities of such movements as the low carbon and circular economy, benign energy generation, fairer trade, green products and sustainable design, greater democratisation and constructive social networking, and the rebuilding of more resilient communities and restored ecosystems, spurred by new thinking and innovation in many sectors beckon as a positive vision of a more sustainable future.

Futurist Paul Raskin (2012: 12) suggests three possible global scenarios: worlds of incremental adjustment ('Conventional Worlds'), worlds of catastrophic discontinuity ('Barbarization') and worlds of progressive transformation ('Great Transitions'). The last of these, sustainable, scenarios depend on 'an enlargement of consciousness', which emphasises 'global citizenship, humanity's place in the wider community of life, and the well-being of future generations' (2012: 13). Raskin suggests that higher education can either drift along with the tides of change, or intentionally choose and contribute to human and global betterment:[1]

> the very plausibility of the Great Transitions depends to a significant degree on whether HE assumes a forceful and proactive role in advancing the necessary shift in culture and knowledge (2012: 14).

So the part played by higher education in helping society shape the future is clearly critically important. It is not just, however, a simple matter of making a stronger marginal or 'add-on' contribution to sustainability – the approach reflected in much of the relevant discourse – although this will be a first step for many institutions. Such is the precarious nature of our times, the challenge for higher education, as for other key institutions, is to reorient itself accordingly, to place sustainability and securing the future at the heart of its *raison d'etre.*

This transition is not simple: universities are complex organisations constantly facing a range of pressures. Even so, this challenge goes beyond 'integrating sustainability' or 'embedding sustainability' into higher education (as is it often termed). Despite decades of debate and work at national and international levels on environmental education, development education, and more latterly, education for sustainability and education for sustainable development (ESD), mainstream educational thinking and practice has still to embrace fully the implications of current socio-economic-ecological trends, let alone explore, critique and inform the urgent changes in thinking, practices and lifestyles that many observers deem necessary to assure a liveable future for all. While such foundational work has been invaluable, the greater challenge now for policy-makers, stakeholders, staffs and students is to integrate *higher education into the wider societal context of sustainability*: that is, to achieve systems that are fully attuned to and alert to the times (Sterling, 2004). In a much quoted phrase, the American writer and academic David Orr points out that the worrying negative global signals are 'not the work of ignorant people (but those) with BAs, BScs, LLBs, MBAs and PhDs' (1994: 7). It is also clear that high income countries, with high levels of educational attainment, have per capita ecological footprints far in excess of middle and low income countries (Oerlemans *et al.,* 2012). Drawing on research that he conducted for the mid-term review of the UN Decade of Education for Sustainable Development, Arjen Wals suggests that 'At present most of our universities are still leading the way in advancing the kind of thinking, teaching and research that only accelerates un-sustainability' (Wals, 2010: 32). This stands in contrast to the view of the UK universities association, Universities UK (UUK), which in 2010 stated:

> Universities already make a significant contribution to the UK's sustainable development strategy and have played an important role in researching the challenges, in developing new approaches to those challenges and also in improving organizational operations.
>
> (UUK, 2010:1)

While this certainly has some truth to it, the 'Statement of Intent' from which this quotation comes is no longer to be seen on the UUK website, and the then chair of the former UUK Sustainable Development Task Group (which wrote the Statement), said in a keynote at a conference in the same year: 'It is a huge challenge to get VCs to see this as core business' (Broadfoot, 2010). Wals' conclusion is no doubt a contentious point of view, but even if it were only partly valid, it raises a fundamental and unavoidable question about the purpose, business and responsibility of universities at this critical point in the twenty-first century.

## Re-thinking the purpose of higher education

That higher education has a critical role to play in developing tomorrow's decision-makers, professionals and citizens is beyond dispute. But the full import of the rapidly changing socio-economic and environmental context within which such roles will be played out lies in the shadows rather than the foreground of higher education discourse, policy-making and practice. There seems to be an implicit assumption that, despite mounting evidence to the contrary, economic futures, energy futures and ecological futures are, by and large, a matter of continuity, and so we can safely continue to educate for a prospective stable socio-economic environment as in the past. (The exception to this assumption is the emphasis on technological futures and the impact of information and communications technology (ICT) on the workplace, where HE has sought to embrace its implications.) At the same time, current concerns about economic recession, and changes in funding of higher education, have raised questions about the purpose of HE, which echo an older debate about the role of universities.

In recent years, the tension between the traditional academic role of universities and the more instrumental role of preparing young people for the workplace and their place in society, has become more marked, particularly – in the UK – through shifts in government policy. In 2008, a White Paper *Innovation Nation* by the then named Department for Innovation, Universities and Skills (DIUS) underlined the economistic, skills-oriented view of the purpose of the university, in contrast to the more liberal conception of the university oriented towards educating for human and social improvement (Gough and Scott, 2008). In the same year, the Higher Education Funding Council for England (HEFCE) commissioned a 'Strategic Review of Sustainable Development in Higher Education in England' which described, for the purpose of the review, the core processes and function of higher education institutions (HEIs), as including:

> To generate advanced knowledge and understanding of the world and of the role of humans and the impacts and implications of human activities within it. HEIs pursue this purpose through Research, and Teaching both of which should lead to in-depth learning.
>
> (Policy Studies Institute *et al.*, 2008: 2)

The tension between these two different conceptions and descriptions of the university's core function – the instrumental and the academic – was thrown into greater relief by further changes in government policy. A key turning point was the Browne report, *Securing a Sustainable Future for Higher Education: An Independent Review of Higher Education Funding and Student Finance* (Browne *et al.*, 2010), which not only further reified an economistic view of the purpose of higher education, but recommended a radical change in funding the sector, moving away from the state-funded grant system towards student loan financing. This thinking was in line with market economy philosophy where universities were to compete as providers for student 'customers' who would ultimately, through market choice, determine which courses and which universities flourished or failed. As Browne put it, 'the money will follow the student' (2010: 4).

While the effects of this radical change in policy are still being played out, critics have pointed to the redefinition of the *purpose* of higher education as being key to the policy's manifestation in practice. Collini (2010: 25), for example, argues that the Browne report 'displays no real interest in universities as places of education; they are conceived of simply as engines of economic prosperity and as agencies for equipping future employees to earn higher salaries'. While some might see Collini's view as too sceptical, the economistic view of HE seems borne out by UUK's futures scenarios study, which states:

> Higher education provides the skills, knowledge and innovation that will help support a productive and successful economy. Universities also equip members of society with the skills, values and knowledge needed to operate on the global stage.
>
> (Collini, 2011: 29)

An earlier report from the New Economics Foundation (NEF) argues that HE's role has narrowed too far towards servicing the market economy and that the purposes of HE need to be reconsidered to advance collective well-being. In NEF's view (Steuer and Marks 2008:12), HE needs to equip its learners 'with the knowledge, skills and understanding to pioneer innovative and creative responses to achieving wider economic, social and environmental well-being'. Their report adds that well-being should be a part of quality assurance (reflecting Sayce, this volume). In a similar vein, M'Gonigle and Starke (2006) are concerned about the increasing corporatisation of higher education and raise basic questions of purpose around the social value and cultural significance of courses studied and research produced.

Hence, as Gough and Scott (2008: xi) outline, there is a fundamental difference of view between those who see the main or even sole function of universities to provide society with the skills base it requires, and those that see universities as not only contributing to the economy, but 'contributing to the intellectual and moral improvement of the human condition'. Similarly, Taylor (2011: 98) contrasts two broad thrusts – one an emphasis on efficiency and effectiveness, delivering graduates who can compete in the global market; and another that values enquiry which allows for vision and imagination; links to the spiritual, emotional and ecological; embraces uncertainty and the possibility of alternatives; and encompasses a plurality of visions. If we then introduce explicitly the sustainability agenda to this picture, the question of the purpose of higher education becomes *broader* still – as sustainability touches on all that constitutes the quality of life for present and future generations. It also becomes more *immediate*, as authoritative reports (as referenced above) are calling for radical and urgent action to address an unprecedented nexus of pressing global and planetary issues. Viewed through the lens of sustainability and well-being, the narrowly instrumental view of higher education – in danger of becoming dominant in current discourse and policy – is potentially subsumed and transformed in an expanded and more holistic view of education. Such a view restores intrinsic learner-centred educational values (the personal dimension) and orients education towards addressing the challenges of sustainability (the global dimension), and has been echoed in the rhetoric of UN initiatives and documentation for some time.

Thus, Agenda 21, emerging from the first Rio Earth Summit (United Nations Conference on Environment and Development) of 1992, states:

> Education, including formal education, public awareness and training should be recognized as a process by which human beings and societies can reach their fullest potential. Education is critical for promoting sustainable development and improving the capacity of the people to address environment and development issues.
>
> (United Nations, 1992: 320)

As outlined in the Introduction to this volume, the sustainability agenda has, to some degree, become a stronger voice in international higher education discourse, if the number of agreements, declarations and accords is any reliable indicator. There is no shortage of mandate and rhetoric at both national and international level, stretching back twenty years and more, with regard to the need for higher education to respond more immediately and fully to the sustainability agenda (see Tilbury, this volume). Yet experience over that period demonstrates that such change is often slow, piecemeal and often simply absent in the framing of policy, practice and discourse. The reasons for this are many, but key is a widespread insufficient appreciation of the profound implications of sustainability for academe. There is a parallel here with sustainable development discourse in wider society: it can be misconstrued as an 'add-on' area of theory and practice, that is, a sectoral interest, which can safely be considered *alongside* but separate from other agendas, *or* it is properly recognised as implying a changed paradigm involving – to a greater or lesser degree – all aspects of social, economic and political organisation. So it is with regard to the relationship between education and sustainability.

In some senses, sustainability *can* be perfectly legitimately regarded and made operational as another agenda alongside others, such as internationalisation, employability and enterprise. Indeed, this is the most common response. Viewed in this way, it is seen at best as having equal status and import, and a discrete agenda, and much good work often ensues. However, this is an accommodative response to sustainability, a first (often only) step for many HEIs. A deeper response calls for a qualitative cultural shift, whereby the richness, complexity, challenge and opportunity of the sustainability agenda, and the probable and possible futures that society faces, set an overarching context for all policy priorities and agendas. Sustainability, while necessarily imprecise, provides an integrative and holistic framework whereby policy coherence in the face of uncertainty can be approached and achieved. Although it requires universities to expand the boundaries of their 'system of interest' and concern (to use a systems term), in so doing, sustainability also offers the chance of renewal of purpose and direction.

To reiterate the challenge: Antony Cortese (2003: 17), the founder of the United States-based Second Nature organisation, which supports sustainability in HE states:

> HE institutions bear a profound moral responsibility to increase the awareness, knowledge, skills and values needed to create a just and sustainable future...

But he goes on:

> why is HE so averse to risk and difficult to change? Because the change sought is a deep cultural shift...

The shift is towards the sustainable university and its nature is considered next.

## Interpreting the sustainable university

Use of the term 'sustainable university' in this book is, of course, entirely intentional. While the word 'sustainable' is undoubtedly overplayed, and often misused and misunderstood, both in debate and literature, it is nevertheless almost unsubstitutable, carrying rich layers of both descriptive and normative meaning. Or it should do: for some, the 'sustainable university' means little more than one that is financially sustainable and will therefore last, at least in funding terms – echoing the diminished sense of sustainability often inferred by the term 'sustainable business'. Indeed, this is the meaning used in the Browne report (see above). For our purposes, I will define 'sustainable university' as shown in Box 1.1.

---

**Box 1.1: The sustainable university**

The sustainable university is one that through its guiding ethos, outlook and aspirations, governance, research, curriculum, community links, campus management, monitoring and *modus operandi* seeks explicitly to explore, develop, contribute to, embody and manifest – critically and reflexively – the kinds of values, concepts and ideas, challenges and approaches that are emerging from the growing global sustainability discourse.

---

This is not prescriptive, there is no blueprint pathway to follow: it is an emergent and evolving wave, giving rise to different forms and initiatives in different universities, as portrayed in this book. There is no threat – as some would have it – to academic freedom, unless the nature of the challenge is misconstrued. The sheer complexity and enormity of the 'sustainability revolution' (Meadows *et al.*, 2005) demands the deep critical debate, insight, cross-disciplinary research and innovation that universities can especially offer. As Raskin suggests:

> The soaring intellectual challenge of creating knowledge, pedagogy, and engagement for the transition would align the twin desiderata of advancing cutting-edge scholarship and human good.
>
> (Raskin, 2012: 15)

This behoves an intentioned shift of culture, a change in landscape: one to be explored, made and critiqued, as much in education as in other fields seeking to respond fully and contribute to the sustainability transition. However, unlike emerging practice towards this transition in other fields – be they economics, health, business, agriculture, law, engineering, science and technology, arts and so on – formal education as such has a unique primary responsibility and formative role to play in relation to society, not least in preparing people for their roles in such fields.

In developing this chapter, and to enrich the discussion, I asked eight leaders involved in advancing this agenda their views about the concept of the sustainable university, in terms of its meaning, characteristics and possibility. Their contributions are shown below:

> Becoming Chancellor of Keele University has made me rethink exactly what is meant today by the notion of 'a sustainable university'. In my mind, the 'sustainable housekeeping' bit is sort-of sorted – on energy efficiency, carbon management, waste minimization and so on. Some universities may only be doing a fraction of what they should be doing, but at least they know what it is they should be doing! Ditto on engaging with local communities and using the purchasing power and influence of a university to create substantial economic multipliers. And there is some, if as yet insufficient evidence, that increasing research efforts are being oriented towards pressing real world issues and challenges. But it's a great deal harder when it comes to what universities are actually teaching, and the way in which they are teaching it. On this matter, the 'best practice' dossier is a lot thinner, and the very robust criticism levelled against HE in the UK – that it reinforces patently defunct world views and expectations amongst students – is easy to substantiate. Addressing this particular 'sustainability deficit' – ensuring that all UK universities are genuinely 'fit for purpose' in a world shaped more and more by the precepts of sustainability – is a challenge for the whole higher education community.
>
> Jonathon Porritt, Founder Director of Forum for the Future
> and Chancellor of Keele University, UK

The sustainable university educates its students to become transformational leaders of a sustainable society through its curriculum, its research, its willingness to serve as a test bed for innovation, its outreach and interactions with the greater community and through behaving sustainably in all of its practices, processes, and deliberations. This means that a sustainable university must first and foremost use its resources in socially responsible ways and must seek to reduce its negative environmental footprint while serving as an exemplar of social justice and environmental stewardship. Most importantly, the sustainable university must help develop a new generation of professionals who embrace, live and promote a sustainable life for all.

> Paul Rowland, Executive Director of the Association for the
> Advancement of Sustainability in Higher Education (AASHE), USA

The sustainable university is one that contributes to, and serves its community and the future challenges of society in multiple ways; through transformational learning, research and active engagements with community members on a regular basis. It is a university where knowledge is not seen only as an object or commodity, but as a richly textured process of changing perspectives; and where reflexivity is valued. In such a place, historically constituted disciplinary boundaries are respected for their integrity and knowledge producing value, but transdisciplinary studies carry equal weighting and cut across and through disciplinary boundaries in formations that create a different kind of knowledge, or knowing. It is essentially a transformative space; where transformational practices are theorized, modelled and imagined. An open space which is not known by its 'ivory towers'; its rigid traditions, or its allegiance to power, but rather by its creativity, and energy for change; a 'hub' of social transformation and social learning for a more sustainable, just and equitable future. At the heart of such a university lies independence of thought, critical debate and social critique, but perhaps more importantly, such critical debate and social critique should feed imagination and re-imagination that is creative, productive, and intellectually rich and stimulating.

> Heila Lotz-Sistika, Murray and Roberts Chair of Environmental
> Education and Sustainability at Rhodes University, South Africa

In my darker moments the dawn of a sustainable university seems as far off as ever and this perplexes me. For where else is it but a university or college that we have the potential to think in new ways and stop making (and teaching) old mistakes, inspire generations with a sense of citizenship and collective responsibility, and regenerate our communities and economy with our access to life-transforming learning, research and engagement. Every college and every university in the land is sitting on a good news story of epic proportions. For our leaders and learners, carbon reduction and resource efficiency are (only) the first steps to a critical new understanding and recognition of the value of our sector. We need to believe that first though.

> Iain Patton, CEO, Environmental Association for
> Universities and Colleges, UK

A sustainable university is a university that contributes to the quality of life and the well-being of the planet through its education, research, management and community outreach. Doing so requires continuous critical scrutiny of its own assumptions, values and practices. Since 'quality of life' and 'well-being of the Planet' are contested and dynamic concepts a sustainable university has a fundamental role to play in recalibrating their meaning as the world changes and new knowledge and insights emerge. Despite progress in recent years, this ideal remains a core challenge for most universities.

> Arjen Wals, Professor and UNESCO Chair of Social Learning and
> Sustainable Development, Wageningen University, The Netherlands

Sustainability is essentially a systemic *worldview* or *epistemic perspective* that presents considerable intellectual and moral challenges to humankind that far transcend the horizons of the prevailing paradigm of scientific reductionism, technological determinism, and economic rationalism. It is an ethos for approaching the world and issues in it from a systemic perspective that includes the integrated embrace of ethical, aesthetic and spiritual dimensions of human activities as well as ecological, economic, political, social and cultural ones.

This provides a context for the development of bio-physical and socio-economic systems that must be designed to operate with responsible concerns for ecological integrity, resource sufficiency and inclusive well-being, even as they attempt to achieve and maintain their productive purposes in the face of the dynamic complexity of the environments in which they are embedded. Such a perspective dictates an emphasis on education as a transdisciplinary process of epistemic transformation: of the facilitated development of high order intellectual and moral competencies that are all too rarely addressed within the academy.

> Richard Bawden, Professor Emeritus University of Western Sydney,
> Adjunct Professor, Michigan State University, and Director of
> the Systemic Development Institute, Sydney

Sustainable universities should be seen as an integral part of the education system that encourages development of new forms of knowledge towards low carbon, resilient and just societies. Such a system encourages diversity of organizational missions and forms which facilitate education and research that holds the interests of communities, nations and humanity at heart. Such universities aspire to:

- facilitate access to higher education for those who were traditionally deprived of it due to their isolation, socio-economic status, ethnicity or gender;
- develop new knowledge and encourage forms of learning appropriate to the characteristics of post-modern society with its complexity, unpredictability and risks;

- cultivate ways of thinking and acting that focus on long-term orientation, dealing with a variety of perspectives and viewpoints, whilst embracing partnerships;
- deal with immediate issues confronting humanity, as well as research and develop long-term scenarios of development;
- and place transdisciplinarity as core, engaging with different forms of knowledge.

Zinaida Fadeeva, Research Fellow at the United Nations University
Institute of Advanced Studies (UNU-IAS)

If a university is to demonstrate its commitment to sustainability, it needs to create appropriate teaching and knowledge exchange programmes to ensure that the knowledge we develop meets societal needs and to be an exemplar in its own right. This requires a systems approach to embed sustainability throughout a university's operations and delivery.

The key challenge is to build sustainability into all aspects of its delivery: governance, research, curriculum, campus, community, culture and reporting mechanisms. Five key pillars could underpin this approach: sustainability, employability, internationalisation, culture and lifelong learning. A specific challenge is to embed sustainability into all students' learning and research experiences. Setting a specific target, eg. 15% of a student's experience, undergraduate and postgraduate, to be delivered through a sustainability lens, will ensure that the agenda is mainstreamed.

To guide the mission, the University's Vice Chancellor needs to explicitly support the agenda and create support frameworks to take it forward; eg. at the University of Wales Trinity Saint David in west Wales, a virtual institute, INSPIRE (Institute for Sustainable Practice, Innovation and Resource Effectiveness), has been created to support internal and external sustainability practice.

Jane Davidson, Director, INSPIRE, Institute for Sustainable Practice,
Innovation, and Resource Effectiveness,
Trinity Saint David University, Wales

These thoughts map out just some of the ingredients that (would) characterise the sustainable university. Taking some keywords from the above – *fit for purpose; transformational learning; reflexivity; transdisciplinarity; critical debate; collective responsibility; engagement; well-being; higher order competencies; long-term orientation; systems approach* – a pattern emerges which indicates the quality of systemic cultural shift that Cortese and others suggest is necessary.

In seeking to clarify this picture, it is important to suspend, temporarily, discussion about the possibility and practicability of implementing this shift (which is, however, returned to below). While common and understandable, such a stance tends to limit the focus of attention to current purposes, policies and structures in terms of how they might be modified to accommodate or bolt-on sustainability, yet often leaving the rest of the system little changed or untouched. Rather, the visionary leader will first look forward to the immediate and longer-term future and consider deeply (Senge *et al.*, 2005) what kinds of education can meet the socio-economic and ecological problematique (Peccei, 1982) that faces societies locally, nationally and globally – and then look back to reappraise current policies and programmes in this light (Holmberg and Robèrt, 2000; Kneale and Maxey, 2012). This backcasting approach is consistent with what Scharmer (2009: 14) refers to as leading, learning and acting 'from the future as it emerges'.

From this perspective, higher education is in the business of 'anticipative education', valuing foresight and insight and fully recognizing the new conditions and discontinuities that face present and future generations, rather than 'retrospective education' looking to reproduce past practices that may no longer be appropriate or valid. In this light, educational institutions need to become less centres of transmission and delivery, and more centres of transformation and critical enquiry; less teaching organisations, more learning organisations critically engaged with real-world issues in their community and region; less discipline based, more inter- and transdisciplinary; less managerial and more participative; less self-contained and self-referential, more engaged with a broad range of stakeholders; less instrumental and reluctant to engage with normative issues, more holistic in purpose and exploring ethical dilemmas and dimensions. The logic here derives from a broad consensus that we are increasingly confronted by a broad range of issues – e.g. food security, energy security, water security, inequity and social coherence, economic vitality and employment, the effects of climate change, public health, degraded ecosystems and so on – characterised by complexity, interrelatedness and indeterminacy (Brown *et al.*, 2010; Homer-Dixon, 2006). These require the critical engagement of academe (Fear *et al.,* 2006) involving research and organisation of knowledge, which necessarily transcends traditional approaches and discipline boundaries, and addresses ethical and normative questions. In this mode, the academy becomes 'a more vigorous partner in the search for answers', affirming 'its historic commitment to what I call the scholarship of engagement' (Boyer, 1996, in Bawden, 2004: 1).

These are indications of the nature and role of the sustainable university. How possible is it to secure a deep response to the challenges and opportunities outlined above? The next section looks at barriers and paradigms, in order to throw some light on how higher education's response might be deepened and accelerated.

## Towards a deep learning response

The responsiveness of universities to these challenges varies, of course, a good deal in both extent and emphasis. There is no doubt that there is a growing movement in which sustainability precepts can potentially inform all aspects of academia including: governance and reporting, the development of relevant and applied research programmes and research centres, greening the estate and sustainable procurement, developing sustainability curriculum programmes and participative pedagogies, and community engagement. A number of drivers and benefits have been evidenced amongst those HEIs who have engaged with sustainability (see Box 1.2).

---

**Box 1.2: Engaging with sustainability: drivers and benefits**

*Drivers*

- public interest and global concern over sustainability-related issues;
- government and funding council mandate and expectations including carbon management and interest in the low carbon economy;
- links to employability and views of employers, particularly around 'green skills';
- the requirements of some professions;
- increasing student demand;
- staff interest;
- competitive league tables and awards;
- links made by senior management between corporate social responsibility and sustainability;
- new 'real-world' research and associated funding opportunities;
- prospect of financial savings made through better environmental practices;
- perceived marketing and recruitment advantage;
- potential to enrich curriculum and enhance quality of teaching and learning;
- a felt obligation by many academics to their students as the next generation;
- interest by many senior management teams in raising their institution's sustainability performance.

*Benefits*

- ability to better integrate other important priority agendas such as internationalisation, enterprise, employability and diversity;
- opportunity to employ innovative interactive pedagogies, and enhance student learning;
- potential for encouraging transformative education and interdisciplinarity;
- increased engagement with students and recognition of the student voice;
- improved student recruitment;
- financial savings;
- improved risk management;
- opportunities to link curriculum to campus and community;
- capitalising on student and staff interest and motivation;
- better environmental management and enhanced campus;
- enhanced engagement with community and external stakeholders and demonstration of social and environmental responsibility.

---

Reviewing progress, Wals and Blewitt (2010) make a historical distinction between 'first and second wave' responses (i.e. the emergence of environmentally-based courses beginning in the 1970s and 80s, and the greening of estates management from the 90s). They then describe a 'third wave response', based on a holistic appreciation of the potential of sustainability, which can be detected only in very recent years. The latter refers to whole institutional change, 'a university's attempt to re-orient teaching, learning, research and university-community relationships in such a way that sustainability becomes an emergent property of its core activities' (Wals and Blewitt, 2010: 56), and while the authors recognise and cite signs of this wave, they also acknowledge that 'the sector is notoriously resistant to change' (2010: 57). Indeed, research shows that there are some very common barriers and inhibitors to sustainability in higher education, and particularly as regards curriculum (Dawe *et al.,* 2005; Velaquez *et al.*, 2005; Holmberg and Samuelsson, 2006; Policy Studies Institute *et al.*, 2008; Jones *et al.*, 2010). Research for HEFCE (Policy Studies Institute *et al.*, 2008: vii) concluded that barriers to curriculum inclusion, where they exist, essentially amount to:

- lack of interest in sustainable development (SD);
- silo or mono-disciplinary thinking and institutional organisation, which militates against the cross-departmental activity that is essential for SD;
- lack of incentives or priority to engage in SD.

Similarly, research for the Higher Education Academy (Dawe *et al.,* 2005) suggested several factors that prevent academics from engaging with SD in the curriculum, such as:

- a crowded curriculum;
- perceived irrelevance;
- limited staff awareness and/or expertise;
- limited institutional commitment;
- limited commitment from external stakeholders;
- seen as too demanding.

A poll conducted by the GUNI network involving the views of some 200 experts in sustainability in HE worldwide came to similar conclusions, citing the following as the top five barriers: difficulties attaining integrative thinking and inter- and transdisciplinarity; sustainable development perceived as an 'add-on'; lack of support and vision by university leaders; lack of understanding of sustainability education; and lack of vision and policy change by governments (Granados-Sánchez *et al.*, 2012).

   All of the above might be characterised into a typology, consisting of: perceptual factors; policy/purpose-related factors; structural factors (governance, compartmentalisation, incentives etc.); and those relating to resources, funding and information. Clearly, all of these have currency and will often apply to differing degrees at many institutions. Yet, however valid, such an analysis does not go far enough to explain adequately the limited 'response-ability' of higher education over recent years. More broadly – and more deeply – the key to understanding the constraints on the full recognition of sustainability in higher education is *cultural*.

This is operative at two levels. First, externally, universities operate inevitably within a certain socio-economic climate, which heavily influences their values, purposes and culture. The advance of the neo-liberal and marketisation agenda in recent years, endorsed by governments (as noted above), has greatly influenced the language and practice of higher education (see Blewitt, this volume). Its attendant emphasis on competition, performance indicators, league tables, employability, business and 'students as customers' dominates earlier discourses of education, which stress more academic and intrinsic values concerning what it means to be educated, developing critical discourse, cultural enrichment and responsible citizenship (Levidow, 2002; Grove, 2011). At the same time, the greater accountability that higher education now needs to demonstrate to society, and the drive towards more interactive pedagogies, arguably can play in favour of HEIs responding more explicitly to sustainability challenges.

Second, at a deeper level, the cultural brake is *paradigmatic*. Higher education still largely reflects the Western intellectual legacy from whence it came, rooted in the memes of the prevalent education epistemology – reductionism, objectivism, materialism, dualism and determinism underlain by a mechanistic metaphor – refracted from the wider cultural milieu and exerting an influence in purpose, policy and provision, as well as in educational discourse. These habits of thought reside in the subterranean layers of the university culture and manifest in the educational landscape above the surface: hierarchical governance, single disciplines, separate departments, abstract and bounded knowledge, belief in value-free knowing and a reluctance to engage with ethical matters in the curriculum, privileging of cognitive/intellectual and technical knowing over affective and practical knowing, prevalence of instrumental rationality, transmissive pedagogy, linearly arranged learning spaces, valuing of analysis over synthesis and an emphasis on first-order or maintenance learning which leaves basic values unexamined and unchanged both individually and institutionally. Of course, there *are* innovative, interdisciplinary and cross-institutional initiatives in universities, notably in research, but these must often contend with a reductive tradition which makes it harder to take root.

To illustrate this further, Figure 1.1 proposes that at both individual and collective scales, it is possible to describe 'levels of knowing' (Sterling, 2003), based on a systems view of thought (Bohm, 1992).

This model of nesting systems suggests that deeper perceptions and conceptions inform, influence and help manifest more immediate ideas and they, in turn, affect more everyday thoughts and actions. At the same time, the downward arrow depicts that experience in the world can, at times, change deeper levels of perception and knowing. A key point arising from this model is

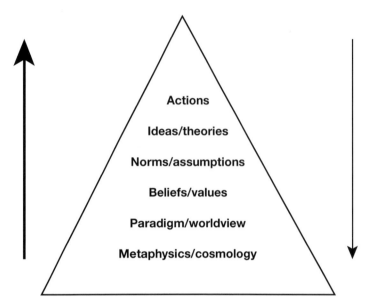

*Figure 1.1* Levels of knowing

that the influence of deeper assumptions often may not be consciously recognised. Assumptions are operative, but may lay largely unexamined (Bawden, 2004). Given that sustainability, as argued above, implies and requires a cultural shift both in society and higher education, this model helps map the 'architecture' of such a shift, which could be envisaged in the model as the whole triangle moving sideways to suggest different sets of knowing (Bateson, 1972), i.e. a changed episteme.

The 'levels of knowing' diagram (Figure 1.1) evokes a parallel model that represents the systemic relationship of the five 'P's of educational institutional culture (Figure 1.2).

As in Figure 1.1, this model suggests that while attention is often focused on the more immediate levels, these are influenced and affected by the deeper levels of orientation. Commonly, an institutional response to sustainability will be to make some changes, say in some courses, at the top end of this spectrum but with little or no effect on the rest of the institution. What lies beneath – that is, *purpose* and *paradigm* – tends to receive less or no attention, as these are part of the consensual 'given'. Yet where change at these deeper levels occurs, it allows (though does not, of course, guarantee) more extensive and coherent change in provision and practice.

The implications of deeper personal and organisational change are further illuminated by Bateson's theory of levels (Bateson, 1972), which has been seminal in organisational learning theory (Watzlawick *et al.,* 1980). Bateson distinguished three orders of learning and change, corresponding with staged

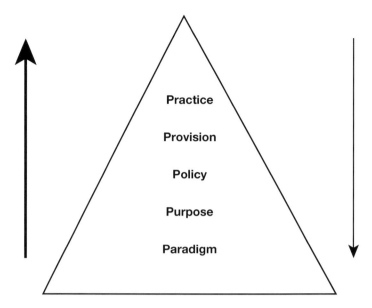

*Figure 1.2* Educational culture: levels of manifestation

increases in learning capacity, and these have been adopted variously by learning and change theorists, particularly in the field of systemic learning and organisational change, such as Argyris and Schön (1996) (single- and double-loop learning), and Ison and Russell (2000) (first-order and second-order change). A key point is that learning can either serve to keep a system stable (i.e. person, group or organisation), or enable it to change to a new state in relation to its environment. Thus Watzlawick *et al.* (1980: 50) distinguish two different types of change: 'one that occurs within a given system which itself remains unchanged, and one whose occurrence changes the system itself'. Most learning that goes on within and outside learning institutions makes no difference at all to individuals' or society's overall paradigm. This is because, in Bateson's model, it is *first order learning* within a consensually accepted framework of assumptions which does not touch on or examine deeper levels of knowing. This is sometimes called basic learning or maintenance learning – adjustments or adaptations are made to keep things stable in the face of change: what Clark calls (1989: 236) 'change within changelessness'. In most cases, this is not in itself a 'bad thing' but a necessary learning response which ensures stability. It becomes maladaptation when first-order change is inappropriate for the scale of the external challenge, in this case the sustainability imperative.

This theory of learning levels helps illustrate the essence of the issue around higher education's limited 'response-ability' to date. As I have commented elsewhere:

There is a double problem here: first, higher education institutions are not primarily reflexive learning systems (learning organisations) but teaching and research systems. Second, higher education is not primarily engaged in the provision of deep learning to students, but in first-order learning: the transmission of information and the development of instrumental skills aligned (increasingly) to the perceived needs of the economy.

(Sterling, 2004: 51)

The common and insufficient understanding of sustainability as requiring an 'add-on' to existing norms reflects an accommodative, first-order response. Where it *is* seen as necessitating an examining, questioning and re-ordering of norms and assumptions, Bateson's second-order learning is reflected. Beyond this, Bateson distinguished a third order of learning involving epistemic or paradigmatic change (affecting the lower layers of levels of knowing, Figure 1.1). This involves 'thinking about and evaluating the foundations of thought itself' (Bawden and Packham, 1993: 6), that is the experience of *seeing our worldview* rather than seeing *with* our worldview so that we can be more open to, and draw upon, other views and possibilities.

Hence, both individual and organisational learning by staff and students can either be *conformative* (first order), or *reformative* (second order), or *transformative* (third order) – a shift towards higher-order learning. While this one sentence is quickly stated, it belies the scale of change involved: a shift of perception from first-order to second-order learning, and particularly from second order to third order, will often involve personal and institutional (and therefore sometimes also a political) resistance on the part of learners, for it poses a significant challenge to existing beliefs and ideas, and interests and involves reconstruction of meaning (Meadows, 2009).

This account of Bateson's theory,[2] while brief, nevertheless provides some insight on the possibilities of, and constraints on, the higher-order learning experience that the crisis of sustainability suggests is necessary, both in terms of worldview change and organisational change (Lyle, 1994; Capra, 2003; Senge *et al.,* 2005; Scharmer, 2009). It also facilitates the mapping of the cultural shift, referred to earlier. *The People's Treaty on Higher Education Towards Sustainable Development* (see Tilbury, this volume) calls for the transformation of higher education. What is missing – and this is surprisingly common in sustainability education discourse – is sufficient mapping of the journey advocated, and by the same token, some insight on why things are as they are. Without deeper analysis and wider discussion of *why* sustainability has made limited progress, the response from educational systems is often likely to be, at most, accommodative first-order change.

By throwing light on the characteristics of our dominant worldview and epistemology, as outlined above, we are in a better position to accept the challenge of transformation. However, taking a whole-systems view of change, we need to consider the cultural milieu that higher education lies within. (This recognises that any transformation in HE needs to be part of a much wider exercise in social learning and reflexivity, and that both HE and society change through a co-evolutionary rather than linear relationship.) As Raskin states:

> The shape of the global future rests with the reflexivity of human consciousness – the capacity to think critically about why we think what we do – and then to think and act differently.
>
> (Raskin, 2008: 469)

So the problem is not *primarily* 'out there', in the many global issues that are increasingly reported, but 'in here', rooted in the underlying beliefs and worldview of the Western mind (Laszlo, 1989; Capra, 1996). There is then, as I have written elsewhere

> a need to recognize the habits of thought and perception that characterize our consciousness, in order to be able to move beyond them ... this will involve an increasing 'second order' knowing or critical reflexivity about the dominant techno-scientific, objectivist, and instrumental rationality that pervades our thinking and exploring a more relational, ecological or participative consciousness appropriate to the deeply interconnected world that we have created. This entails a shift of emphasis from relationships largely based on separation, control, manipulation, individualism and excessive competition towards those based on participation, appreciation, self-organisation, equity, justice, sufficiency and community.
>
> (Sterling, 2009: 108)

This shift – second order, and for some, third order – is reflected in calls for a more holistic, integrative, co-evolutionary, participative, 'living systems', ecological worldview: in sum, an essentially *relational* worldview or sensibility, appropriate to a precarious world characterised by dynamic feedback, complexity and consequence (Reason and Bradbury, 2001; Capra, 2003, Senge *et al.,* 2005; Sterling, 2007; Raskin, 2012). These ideas lend some richness and insight in understanding the 'staged change' process of higher education institutions towards sustainability, from fragmentation to integration (a process encouraged in the 'Learning in Future Environments' (LiFE) index: www.thelifeindex.org.uk). Similarly, HEFCE's strategic review suggests a useful four-phase model of progressive integration of sustainable development in an HEI, the last stage being 'full commitment' which 'sees sustainable development as absolutely integral to the institution' (Policy Studies Institute *et al.*, 2008: 64).

The initial response, which as noted above is all too common, is a nil response. This may be through ignorance of the challenge of sustainability, or misunderstanding or denial of its import. A first-order learning response is adaptive and *accommodative*, a 'bolt-on' of sustainability ideas to the existing system, which otherwise remains unchanged. This is often characterised by 'education about sustainability', which typically occurs in likely subjects, may be found in isolated parts of the university, and may be contradicted by other formal or hidden curriculum messages. While a long way from an organisational shift in culture, this common initial response can sometimes open the door to and spark deeper change, as is demonstrated in several chapters in this volume.

Beyond this, a second-order learning response is *reformative*, a 'build-in' process whereby questioning and reformulating some policies and programmes, and revising some guiding institutional norms in line with sustainability ideas and principles, leads to quite significant institutional change. The cultural shift here is more towards 'education for sustainability' and learning for change.

A third-order learning response involves a deep questioning of educational paradigms, and therefore also purposes, policies and programmes, and a *transformative* redesign process that involves 'learning as change' throughout the educational community. Clearly, this is most difficult to achieve as it involves significant re-thinking of long established norms and structures, yet this kind of fundamental shift is called for by many who critique the current disconnect between higher education and the contextual problematique in wider society, outlined at the outset of this chapter. To illustrate further, this stage

> requires in-depth engagement from within and across a range of disciplinary frameworks. It also requires a transformation in teaching epistemologies, away from transfer oriented epistemologies of certainty, to more complex teaching epistemologies that recognise uncertainty and risk. It also requires a re-orientation of community engagement, from models of re-active responsiveness to existing community need and interests, to more pro-active, preventive engagement with potential future risks and development of competences for adaptive management and risk prevention.
>
> (Lotz-Sisitka, 2010: 113)

The picture is complicated by the fact that different individuals and groups in the same university can be at different stages of learning response, although this can help the organisational change process where some groups act as a catalyst. Examples of universities that have attempted whole institutional change in response to the new conditions of the twenty-first century include Aberdeen, Melbourne, Leuphana Universität Lüneburg, Chalmers (Sweden), and Universiti Sains Malaysia, the latter three specifically focusing on sustainability as a guiding principle (Razak *et al.,* 2010; see Ryan, 2012 for further examples). One interesting illustrative experiment dates from some years back when the AFANet European network of agricultural educational institutions sought to explore the implications of sustainability by developing a number of projects to integrate sustainability concepts and practices between 1997 and 2000. Their experience is summarised in Box 1.3.

**Box 1.3: Six lessons about sustainability and transformative learning from AFANet**

1 **Integrating sustainability presupposes the re-thinking of institutional missions**
The integration of sustainability will never lead to anything fundamentally new if the institution is not prepared to re-think its academic mission.

2 **It is no use crying over vague definitions**
Sustainability is a non-prescriptive concept, which needs to become meaningful in a specific context. Its non-specific, imprecise nature can be seen as an advantage in stimulating dialogue on meaning and implications for curriculum, pedagogy and so on.

3 **Sustainability is as complex as life itself**
It is related to the social, economic, cultural, ethical and spiritual domain of our existence. It differs over time and space and it can be discussed at different levels of aggregation and viewed through different windows. It does not lend itself to unilateral, linear planning or a reductionist scientific paradigm and thus involves the systemic integration between theory and practice into systemic praxis.

4 **Teaching about sustainability requires the transformation of mental models**
It includes deep debate about normative, ethical and spiritual convictions and directly relates to questions about the destination of humankind and human responsibility. In this way, it differs from a modernist and positivistic way of thinking.

5 **There is no universal remedy for programmatic reconstruction**
The possibility of innovation depends on the cultural and academic context. There is then, no panacea for curricular reform, no blueprint for change in education towards sustainability. Therefore, change must be based on an inclusive, participative approach.

6 **Programming sustainability demands a re-thinking of teaching and learning**
Reorientation requires ample opportunity for staff members and students to embark on new ways of teaching and learning, the opportunity to re-learn their way of teaching and learning, and to re-think their mutual relationships.

(Based on and adapted from Wals and Bawden, 2000: 20)

In sum, the notion of the sustainable university presents a challenge which is both pressing and difficult to realise because it involves deep learning across whole educational communities and amongst policy-makers.

More than ten years ago, I first outlined the basis and possibility of a changed educational paradigm or culture, based on holistic/ecological or systemic thinking (Sterling, 2001, 2003, 2009), which I deliberately named '*sustainable education*', to distinguish it from 'education for sustainable development (ESD)'. This might seem an unnecessary distinction to some, but I would argue that in seeking paradigm change, the 'education for sustainable development' discourse is both helpful *and* unhelpful: helpful because it has over the last twenty years developed significant debate and a degree of consensus around the characteristics of education oriented toward sustainability (see Box 1.4); unhelpful in the sense that boundaries tend to be placed – sometimes by advocates but often by others – around what constitutes and what does not constitute ESD. This carries the

---

**Box 1.4: Essential characteristics of education for sustainable development**

Education for Sustainable Development:

- is based on the principles and values that underlie sustainable development;
- deals with the well-being of all three realms of sustainability – environment, society and economy;
- promotes lifelong learning;
- is locally relevant and culturally appropriate;
- is based on local needs, perceptions and conditions, but acknowledges that fulfilling local needs often has international effects and consequences;
- engages formal, non-formal and informal education;
- accommodates the evolving nature of the concept of sustainability;
- addresses content, taking into account context, global issues and local priorities;
- builds civil capacity for community-based decision-making, social tolerance, environmental stewardship, adaptable workforce and quality of life;
- is interdisciplinary: no one discipline can claim ESD as its own, but all disciplines can contribute to ESD;
- uses a variety of pedagogical techniques that promote participatory learning and higher-order thinking.

(UNESCO, 2005)

attendant danger that it is put into a box both mentally and in practice, alongside – but not necessarily relating to – other education-for-change movements whether development education, peace education, health education, human rights education, futures education, global citizenship education, education for all, or indeed any current discipline or subject seeking to respond to sustainability. So there remains confusion in the higher education sector about the qualitative difference between, on the one hand, 'embedding sustainable development in education' – most often an accommodatory add-on response that does not necessarily impinge on or challenge existing norms – and, on the other, a reformative or transformative change as a more holistic response involving cultural change and whole institutional shifts (Sterling and Scott, 2008).

Intended to help evoke the latter response, the 'sustainable education' label invites people to shift their sole focus from 'how do we educate for sustainability or for sustainable development?' towards deep attention to *education* itself – its paradigms, policies, purposes and practices, and its *adequacy* for the age we find ourselves in. I saw this shift of focus as

> one which develops and embodies the theory and practice of sustainability in a way which is critically aware. It is therefore a transformative paradigm which values, sustains and realises human potential *in relation to* the need to attain and sustain social, economic and ecological wellbeing, recognising that they must be part of the same dynamic.
>
> (Sterling, 2001: 22)

Sustainable education implies four descriptors: educational thinking, policy and practice that are sustaining, tenable, healthy and durable:

- *Sustaining*: it helps sustain people, communities and ecosystems.
- *Tenable*: it is ethically defensible, working with integrity, justice, respect and inclusiveness.
- *Healthy*: it is itself an adaptive, viable system, embodying and nurturing healthy relationships and emergence at different system levels.
- *Durable*: it works well enough in practice to be able to keep doing it.

John Blewitt (2012) has argued that this paradigm shift is further away than ever. He might be right, but he may also be wrong (or perhaps partly right, partly wrong). The economistic and neoliberal forces he cites, noted earlier in this chapter, exert a powerful and limiting influence on the scope and accepted purposes of higher education. At the same time, there is a discernible awakening among key stakeholders – including students and enlightened business – to the severity of the contemporary socio-economic and ecological conditions we face, to the extent that there may be the possibility of new dialogue and synergy between the instrumental, skills-based, business-facing agenda, and the holistic, sustainability-oriented view of education, where the former is broadened and the latter sharpened. For example, in leading-edge thinking in economics and business, there is an increasing recognition that significant change is upon us, whether related to the effects of climate change, resource depletion, or peak oil, to name some of the critical factors coming into play. As the United Nations Secretary-General's High-level Panel on Global Sustainability states:

> To achieve sustainability, a transformation of the global economy is required. Tinkering at the edges will not do the job.
>
> (UNSGHP, 2012: 48)

In the UK, the Ellen MacArthur Foundation (2012) is supporting the systemic vision of the circular sustainable economy reflecting growing international engagement with concepts such as steady state, degrowth and cradle-to-cradle economics (Daly and Cobb, 1990; Seyfang, 2010; McDonough and Braungart, 2002; Jackson, 2009). Again, the importance of education in supporting this radical shift is acknowledged. All this raises a question as to how far students are being prepared for, and equipped to handle, the probability of very radically changed economic conditions and patterns.

The future that graduates face will require them to be resilient, resourceful, enterprising, sustainability literate, adaptable, imaginative, creative, caring, able to think critically and systemically, and both self-reliant and collaborative learners. These are the kind of qualities, competencies and capabilities that some universities are beginning to reflect in their 'graduate attributes' statements, but sustainability adds immediacy, purpose, application and direction.

While it might be unfair to pick out one exemplar, the University of British Columbia (UBC) stands out as an illustration of how a mission statement can be presented in this light:

> The graduates of UBC will have developed strong analytical, problem-solving and critical thinking abilities; they will have excellent research and communication skills; they will be knowledgeable, flexible, and innovative. As responsible members of society, the graduates of UBC will value diversity, work with and for their communities, and be agents for positive change. They will acknowledge their obligations as global citizens, and strive to secure a sustainable and equitable future for all.
>
> (UBC, 2012)

Further, the UBC Sustainability Initiative's Teaching and Learning Office statement *Transforming Sustainability Education at UBC: Desired Student Attributes and Pathways for Implementation* (USI, 2011), sets out a bold and detailed vision for the undergraduate curriculum around four key student attributes: holistic systems thinking, sustainability knowledge, awareness and integration, and acting for positive change.

The experience of my university, Plymouth, where some of the ideas discussed above have been applied, adapted and developed is briefly reviewed next to give an illustrative example of attempts to transform a university, and to provide some insight into our approach that might be of assistance to others in this journey – alongside other chapters in this volume. To do this necessitates giving some background story.

## Towards the sustainable university: a view of Plymouth

As noted in the Preface, Plymouth University has much to commend it in terms of its sustainability record. It has emerged as overall top performer in the whole UK sector in the People and Planet Green League in the five years since its inception in 2007, won the Green Gown Award in 2011 for 'Continuous Improvement – Whole Institutional Change', and its students union, UPSU, won gold status in 2012 in the National Union of Students' Green Impact scheme. But our staff, students and the wider academic world would be mistaken in thinking that Plymouth is now a sustainable university in its full sense. Indeed, the danger of such awards is in giving the sense that, first there is a final end point or state, and second, that it has been achieved, and therefore there is little more to be done – a concern that was aired by some winners at the Green League Award Ceremony held in London in June 2012. Rather, a long and radical journey is involved, and Plymouth is simply further along the road than many others, while some HEIs – particularly internationally – are ahead of Plymouth. Our particular journey thus far is now outlined.

In 2005, Plymouth won four significant funding awards from HEFCE to set up Centres for Excellence in Teaching and Learning (CETLs). One of these bids was to establish the Centre for Sustainable Futures (CSF), one of only two CETLs nationally having a specific sustainability focus (see Taylor, this volume, for reference to the Kingston CETL). The mandate that CSF set out, and which was endorsed by senior management, was

> [t]o transform the University of Plymouth from an institution characterised by significant areas of excellence in Education for Sustainable Development (ESD) to an institution modelling university-wide excellence and, hence, able to make a major contribution to ESD regionally, nationally and internationally.
>
> (Dyer and Selby, 2004: 1)

From the start, the vision was to develop a strategy and activities that could transform the university towards a state where sustainability permeated the curricula, physical campus, and the whole institutional culture, as well as influencing relations with its immediate environs, and the wider region, and contribute to similar work across the sector nationally and internationally. Importantly, the core team of eight staff shared this bold vision, and were also informed philosophically by holistic and ecological thinking, and by organisational change theory, underpinning a belief in the need for a systemic approach to institutional change (Argyris and Schön, 1996; Banathy, 1991; Bawden, 1997; Meadows, 2002; Meadows, 2009).

Early on, we developed what became known as the '4C' model to represent a *gestalt* view of the whole (which has since been taken up by a number of institutions nationally and internationally). With this framework in mind, the CETL-funded years of CSF (2005–10) were built upon an holistic '4C' approach, seeing *Curriculum,* learning and teaching (and related research), *Campus* change, and *Community* engagement as mutually embedded and enhancing spheres and, as such, powerful contributors to the student learning experience, all encircled and in relation to a fourth C, *Culture,* as reflected in institutional values, policies and practices (Selby, 2009) (see Figure 1.3).

With the four interrelated foci of the C model in mind, CSF's methodology and working style emerged – summarised by six more Cs. Figure 1.3 indicates that the *process* we used was primarily about *connection* – particularly connecting isolated individuals and parts of the university that otherwise had little or no contact in order to develop working links and synergies; about *communication* of information, experience, support, and feedback to combat isolation and build a sense of being part of a wider movement; and about facilitating *collaboration* to inspire joint working and new initiatives. In many cases, the existence of CSF allowed staff to pursue their own enthusiasms and network more widely, knowing they were part of a bigger movement – which grew over time.

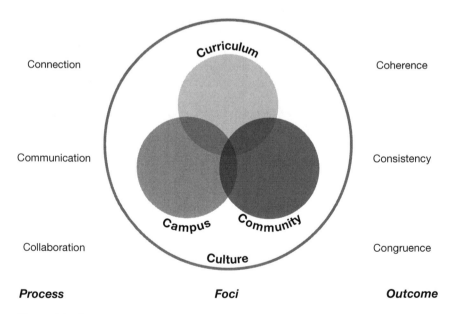

*Figure 1.3* The 'C' model at Plymouth University

The *outcomes* we sought were *coherence,* giving a sense to the university community that the initiative was substantial, made sense conceptually and intellectually and had added value; *consistency* in terms of cross-institutional working and demonstrable synergies across the 4Cs; and *congruency* so that the results of the work were in line with what we understood to be the requirements of a more sustainable university. One 'C' we were *not* seeking was 'conformity' or homogeneity, and following organisational change theory we sought to facilitate diverse and creative responses to sustainability at all levels of the university system. Further, we made a distinction between what we saw as the 'systematic' and 'systemic' aspects of the university: the former being the policies, strategies, rules, procedures, assessment, evaluation, structures etc. where CSF adopted a *catalytic* role, for example in developing through consultation the university's first holistic sustainability strategy (which was approved in 2008). The latter was the 'glue' of the organisation, the collegiality, social learning and exchange, informal networks, ethos, self-organisation, where CSF adopted an *enabler* or facilitative role. It became clear that both aspects – the 'hardware' and the 'software' – were important, and that we needed to work on both at the same time. We also, early on, made the important distinction (see Introduction, this volume) between *designed learning,* being the formal curriculum, and *organisational learning,* sufficient progress in the latter being absolutely critical to facilitating significant progress in the former.

In essence, CSF staff and colleagues attempted to develop and grow a sustainability-oriented 'critical, connective, and collective intelligence', through an open and invitational approach; to make horizontal connections and communication between faculties and between academic and support staff and initiatives; and to plan and execute initiatives with an eye to their systemic effects and potential, monitoring progress periodically through developing and updating a whole institutional sustainability SWOT analysis (Strengths, Weaknesses, Opportunities, Threats).

A key part of this was a Centre Fellow scheme involving (eventually) nearly 50 academic staff working across all schools and faculties, spearheading curriculum change, associated research, and organising sustainability-related CPD events for both academic and administrative staff.

Three years into the CETL, CSF decided to self-appraise progress. Based on international literature we came up with sixteen criteria:

1   sustainability vision – policy statement in corporate plan;
2   whole institution strategy and action plan;
3   senior manager with known responsibility for implementation;
4   senior executive committee;
5   regular sustainability and environmental auditing;
6   sustainability applied to all aspects of campus operation;
7   ethical investment policy;
8   excellent internal communication;
9   excellent external communication of sustainability message;

10    holistic perception and management of 4 Cs;
11    embedding sustainability in formal and informal learning of students;
12    sustainability principles and pedagogy in learning and teaching policy;
13    university sustainability research centre and research strategy;
14    culture of organisational learning and improvement;
15    concern for well-being of the whole community as well as achievement;
16    engaged with local community.

Our evaluation showed that the university was making good progress on most counts. Without listing and describing achievements, the key point about the 2005–10 period is that it shifted the university sufficiently that sustainability became widely recognised by senior management, staff and students as one of Plymouth's platforms and part of its identity. Very importantly, it laid the foundations for the post-CETL stage, and cemented commitment where it mattered.

Briefly, from 2010 on, a revised whole institution sustainability strategy was put in place; a revised teaching and learning strategy included sustainability as one of the key themes; a new Office of Procurement and Sustainability was established; a new research institute (Institute for Sustainability Solutions Research, ISSR) was launched – complementing the existing Marine Institute; a pedagogic research institute (PedRIO) was launched, with sustainability education research as a key strength; a Sustainability Executive and a cross-university Sustainability Advisory Group was initiated; the ISO 140001 environmental management standard was secured; a sustainability curriculum review was undertaken; a comprehensive Sustainability Report was developed (OVC, 2011) involving a broad range of stakeholders, and many other initiatives put in place. Sustainability synergies have been created across the university with research-enriching operational strategies whether carbon management, food, travel or water, and there has been close engagement with the student community wherever possible, recently enhanced by a key 'Students as Partners' initiative. Further, post-CETL, CSF has been funded by the university to continue its work, specialising more on curriculum support and pedagogic research related to sustainability education, under the auspices of PedRIO. This includes working closely with specific schools across the university.

Looking back, and comparing the situation with 2005, it is clear that an important degree of systemic change has been achieved. Certainly, we can claim that some individuals have experienced a degree of second-order change as a result of their involvement in this work, which has affected their whole outlook on teaching and learning. It would be harder to make transformational claims about the university as a whole, but there *have* been sufficient shifts in the systematic and systemic aspects of university operation and life to believe that sustainability is firmly embedded. The result of seven years' work has been to build a broad sustainability community among staffs and students (for example the sustainability institute ISSR has over 200 academics involved in sustainability research), which in different ways affects the whole institution. At the same time, external recognition provides a vital feedback loop, which helps secure further progress. But despite Plymouth's national and international reputation, this work only goes so far. There are parts of the university unaffected by and probably unaware of this work. Given the global challenges reviewed briefly in this chapter, there is much more to be done. In addition, progress cannot be taken for granted but must be striven for; staff and student populations change, and universities are under such pressures that progress on sustainability can slip backwards.

## Conclusion

In this chapter, I have sought to make a case for higher education to lift its attention and expand its sphere of concern and interest to embrace the critical issues and conditions that will face graduates and their offspring throughout their lives. These are an unprecedented set of life-affecting trends and factors to which universities could make a positive, radical difference. Conversely, they will have a net negative effect if this 'different world' is not taken into full account in planning, policy and programmes, and if the purposes of higher education are not re-thought accordingly. But assuming so, perhaps the next Browne-type review will change one word in its title, and be known as 'Securing a Sustainable Future *through* Higher Education'.

## Notes

1    For further information on the Great Transition Initiative see www.GTinitiative.org.
2    For more detail on Bateson's learning levels, see Sterling, 2003, 2004.

## References

Argyris, C and Schön, D (1996) *Organisational Learning II*, New York: Addison Wesley
Banathy, B (1991) *Systems Design of Education*, New Jersey: Educational Technology Publications
Bateson, G (1972) *Steps to an Ecology of Mind*, San Francisco: Chandler
Bawden, R (1997) 'Leadership for systemic development', in Centre for Systemic Development, *Resource Manual for Leadership and Change*, Hawkesbury: University of Western Sydney
Bawden, R (2004) 'Engagement, reflexive scholarship, and the learning turn within the academy', Paper presented at the ALARPM conference, South Africa
Bawden, R and Packham, R (1993) 'Systemic praxis in the education of the agricultural systems practitioner', *Systems Practice*, 6: 7–19
Ben-Ze'ev, A (2000) *The Subtlety of Emotions*, Cambridge, MA: Massachusetts Institute of Technology
Blewitt, J (2012) *Radicalising Education for Sustainability*, A Schumacher Institute Challenge Paper, Bristol: Schumacher Institute
Bohm, D (1992) *Thought as a System*, London: Routledge

Boyer, E L (1996) 'The scholarship of engagement', *Journal of Public Service and Outreach,* 1: 11, in Bawden, R (2004) 'Engagement, reflexive scholarship, and the learning turn within the academy', Paper presented at the ALARPM conference, South Africa

Broadfoot, P (2010) 'Sustainability and responsibility in higher education institutions', Conference presentation, *Higher Education and the Sustainability Challenge,* 22 June 2010, Grosvenor Hotel, London

Brown, V, Harris, J and Russell J (2010) *Tackling Wicked Problems: Through the Trandisciplinary Imagination,* London: Earthscan

Browne, J (chair) (2010*) Securing a Sustainable Future for Higher Education: An Independent Review of Higher Education Funding and Student Finance.* Available at www.bis.gov.uk/assets/biscore/corporate/docs/s/10-1208-securing-sustainable-higher-education-browne-report.pdf (accessed 25 May 2012)

Capra, F (1996) *The Web of Life,* London: Harper Collins   Capra, F (2003) *The Hidden Connections,* London: Flamingo

Clark, M (1989) *Ariadne's Thread: The Search for New Ways of Thinking,* Basingstoke: Macmillan

Collini, S (2010) 'Browne's gamble', *London Review of Books,* 32 (21), 4 November 2010: 23–5

Cortese, A D (2003) 'The critical role of higher education in creating a sustainable future', *Planning for Higher Education,* 31 (3): 15–22

Diamond, J (2005) *Collapse: How Societies Choose to Fail or Survive,* London: Allen Lane

Daly, H and Cobb, J (1990) *For the Common Good,* London: Greenprint Press

Dawe, G, Jucker, R and Martin, S (2005) *Sustainable Development in Higher Education: Current Practice and Future Developments.* York: Higher Education Academy. Available at: www.heacademy.ac.uk/resources/detail/sustainability/dawe_report_2005 (accessed 25 May 2012)

Dyer, A and Selby, D (2004) *Centre for Excellence in Teaching and Learning Education for Sustainable Development: Stage Two,* Plymouth: University of Plymouth

Ellen MacArthur Foundation (2012) *Towards the Circular Economy: Economic and Business Rationale for an Accelerated Transition,* Cowes: Ellen MacArthur Foundation, www.thecirculareconomy.org

Fear, F, Rosaen, C, Bawden, R, and Foster-Fishman, P (2006) *Coming to Critical Engagement,* Lanham, Maryland: University Press of America

Gough, S and Scott, W (2008) *Higher Education and Sustainable Development: Paradox and Possibility,* London: Routledge

Granados-Sánchez, J, Wals, A, Ferrer-Balas, D, Waas, T, Imaz, M, Nortier, S, Svanstrom, M, Van't Land, H and Arriaga, G (2012) 'Sustainability in higher education: moving from understanding to action, breaking barriers for transformation', in Global Universities Network for Innovation (GUNI) (2012) *Higher Education's Commitment to Sustainability: From Understanding to Action,* World in Higher Education Series No. 4, Barcelona, Spain: GUNI

Grove, J (2011) 'Sector must reject neoliberal business-speak, event hears', *Times Higher Education,* 1 December 2011. Available at www.timeshighereducation.co.uk/story.asp?storycode=418295 (accessed 24 May 2012)

Holmberg, J, and Robèrt, K-H (2000) 'Backcasting from non-overlapping sustainability principles: a framework for strategic planning', *International Journal of Sustainable Development and World Ecology,* 7: 291–308

Holmberg, J and Samuelsson, B (eds) (2006) *Drivers and Barriers for Implementing Sustainable Development in Higher Education,* Göteborg Workshop, 7–9 December *2005* (61–67), Paris: UNESCO (Education for Sustainable Development in Action; Technical Paper No. 3). Available at: http://unesdoc.unesco.org/images/0014/001484/148466E.pdf (accessed 23 May 2012)

Homer-Dixon, T (2006) *The Upside of Down: Catastrophe, Creativity and the Renewal of Civilisation,* London: Souvenir Press

Homer-Dixon, T (2012) *Our Panarchic Future,* Washington DC: Worldwatch Institute. Available at: www.worldwatch.org/node/6008 (accessed 26 June 2012)

Jackson, T (2009) *Prosperity Without Growth: The Transition to a Sustainable Economy,* London: Earthscan

Kneale, P and Maxey, L (2012) 'Creating the Life You Want: Lifelong Professional Development for Geographers', in Solem, M, Foote, K and Monk, J (eds), *Practising Geography: Careers for Enhancing Society and the Environment,* Upper Saddle River, NJ: Prentice-Hall

International Council for Science (2012) *State of the Planet Declaration,* Planet Under Presssure: New Knowledge Towards Solutions, 26–29 March, London. Available at: www.planetunderpressure2012.net/pdf/state_of_planet_declaration.pdf (accessed 25 May 2012)

Ison, R and Russell, D (2000) *Agricultural Extension and Rural Development: Breaking Out of Traditions, A Second-order Systems Perspective,* Cambridge: Cambridge University Press

Jones, P, Selby, D and Sterling, S (2010) 'Introduction', in Jones, P, Selby, D and Sterling, S (2010) *Sustainability Education: Perspectives and Practice across Higher Education.* London: Earthscan

Laszlo, E (1989) *The Inner Limits of Mankind,* London: One World

Levidow, l (2002) 'Marketizing Higher Education: Neoliberal Strategies and Counter-Strategies', in Robins, K and Webster, F (eds) *The Virtual University? Knowledge, Markets and Management,* Oxford: Oxford University Press, pp. 227–48

Lotz-Sisitka, H (2010) 'Knowledge questions associated with the public health and climate change relation: Some implications for universities in Southern Africa', in Kotecha, P, *Climate Change, Adaptation and Higher Education: Securing Our Future,* Sarua Leadership Dialogue Series, Volume 2, no. 4, Wits, South Africa: South African Regional Universities Association

Lyle, J (1994) *Regenerative Design for Sustainable Development,* New York: John Wiley

McDonough, W and Braungart, M (2002) *Cradle to Cradle: Remaking the Way we Make Things,* San Francisco: North Point Press

Meadows, D (2002) 'Dancing with Systems', *The Systems Thinker,* 13 (2) (March 2002). Available at: www.sustainer.org/pubs/Dancing.html (accessed 25 June 2012)

Meadows, D (2009) *Thinking in Systems: A Primer,* London: Earthscan

Meadows, D H, Meadows, D L, and Randers, J (2005) *Limits to Growth: The 30-year Update,* London, Earthscan

M'Gonigle, M and Starke, J (2006) *Planet U. Sustaining the World, Reinventing the University,* New Society Publishers: British Columbia

NEF (2008) *Triple Crunch –Joined up Solutions to Financial Chaos, Oil Decline and Climate Change to Transform the Economy,* London: New Economics Foundation

OECD (2010) *OECD launches first global assessment of higher education learning outcomes.* Available at: http://globalhighered.wordpress.com/2010/01/28/oecd-launches-first-global-assessment/ (accessed 24 May 2012)

Oerlemans, N, McLellan, R and Grooten, M (eds) (2012) *WWF Living Planet Report 2012*, Gland, Switzerland: WWF. Available at: http://assets.wwf.org.uk/downloads/lpr_2012_summary_booklet_final_7may2012.pdf (accessed 15 June 2012)

Orr, D (1994) *Earth in Mind: On Education, Environment and the Human Prospect*, Washington: Island Press

OVC (2011) *Sustainability Report 2011*, Office of the Vice-Chancellor, Plymouth: Plymouth University. Available at www1.plymouth.ac.uk/sustainability/environmentandsocialresponsibilities/Documents/University%20of%20Plymouth%20Sustainability%20Report%202011%20v.13.2%20Final.pdf

Peccei, A (1982) *One Hundred Pages for the Future*, London: Futura

Policy Studies Institute / CREE / PA Consulting (2008) *Strategic Review of Sustainable Development in Higher Education in England*, London: Policy Studies Institute. Available at: www.hefce.ac.uk/data/year/2008/hefcestrategicreviewofsustainabledevelopmentinhighereducationinengland/ (accessed 24 May 2012)

Raskin, P (2008) 'World lines: A framework for exploring global pathways', *Ecological Economics*, 65, 461–70

Raskin, P (2012) 'Higher education in an unsettled world: handmaiden or pathmaker?', in Global Universities Network for Innovation (GUNI) (2012) *Higher Education's Commitment to Sustainability: From Understanding to Action*, World in Higher Education Series No. 4, Barcelona, Spain: GUNI

Razak, D, Hamid, Z, Sanusi, Z, and Koshy, K (2010) 'Transforming higher education for a sustainable tomorrow: a case of learning by doing at Universiti Sains Malaysia', in Witthaus, M, McCandless, K and Lambert, R (eds) *Tomorrow Today*, Leicester: Tudor Rose on behalf of UNESCO

Reason, P and Bradbury, H (2001) 'Introduction: Inquiry and Participation in a Search of a World Worthy of Human Aspiration', in Reason, P and Bradbury, H (eds), *Handbook of Action Research: Participative Practice and Enquiry*, London: Sage Publications, pp. 1–14

Royal Society (2012) *People and the Planet*, The Royal Society Science Policy Centre Report 01/12, London: The Royal Society. Available at: http://royalsociety.org/policy/projects/people-planet/report/ (accessed 1 June 2012)

Ryan, A (2012) *Education for Sustainable Development and Holistic Curriculum Change: A Review and Guide*, York: Higher Education Academy. Available at: www.heacademy.ac.uk/assets/documents/esd/ESDandHolisticCurriculumChange_review_and_guide.pdf (accessed 26 June 2012)

Ryan, A (2012) *Education for Sustainable Development and Holistic Curriculum Change: A Review and Guide*, York: Higher Education Academy. Available at: www.heacademy.ac.uk/assets/documents/esd/ESD_Artwork_050412_1324.pdf (accessed 27 June 2012)

Scharmer, O (2009) *Theory U: Leading from the Future as it Emerges: The Social Technology of Presencing*, San Francisco: Berrett-Koehler Publishers

Selby, D (2009) 'Towards a sustainability university', *Journal of Education for Sustainable Development*, 3 (1): 103–6

Seyfang, G (2010) 'Community action for sustainable housing: Building a low-carbon future', *Energy Policy*, 38: 7624–33

Senge, P, Scharmer, O, Jaworski, J and Flowers, B, (2005) *Presence: Exploring Profound Change in People, Organisations, and Society*, London: Nicholas Brearley.

Sterling, S (2001) *Sustainable Education: Re-Visioning Learning and Change*, Schumacher Society Briefing no. 6, Dartington: Green Books

Sterling, S (2003) *Whole Systems Thinking as a Basis for Paradigm Change in Education: Explorations in the Context of Sustainability*. PhD diss., Centre for Research in Education and the Environment, University of Bath. Available at: www.bath.ac.uk/cree/sterling/sterlingthesis.pdf (accessed 15 June 2012)

Sterling, S (2004) 'Higher education, sustainability and the role of systemic learning', in Corcoran, P B and Wals, A E J (eds), *Higher Education and the Challenge of Sustainability: Contestation, Critique, Practice, and Promise*, Netherlands: Kluwer Academic

Sterling, S (2007) 'Riding the storm: towards a connective cultural consciousness', in Wals, A (ed.) *Social Learning Toward a More Sustainable World: Principles, Perspectives, and Praxis*, Netherlands: Wageningen Academic

Sterling, S (2009) 'Sustainable Education', in Gray, D, Colucci-Gray, L and Camino, E, *Science, Society and Sustainability: Education and Empowerment for an Uncertain World,* Abingdon: Routledge, pp. 105–88

Sterling, S and Scott, W (2008) 'Higher education and ESD in England: A critical commentary on recent initiatives', *Environmental Education Research,* 14 (4): 386–98

Steuer, N and Marks, N (2008) *University Challenge: Towards a Well-being Approach to Quality in Higher Education,* London: New Economics Foundation

Taylor, P (2011) *Higher Education Curricula for Human and Social Development.* Available at: http://upcommons.upc.edu/revistes/bitstream/2099/8118/1/taylor.pdf (accessed 15 June 2012)

UBC (2012*) Documentation: Guide to Teaching for New Faculty at UBC/Welcome to Teaching at UBC.* Available at: http://wiki.ubc.ca/Documentation:Guide_to_Teaching_for_New_Faculty_at_UBC/Welcome_to_Teaching_at_UBC (accessed 28 June 2012)

UNEP (2012) *21 Issues for the 21st Century: Result of the UNEP Foresight Process on Emerging Environmental Issues,* Alcamo, J and Leonard, S (eds.), Nairobi, Kenya: United Nations Environment Programme (UNEP). Available at: www.unep.org/publications/ebooks/foresightreport/ (accessed 10 May 2012)

UNESCO (2005) *International Implementation Scheme (IIS) for the DESD.* Paris: UNESCO. Available at: http://unesdoc.unesco.org/images/0014/001486/148654e.pdf (accessed 25 June 2012)

United Nations Population Fund, (UNFPA) (2011) *State of World Population 2011: People and Possibilities in a World of 7 Billion*, New York: UNFPA. Available at: http://foweb.unfpa.org/SWP2011/reports/EN-SWOP2011-FINAL.pdf (accessed 12 May 2012)

United Nations Secretary-General's High-level Panel on Global Sustainability (2012) *Resilient People, Resilient Planet: A Future Worth Choosing,* New York: United Nations. Available at: www.un.org/gsp/report/ (accessed 15 June 2012)

United Nations (1992) 'Promoting Education, Public Awareness and Training', Chapter 36, *Agenda 21,* UN Department of Economic and Social Affairs, Division for Sustainable Development. Available at: www.un.org/esa/dsd/agenda21/res_agenda21_36.shtml (accessed 14 June 2012)

USI Teaching and Learning Office (2011) *Transforming Sustainability Education at UBC: Desired Student Attributes and Pathways for Implementation.* Available at: www.sustain.ubc.ca/sites/sustain.ubc.ca/files/uploads/pdfs/Sustainability%20Attributes_August%202011_FINAL%20%282%29.pdf (accessed 28 June 2012)

UUK Sustainable Development Task Group (2010) *A University Leaders' Statement of Intent on Sustainable Development,* London: UUK. Available at: www.eauc.org.uk/ universities_uk_statement_of_intent (accessed 16 May 2012)

UUK (2011) *Futures for Higher Education: Analysing Trends. Higher Education: Meeting the Challenges of the 21st Century.* London: UUK. Available at: www.universitiesuk.ac.uk/Publications/Documents/FuturesForHigherEducation.pdf (accessed 16 May 2012)

Velaquez, L, Munguia, N and Sanchez, M (2005) 'Deterring sustainability in higher education', *International Journal of Sustainability in Higher Education,* 4 (6): 383–91

Wals, A (2010) 'Message in a bottle: learning our way out of unsustainability', Inaugural lecture, Wageningen University, 27 May 2010. Available at: www.lne.be/themas/ natuur-en-milieueducatie/algemeen/edo/docs/inaugurele-rede-prof.-dr.-ir.-arjen-wals (accessed 15 May 2012)

Wals, A and Bawden, R (2000) *Integrating Sustainability into Agricultural Education – Dealing with Complexity, Uncertainty, and Diverging Worldviews,* Interuniversity Conference for Agricultural and Related Sciences in Europe (ICA), University of Gent, Gent

Wals, A and Blewitt, J (2010) 'Third-wave sustainability in higher education', in Jones, P, Selby, D and Sterling, S (2010) *Sustainability Education: Perspectives and Practice Across Higher Education,* London: Earthscan

Watzlawick, P, Weakland, J H and Fisch, R, (1980) 'Change', in Lockett, M and Spear, R (eds) (1980) *Organizations as Systems,* Milton Keynes: The Open University Press

WEF (World Economic Forum) (2012) *Global Risks 2012,* Seventh Edition, An Initiative of the Risk Response Network, WEF. Available at: www3.weforum.org/docs/WEF_ GlobalRisks_Report_2012.pdf (accessed 25 May 2012)

# 2 EfS

## Contesting the market model of higher education

*John Blewitt*

## Introduction

For a moment, it seemed, another world might just be possible. The *greed is good, no such thing as society, the business of business is business* lie, *loadsamoney* culture and end-of-history proclamations that had enabled the ideology of neoliberalism to pervade virtually every aspect of public policy, every social and economic practice, every major private and public institution, seemed to be at an end. Neoliberalism, understood as the relentless and fundamentalist promotion of market forces as morally and practically essential to capitalist economic development, had faltered – perhaps fatally. In 2008 capitalism was not only facing yet another of its periodic cyclical crises, necessary for its restructuring and renewal, but was actually confronting a potential meltdown of its values, its rationale and its own propaganda. Richard A. Posner, a leading and highly respected figure among advocates of the free market as well as being a US federal judge and Professor of Law at the University of Chicago, publicly argued that the crisis was the logical outcome of unregulated market capitalism (Posner, 2010, 2011). The prospect of endless wealth had been transformed to one of disaster and catastrophe. The financial scams that had engulfed Enron, the speculative real estate bubble, the reckless expansion of consumer credit and the global fall-out following the collapse of Lehman Brothers required a rethink, retrieval and re-evaluation of ideas, theories and propositions that had earlier been gleefully thrown into the trashcan of history. The financial storms had blown off the lid and the opening decade of the twenty-first century saw the work of Keynes and, indeed, Marx being taken seriously once again. The iconic status of the great business gurus such as Peter Drucker and Milton Friedman all came under a close, intense and critical scrutiny ... at least for a moment.

Even the Green movement was seen by some sections of what C W Mills (1956) identified as the military-corporate-government-academic power elite, to be offering an analytic and a practical political alternative that might just be credible if the worst came to the worst. Drawing on some ideological constructions of the 1930s, the Green New Deal articulated a system reform that privileged environmental sustainability, ecological economics and participatory and, in some variants, direct democracy rather than corporate defined business as usual (Green New Deal Group, 2008). Education for Sustainability (EfS) would be an important element in shaping this expectant reality. However, the moment soon passed. The Green New Deal lacked ideological purchase, and actually existing, that is to say, institutionalised EfS continued to demonstrate a weakness that came with decades of accommodation, compromise and collusion. EfS had played its part in making higher education an auditable commodity, and had been party to the commercialisation of research, the scramble for private sector partnerships, corporate sponsorship and new ways of competing effectively in the global marketplace. (Self-interested individualism played a part, too.) In the struggle to be taken seriously by institutional managers, policy-makers and business leaders, EfS was seduced by the illusory prospect of making a real difference. Thus, in collaborating with the myth of efficiency, quantification, targets, performance indicators, attributes, strategies, action plans, work plans, outputs, outcomes, specifications, institutionalised visions, missions and so on has meant that EfS has become an instrument of a managerialist culture and governmentality that has compromised its radical potentialities. Additionally, in aligning itself with the green skills and employability agenda EfS has so far failed in practice to rearticulate the principle aims of the capitalist enterprise from increasing market share and profitability to fashioning an eco-centric economy, generating full and worthwhile employment, social equality and environmental justice. This would indeed require systemic change – a paradigm shift not only in education but in the nature of capitalism itself.

## Capitalism now

The EfS critique of modern industrial society has evolved very broadly, articulating notions of sustainability and sustainable development that rarely secure consensus and sometimes not even agreement. This is evident in the huge academic publishing output, the continuous revision of declarations, policies, statements, protocols and definitions, and the regular conferences and congresses on climate change, habitat destruction, technology, education or the economy that occur globally and that usually result in little more than an agreement that at some future date delegates will fly to somewhere else. In the meantime, the neoliberal shock therapy administered to the former communist states in the 1990s, the credit boom, and the modernization of the People's Republic of China has ensured the capitalist economy has rumbled globally on. These states provided investment opportunities that absorbed the capital surpluses generated by previous economic activity in the West, which now required new capital outlets to ensure future, necessary economic growth. Following the financial collapse of 2007–08, the now indebted Western countries, particularly the US whose debts are being serviced by the Chinese, are increasingly dependent on the continuing economic growth and financial stability of China, India, Brazil and Russia. Moreover, the continuing political subordination of Middle Eastern oil-producing nations to the needs and imperatives of the major oil corporations is essential if the Western way of life is to remain the envy of the world.

Sustainability practitioners and educators have, of course, vigorously engaged with the realities of climate change, oil dependency, social and gender inequality, habitat destruction, species extinction, consumer materialism, war and poverty but there has also been a tendency to accept the capitalist mode of production as a given, albeit one in need of institutional reform and a course of ethical responsibility. There has been some modest progress in developing new economic instruments, green skill strategies and organisational adaptations but this has only gone as far as neoliberal market policy and managerialist practice has allowed. It has worked within the paradigm it wants to shift and in so doing helped to sustain it. New specialist courses have appeared, sustainability issues have been addressed in a range of professional and academic programmes, and research, particularly on energy and climate change, have offered decision-makers important but not necessarily attractive policy options. A form of repressive tolerance, that is a tolerance and acceptability determined and defined by institutionalised inequality and legalised suppression (Marcuse, 1969), has nurtured a pragmatic self-censorship, a holding back and a compliance to the power of Capital. Consequently, capitalism's historical contingency and internal contradictions have been overtly contested in an oppositional manner by a relatively small number of EfS practitioners despite the logic and persuasive power of the arguments presented by Foster (1994), Meiksins Wood (2002), Harvey (2011) and the like. These analysts are way outside the reformist frame of mind of most EfS practitioners who assume that institutional higher education and dominant corporate practices can be reformed meaningfully and effectively from inside the tent. To believe this is perhaps a terrible act of self-deception. David Harvey writes:

The risk and uncertainty we now experience acquires its scale, complexity, and far reaching implications by virtue of processes that have produced the massive industrial, technological, urban, demographic lifestyle and intellectual transformations that we have witnessed in the latter half of the twentieth century. In this, a relatively small number of key institutions, such as the modern state and its adjuncts, multinational firms and finance capital, and "big" science and technology, have played a dominant and guiding role. For all the inner diversity, some sort of hegemonic economistic-engineering discourse has also come to dominate discussion of environmental questions, commodifying everything and subjecting almost all transactions (including those connected to the production of knowledge) to the singular logic of commercial profitability and the cost benefit calculus. The production of our environmental difficulties, both for the working class, the marginalized and the impoverished (many of whom have had their resource base stripped from under them by a rapacious commercialism) as well as for some segments of capital and the rich and the affluent, is broadly the result of this hegemonic class project (and its reigning neoliberal philosophy).

(Harvey, 1998: 23)

Thus, Paul Hawken *et al.* (1999) and Jonathan Porritt (2005) have identified what capitalism should do if the world really mattered. They have argued for a 'natural capitalism' and Porritt has tried hard to persuade capitalists to be ecologically sustainable; but his work and career are an interesting example of an optimism of the will gradually succumbing to a pessimism of the intellect when facing the obduracy of capitalist realism. As Chair of the UK's Sustainable Development Commission, Porritt had the brief, but not the power, to make a difference. His book, *Capitalism As If The World Matters* (Porritt, 2005), eloquently showed how the globe's dominant mode of production has been consuming the planet and how, without too much of a shock to the system, things could be different. His resignation as Chair of the Commission and the failure of the Green New Deal to become counter hegemonic at the moment when neoliberalism had palpably failed, showed just how resilient neoliberal capitalism is.

This should come as no surprise. Neoliberalism has informed and shaped the very fabric of our culture – our entertainment, our politics, the way we live, learn and work. Business leaders have been transformed from once being perceived as predators into icons, saviours and celebrities with popular television programmes such as *The Apprentice, Dragon's Den* and even *Big Brother* becoming models for an economically relevant public pedagogy (Couldry, 2010). For David Harvey (2011) and Colin Crouch (2011) the ideology of free market capitalism will be with us for some time. An osmotic relationship exists between the major corporations, the institutions of national government and the international agencies of global finance capital. Political freedom is routinely identified with spurious notions of economic choice grounded in the cultural practices of consumer materialism, possessive individualism, capital accumulation and the supportive buttresses of corporate-owned mass media and a complicit and (increasingly) privatised public education. For Crouch (2011), the lack of an effective counter-hegemonic culture means that the best that can be hoped for is an enhanced role for corporate social and environmental responsibility and a vibrant, critical, civil society that may demand some kind of accountability and develop the capacity to provide those public and social goods ignored by the private sector in the drive for profit. For higher education this will be reflected in a few warm words hidden away in strategy documents, a couple more modules on business ethics and a beady eye on whether anyone takes the green league tables seriously.

The dialectical relations between the system's various spheres of activity – the labour process, intellectual production, the socially constructed environment, revolutionary technologies, organisational forms and social relations – kept the wheels turning historically, creating a compound growth rate of around 3 per cent (Harvey, 2011). Planned product obsolescence, technological innovation and new exploitable markets have kept the system going but without new opportunities for productive investment this growth rate may not be achieved so easily and the system and its elites, who have benefitted so spectacularly from the disposal of public assets will be in trouble. If, as EfS practitioners say, the right and proper goal for public policy is greater social and human well-being, then what is required is more equality and fairness, not more market-driven inequality, conflict, uncertainty and insecurity (Wilkinson and Pickett, 2010). However, as history shows, capitalism is immensely resourceful, being able through innovation and technological invention to survive crises and emerge, if not stronger, then certainly fit enough to fight another day. Nonetheless, there are systemic, social and ecological limits to growth as Foster *et al.* (2010) have shown and as Harvey has so clearly analysed. An ecological rift exists between the relations of capitalist production (and consumption) and nature, which, conceivably, no amount of entrepreneurial invention, technological innovation, expropriation or exploitation will gainsay. A compelling logic flows from this fact for the Jevons paradox has returned to plague natural capitalists like Hawken, the Lovins and other advocates of eco-efficiency and ecological modernisation: increased resource efficiency will lead to ever-more resource use as production and consumption increase to satisfy the systemic demand for continuous growth. Instead, we must produce and consume less stuff. We must contest and abandon the gods of neoliberalism and economic growth, recognising that in a globally connected world relations of exploitation have both proliferated and deepened. Higher education has played its part in all this, too.

## The corporatisation of higher education

Over 40 years ago the English socialist historian E P Thompson reflected critically on his experience working for the new business-friendly University of Warwick. He asked:

> Is it inevitable that the university will be reduced to the function of providing, with increasingly authoritarian efficiency, pre-packed intellectual commodities which meet the requirements of management? Or can we by our efforts transform it into a centre of free discussion and action, tolerating and even encouraging "subversive" thought and activity, for a dynamic renewal of the whole society within which it operates?
>
> (Thompson, 1970: 166)

Thompson's question is more valid and urgent now than ever, particularly as the dominant value syntax is one where neoliberal clichés have become institutionalised and embedded in the practices, procedures, strategies and policies governing higher education. Neoliberal assumptions have taken on the aura of physical laws: *the private sector is more efficient than the public*; or *only the market determines the value of something*; or *there is no alternative*. When such a state of affairs becomes dominant, 'a value system or ethic becomes a *program*' and 'its assumed system of worth rules out thought beyond it' (McMurtry, 1998: 15). 'Freedom' is now thought of in terms of 'market freedom'. Higher education's major purpose is to serve the needs of the economy with 'employers' becoming a metonym for the capitalist mode of production and the exploitative labour process.

Many critical and sustainability educators have questioned these assumptions, exploring the way language use shapes human perception, professional consciousness and institutional practice that 'prevent today's students from becoming aware of that cultural commons – that is, the community-centered alternatives to a consumer-dependent and ecologically destructive form of existence' (Bowers, 2008: 3). However, the market model is robust and the language of education has too readily incorporated metaphors from the worlds of business and the military: *products, customers, clients, institutional brands, mission statements, targets, markets, competition, vision, quality management, cohorts, value for money, units of resource.* It seems the student experience is understood as basically another element of the wider-experience economy and cultural brandscape. For Pine and Gilmore (1999), businesses market and sell transformative experiences that are in various ways commoditised in the form of a good or service – a latte at Starbucks or the carrying of a Mulberry handbag. Universities now sell a 'student experience': a learning experience and a work experience with the promise of more commoditised experiences to come. This has prompted McMurtry (1991) to suggest a set of contrasts between education as it should be and the capitalist market as it is.

Although Barrett (2000) has pointed out that McMurtry's schema is not without its flaws, the general argument is well made. For some time, higher education in the United States has been increasingly marketised and internally transformed as its future prosperity has been tied ever more tightly to the development needs of the major capitalist corporations to maintain a dynamic for

*Table 2.1* Contrasting education and the market

|  | *The market* | *Education* |
|---|---|---|
| *Goals* | To maximise money-profits | To advance and disseminate shared knowledge |
| *Motivations* | To satisfy the wants of anyone who has the money to purchase what is wanted | To satisfy the desire for knowledge of anyone who seeks it, whether they have the money or not |
| *Methods* | The buying and selling of ready-made products at whatever price can be had | Never to buy and sell it as a good to anyone but to require all who would have it to fulfill its requirements autonomously |
| *Standards of excellence* | (i) To successfully sell the product line<br>(ii) How problem-free the product is to the buyer | (i) How inclusively it takes into account others' interests and avoids bias<br>(ii) How deep and broad the problems it poses are to one who has it |

continual growth (Slaughter and Leslie, 1997; Rhoades and Slaughter, 1997; Slaughter and Rhoades, 2004). In *Global Education Inc.*, Ball (2012) identifies education as a global phenomenon characterised by a 'neoliberal imaginary' that articulates 'best practice' as something that is naturally commoditised, bought and sold for profit, and where issues relating to education development, quality and inequality are addressed by applying market solutions. With recent steps being taken towards the confused but nonetheless *de facto* privatisation of UK universities by the Conservative–Liberal Democrat Coalition Government, an increased emphasis has been placed on the importance of employable graduates, exploitable research and enterprise by corporate stakeholders and politicians in an environment where student-customers look for work in an increasingly jobless and depressed global economy. Arguably, with the drift towards privatisation, many of the values, concepts and practices of the 'corporate university' have gained purchase within traditional universities as their roles become predominantly about selling qualifications. Thus, in this context, the 'corporate university' should not be understood exclusively in terms of its current manifestations as with the *Virtual University* (BAE Systems), the *Hamburger University* (MacDonald's) or the *Academy of Business* (Ernst and Young) but more as a concept, an ideal or a neoliberalist end-in-view. The 'corporate university' is, therefore, not so much a distinct physical place where organised learning occurs (Meister, 1998) but as a space for higher learning that is increasingly about realising the objectives and priorities of corporations, the engines of economic growth, rather than satisfying an individual's broader educational needs. In other words, the ethos of corporate 'universities' is becoming infused into an actually existing university system, transforming it into a strategic resource for developing and educating employees, customers and suppliers and for realising the financial strategies of business and the growth economy. Consequently, many universities in the UK and the US are helping Lloyd Blankfein, CEO of the highly profitable Goldman Sachs, do 'God's work' (Arlidge, 2009) and Charles Ferguson's 2010 documentary *Inside Job* has graphically demonstrated how a complex of top academics, policy gurus and politicians has enabled Goldman to be a controlling figure in various governments' neoliberal economic and financial policy development and implementation. In 2012 the finance company launched a scheme, *10,000 Small Businesses*, to work in partnership with five top UK business schools to provide 'high-quality, practical support to the owners and leaders of established small businesses' (www. goldmansachs.com/citizenship/10000-small-businesses/index.html). This can only add a something extra to the competitive market position of those schools in an ideological environment fixated on business development. In order to bring this about, university collegiality is being challenged by an instrumental and autocratic managerialism, for only such a system is deemed to be efficient and fit for purpose. For Prince and Beaver:

a corporate university charged with leading and managing an organisation's knowledge and learning initiatives needs to be at the very heart of the enterprise and its decision making. To be successful in engineering strategic change it has to be proactive, innovative and professionally managed.

(Prince and Beaver, 2001: 24)

The private university college BPP in the UK, and many traditional universities, offer very similar variations to this thematic in their own specific mission statements. The 'enterprise' or the 'business-friendly' university is now a common phrase among many traditional universities, accompanying the usual rhetoric about the competitive economy, skills development, knowledge and, of course, sustainability.

The organisational visions of many universities are attempts at positioning themselves to be the preferred outsourcing solution for corporate research, development and training (Boher, 2010). Of course, there has always been an economic imperative to education generally, and technical education specifically has been linked directly to economic performance; but the current instrumentalism is arguably superseding any notion of education being a good in itself. Higher education's value resides in what it can provide economically and materially, and like the value-added tax can only be realised through the system of market exchange. However, as Blass (2005) correctly observes, all this does make a great deal of sense for the majority of undergraduate and postgraduate students who are primarily studying to improve their career prospects, and they are the major source of income for most traditional universities. But this is more than simply mission drift or pragmatic accommodation. As one critic writing in the *New York Times Magazine* has noted:

> Universities create jobs, develop new therapies and technologies and train America's young people for the modern knowledge economy. All this is true. But comparable claims could be made for a pharmaceutical company. What makes the modern university different from any other corporation? There is more and more reason to think: less and less.
>
> (Delbanco, 2007)

In the UK, a major review of higher education was conducted by the former CEO of BP, John Browne, and although not all of the report's recommendations were accepted new modes of university governance, funding, teaching and research evaluation have been introduced (Collini, 2011). These higher education reforms have been presented by government and business as modernising, necessary, inevitable, efficient and relevant, and although higher education may not yet be a totally unreconstructed cheerleader for corporate capitalism the connection between knowledge and power, learning and the economic needs of the private corporations has certainly been 'refreshed'. This is most clearly illustrated in the business schools. Durham Business School, for example, not only carries endorsements from PriceWaterHouseCoopers, Accenture, Barclays and KPMG but displays the corporations' logos on its website despite the fact that some have questioned whether the financial collapse of 2008 can in some way be considered a failure of business education (Corbyn, 2008). Indeed, in less prestigious university environments the employer-led foundation degrees have been specifically designed to meet industry's workforce development needs. The UK case brings the issues into high definition and sharp focus. Unsurprisingly, many academic staff in Europe and the USA are distinctly unhappy and uncomfortable with higher education's neoliberal corporate turn. American sociologist Stanley Aronowitz (2000) argues in *The Knowledge Factory* that the vocationalisation of higher education is short-sighted and misdirected. Students are not served well by learning too specialist a curriculum or being umbilically connected to the workplace. Students are not simply future workers but persons and global citizens who need to acquire certain intellectual habits, a cultural breadth and a predisposition to learn throughout life. The world changes quickly and geopolitical and economic power structures are shifting eastwards. The concept of career is quickly becoming anachronistic. We are in the age of the portfolio, part-time worker with many professional working lives becoming a succession of discrete jobs rather than a coherent lifetime journey (Handy, 1995; Sennett, 1998). 'The learner who really understands the economy knows how fragile is the concept of career' writes Aronowitz (2000: 161). Economic recoveries in recent years have essentially been jobless and those jobs that remain in the private and public sector, or will be created in the future, are often insecure, stressful and alienating. Corporatised higher education has therefore been charged to produce obedient neoliberal subjects or, as Madeline Bunting (2005) puts it in her study of the labour process, 'willing slaves'. Economic value takes precedence over social value, welfare and well-being and so, to cope with the likely psychological fall-out from this, a therapeutic educational/cultural practice has emerged. This practice has undoubtedly enabled educators to understand the affective conditions of effective learning and especially the acquisition of abstract knowledge; but it is also complicit in fashioning neoliberal subjectivities. As Sarah Amsler remarks:

> The irony of this is that some of the most affectively-oriented educational practices in fact prepare students to adapt to the very forms of life, political constraints and mis-educative ideologies that produce emotional suffering in the first place.
>
> (Amsler, 2011: 58)

There are other worries and dangers. It is increasingly evident that capitalism can proceed quite effectively without representative, let alone participatory, liberal political democracy. The interwar fascist experience demonstrated this decades ago (Poulantzas, 1979) and western crony capitalism (Foulkes, 2006) and China's present economic expansion (Reich, 2006) suggest that public accountability need not extend much beyond the sensitivities of the financial markets or consumer choice. If this is so, the university may no longer remain a centre for democratic, free and open critique. Indeed, it may even be in their collective corporate interests not to. For example, many higher education institutions have become increasingly involved with arms-producing companies raising important concerns not only about the instrumental commercialisation of higher education, human rights and ethics but also its incipient militarisation, particularly in the areas of business, science and engineering (Langley, 2006, 2009; Stavrianakis, 2009). As public funding declines further, many universities in the UK and elsewhere will become increasingly dependent on military-related money, which will be very hard, if not impossible, to replace if university managers and administrators believe the business risks are too great and consequently do not wish to do so. Faculty and student criticism, and especially actual physical protest, are bad for the higher education business, so accompanying this there has evolved a culture of secrecy, a lack of executive transparency. At the same time, this has been accompanied by audit procedures resembling Foucauldian surveillance, together with coercive accountability (Shore and Wright, 2000). The critical educator Henry A. Giroux writes:

> More striking still is the slow death of the university as a center of critique, vital source of civic education and crucial public good. Or, to put it more specifically, the consequence of such dramatic transformations has resulted in the near death of the university as a democratic public sphere .... As an adjunct of the academic-military-industrial complex, higher education has nothing to say about teaching students how to think for themselves in a democracy, how to think critically and engage with others and how to address through the prism of democratic values the relationship between themselves and the larger world. We need a permanent revolution around the meaning and purpose of higher education, one in which academics are more than willing to move beyond the language of critique and a discourse of both moral and political outrage, however necessary to a sustained individual and collective defense of the university as a vital public sphere central to democracy itself .... Central to any viable, democratic view of higher education is the necessity to challenge the notion that the only value of education is to drive economic progress and transformation in the interest of national prosperity.
>
> (Giroux, 2011: XX)

## Contesting neoliberal higher education

There has been some resistance to the neoliberal corporatisation within the higher education establishment but much has been often politically quite modest, reasonable and moderate (Hotson, 2011; Holmwood, 2011). Bailey and Freedman (2011), in their *The Assault on Universities*, do go further but their aim is to return to the social democratic status quo that was overturned some years ago. For a return to occur, there needs to be fundamental changes to the economic and political values dominating contemporary society. Grumbling within the academy or attempting to reason with the irrationalities of policy-makers, business gurus and other 'leaders' is unlikely to lead to radical systemic change and, of course, by no means everyone in the academy wants it. More to the point perhaps, the authors of *Better Questions on the University* (The Provisional University, 2010) conceptualise the university as an apparatus coordinating a system of partially autonomous machines each with their own specific logic – one machine for issuing qualifications, another machine for giving instructions and setting assessments, a machine for establishing standards and a machine for depoliticising the student experience (just think of the emphasis on employability). The political-pedagogic drive towards socio-economic egalitarianism or radical political and collective subjectivities has declined over the last 40 years for the academy has been constrained in its capacity to offer, or constitute, strong alternatives to the commodification of knowledge and the dominance of the market. Thus, the paradigm shift towards a sustainable education that Stephen Sterling (2001) called for over a decade ago is today further away than ever. Sustainability educators are now confronted with a serious challenge, not least because, as systems thinkers working within the capitalist system, compromise, accommodation and incorporation have become their means of survival. The issue here is not so much that systemic thinking is mistaken but that it has induced EfS practitioners to be insufficiently political in their contestations of dominant institutional practices, neoliberal ideologies and the basic value assumptions of the free market. As Herbert Marcuse wrote in *One Dimensional Man*, 'man [sic] and nature are fungible objects of organization' and 'the web of domination has become the web of reason' (1972: 137–8). Thus sustainability educators, unlike Marcuse, pay too little attention to the dialectical realities of political power and the capabilities necessary to fashion a political praxis of educative liberation and ideological emancipation. Organisational constraints and internalised limitations need to be overcome. New, as well as subaltern, voices need to be heard if meaningful democracy is to be renewed and if people are to be more than the agents of profit-making, injustice and inequality (Couldry, 2009, 2010). Market-driven media, like formal higher education, offer few, if any, alternatives to the dominant neoliberal worldview and are not well positioned to offer effective resistance.

However, a critical pedagogy is emerging outside the academy, aided and abetted by a number of critical educators working within the system. The 'Reimagining The University' issue of the *Roundhouse Journal* (2011) sees the currently historical moment as offering certain cracks within the existing system that can be widened through combining a refusal to endorse the university as 'a factory' for the knowledge economy and by extrapolating the concept of open source or self-organising open schools as a way of recalibrating capitalist social relations, repoliticising higher learning and successfully contesting the 'university-as-fetish'. The experience and compass of higher learning needs to be repurposed to promote democratic association, radical dialogue and critical deliberation, and for Hall (2011), the Occupy movement offered a temporary countercultural space for this to emerge more fully. Other alternative structures such as Dougald Hine's *Dark Mountain Project* with its manifesto, challengingly titled *Uncivilisation* (Dark Mountain, 2009), offer similar possibilities whereby the cultural politics of education may be reconfigured to enable the adventure of social and educational creativity. With its intellectual debts to Paulo Freire (1996) and Ivan Illich (1973), a critical space of and for deinstitutionalised learning is perhaps emerging. The idea of deschooling society, of creating webs of informal and formal learning for sustainability and radical contestation, informed my own attempt at refashioning a lifelong learning, beyond the walls of the university but somehow enabled by it (Blewitt, 2010). There are others such as the Social Science Centre at Lincoln, a not-for-profit higher education cooperative (http://socialsciencecentre.org.uk), the 'autonomous university' promoted by the Provisional University collective (http://provisionaluniversity. tumblr.com/about), the Really Open University established in Leeds and, in North America, the Edu-factory Collective, which promotes a Global Autonomous University (Edu-factory Collective, 2009). Green economist Mary Scott Cato (Cato, 2012) has argued that a new parallel economy could be developed within the existing higher education framework in the UK based on a local currency and time banking scheme.

Other educators have been inspired in their critical scholarship and pedagogy by Guy Debord and the Situationists of the 1950s and 1960s whose practices of *détournement* and *dérive* sought both to divert and subvert established priorities, discourses, organisational structures and everyday behaviours (Welsh *et al.*, 2010). The Canadian radical educator Richard Day (2004) has drawn on the anarchist tradition, and particularly the work of the relatively unknown thinker and activist Gustav Landauer, perceiving within new social movements the kernel of an alternative pedagogy and processes of social renewal informed by a logic of affinity. The editors of a recent collection of essays, *Utopian Pedagogy: Radical Experiments against Neoliberal Globalization* (Coté et al., 2007a) included theoretical pieces and discussions of actual radical/utopian education actions. It is a valuable handbook for activist academics; but there is paradox here, if not irony, in that many of the authors, editors and publishers work within the education system they attack and wish to see reformed or overturned. This shows there is still some space for dissenting academics to be progenitors of alternatives if they are courageous enough to act. Historically, academics have usually not been fully controllable, and although organisational and material pressures on them are ratcheting up constantly, many do have close links and closer affinities to protest movements and oppositional groups. The task then for these dissenters is to increasingly become public intellectuals inventing, building, experimenting and creating alternative mental conceptions and practices that challenge the morally and financially bankrupt culture of neoliberalism. Coté *et al.* perceive

> academic dissent as a laboratory of experiment in non-capitalist, non-statist futures, [which] seeks to produce encounters in the name of exploring a potential politics of solidarity across all the divisions – by race, class,

gender, sexuality, ability, age – that are crucial to the continued functioning of the systems of states and corporations. The persona of *academicus affinitas* is, therefore, a style of work that responds to a transformative impulse, that prefers open experimentation to rule-based procedures, chooses the politics of the act over a politics of demand, pursues inventions rather than reforms, respects heteronymous systems of difference rather than universalizing hegemonic formations, and is committed to the task of minimizing the operation of power as domination in every situation.

(Coté *et al.*, 2007b: 325)

## Conclusion

Clearly, another world is possible, but the problem is that the social relations reproducing neoliberalism, combined with the bursting of the financial debt bubble in 2008, have left many people, including many academic faculty, to see only catastrophe as the alternative to an imagined future of perpetual money-wealth (Graeber, 2002, 2011). At the time of writing, a new legitimation crisis is clearly upon us, perhaps best voiced by the non-leaders of the Occupy movement who refused to present a programme of reforms or demands in the same old way. There is a lesson here for EfS practitioners; for the legitimacy of institutionalised higher education is likely to suffer by association given its own articulation of neoliberalist ideology and corporate business practice.

The task of EfS and other intellectual cultural workers, therefore, is to build a new legitimacy by helping to build a genuine alternative to what we have at present. As Žižek writes:

One can sincerely fight to preserve the environment, defend a broader notion of intellectual property, or oppose the copyrighting of genes, without ever confronting the antagonism between the Included and Excluded ... In short, without the antagonism between the Included and the Excluded, we may find ourselves in a world in which Bill Gates is the greatest humanitarian battling against poverty and disease, and Rupert Murdoch the greatest environmentalist mobilising hundreds of millions through his media empire.

(Žižek, 2009a: 98)

An emancipatory politics and educative practice will not emerge from a single social agent or from another accredited course on sustainable development, but may do so from a combination of different agents and agencies ... inside and outside the tent. David Schweickart, the author of *After Capitalism* (2002), sees the answer in increased democratisation of the labour process and investment. Necessary to this, as Harvey argues, is a more nuanced understanding of the interplay between environmental transformations and social relations 'of which the class dimension is fundamental because that is what capitalism is all about' (1998: 30). Social class needs to become central to many EfS debates as the global labour force, unemployment and the ratio of wage labour to capital is larger today than it has ever been (Freeman, 2005) at a time when capital is not working and corporate social responsibility is still too often just a veneer. Much more needs to be done if global equity, social and environmental justice, and ecological sustainability is to be achieved and at the root of much of this is inequality in the workplace and in the wider society, North and South. A strong infusion of critical and visionary dialectics (Ollman, 2003), critical pedagogy and oppositional politics could invigorate and radicalise EfS. New learning configurations and opportunities are emerging and older cooperative, democratic and egalitarian conceptions of the ideal university are taking on a fresh complexion and attractiveness as movements outside the academy capture the imagination. Calculation kills, inspiration inspires. Subjectivity rather than objectivity may be the key to the effective governance of the sustainable university and the free development of learning that will shape a more democratic future. But, this is clearly antithetical to the corporatisation of traditional universities, our democratic polities, our governing elites and our unsustainable economic practices. Radical and critical educators, particularly in the global EfS community, should see the coming years as an opportunity rather than a threat. There are plenty of examples of EfS projects and curriculum developments, exhaustively documented in many books and journals, that could help to open up the cracks within the existing system (Corcoran and Wals, 2004; Jones *et al.*, 2010). If anything is to be salvaged from Global Education Inc., it is the unrealised possibilities and potentialities of not just thinking critically or culturally but of actually refusing and transforming capitalist social relations and means of production. A new radical praxis needs to emerge.

So, if ever there was a time for a real paradigmatic change in society and in the academy, it is now. Neoliberalism has failed, even in its own terms. There is an alternative. History has not ended. The impossible does happen. Alternative structures are taking to the streets (Žižek, 2009b). The big question, of course, is whether these assist and encourage those within the academy to create a free, open, public and sustainable university – a lifelong university without walls. Only by trying shall we know.

## Higher education study and sustainability: a comment from students at Aston University, UK

Surveys commissioned by the UN Global Compact, conducted in 2007 and 2010 (Oppenheim *et al.*, 2007; Lacy *et al.*, 2010), reveal a significant shift in CEO's consideration of sustainability issues. In the latest figures, 93 per cent of them thought sustainability will become a critical business driver for the success of their company in the years to come. Interestingly, the academic community's engagement was repeatedly identified as fundamental to the development of the mindset and skills required if future managers are to embody a change toward a new economy. In 2007, less than 8 per cent considered that their own organisations or business schools had shown enough interest or had allocated enough means to develop such capabilities. In 2010, one CEO out of four believed that the lack of knowledge of sustainability issues among current senior and middle managers was one of the three main barriers to implementing an integrated and strategic company-wide approach to sustainability.

How should sustainability knowledge be taught to students in the academic community? Sustainability not only needs to be integrated into business courses, but also into politics and engineering education in order to make sense and have an effective impact on society as a whole. Business schools, according to Goshal (2007), play an important role in the formation of future business leaders and consequently the inclusion of environmental and social concerns in the decision-making process contributes to sustainability.

Further, Goshal (2007) points out that there is a need to change the models used to analyse business situations and that sustainability should not be set apart as a specific discipline. Sustainability must not be considered as 'a compartmentalised add-on to business' (Pinnington *et al.*, 2007); it should be a fully included dimension of business education programmes. In their paper, Atkinson and Wade (2011) highlight the importance of sustainability education in political science. They argue that sustainability is about social change; therefore political sciences must provide the skills and routes to bring about the necessary change.

From a personal standpoint, the Social Responsibility and Sustainability courses taught at Aston University provided us with significant analytical skills for complex problem-solving involving practical exercises and real-world examples. The issues addressed involved dealing with uncertainties, using both broad and local perspectives and non-traditional approaches to pinpoint the most appropriate solution. Additionally, the interdisciplinary nature of sustainability provided us with the possibility to choose and follow different career paths and easily move into new opportunities if necessary.

Practically, there is a vital need to teach students the capacity to think and act beyond the traditional theoretical models with different lights focused on the ethical and sustainable. However, to be meaningful sustainability still needs to be designed and applied to each sector, company and project. Therefore, universities should take the responsibility to incorporate sustainability issues within each degree as part of its core syllabus. In particular, ethical and strategic business sustainability classes should be included in any business, politics and engineering course. It appears fundamental to us that basic marketing, finance and economic modules are redefined in order to include the central issue of sustainability in the underlying principles upon which business is run.

In a nutshell, sustainability education equips students with the ability to deal with uncertainties along with being more adaptable and flexible in their thinking and future decision-making. However, the timescale and expected impact should be recognised, as today's chief executives were on the bench of schools around 30 years ago. This means that if this type of education starts to develop now it will only reach future executives around 2040/2050. Since a shift to a new sustainable economy is needed sooner than this, organisations may have to invest in enhanced training in order to integrate sustainability into their strategy and daily operations. Companies have an urgent responsibility to educate their employees throughout their professional life as well. Studying sustainability is a necessity for each of us to learn how to take our responsibilities as human beings personally but also for humanity as a whole.

*Elena Schad, Adrian Poher, Jung Park, Sirapop Kaewthipharat, Mathilde Colin, Sudipa Bhattacharjee and David Aboulkheir*

## References

Atkinson, H and Wade, R (2011) *Education for Sustainability (ESD) and Political Science.* Paper presented to the annual conference of the Political Studies Association, London. Available at: www.psa.ac.uk/journals/pdf/5/2011/183_123.pdf (accessed 18 February 2012)

Goshal, S (2007) 'Bad management theories are destroying good management practices', *Academy of Management Learning and Education*, 4 (1): 75–91. Available at: www.aom.pace.edu/amle/AMLEVolume4Issue1pp75-91.pdf (accessed 18 February 2012)

Lacy, P, Cooper, T, Hayward, R and Neuberger, L (2010) *A New Era of Sustainability, UN Global Compact-Accenture CEO Study 2010.* Available at: www.unglobalcompact.org/docs/news_events/8.1/UNGC_Accenture_CEO_Study_2010.pdf (accessed 18 February 2012)

Oppenheim, J, Bonini, S, Bielak, D, Keh, T and Lacy, P (2007) *Shaping the New Rules of Competition: UN Global Compact Participant Mirror.* Available at: www.unglobalcompact.org/docs/news_events/8.1/McKinsey.pdf (accessed 18 February 2012)

Pinnington, A H, Macklin, R and Campbell, T (eds) (2007) *Human Resource Management: Ethics and Employment.* Oxford and New York: Oxford University Press. Available at: http://people.brandeis.edu/~molinsky/documents/Human%20Resource%20Management%20Ethics%20and%20Employment.pdf (accessed 20 February 2012)

# References

Arlidge, J (2009) 'I'm doing "God's work". Meet Mr Goldman Sachs', *Sunday Times Magazine*, 8 November

Amsler, S (2011) 'From "therapeutic" to political education: the centrality of affective sensibility in critical pedagogy', *Critical Studies in Education*, 52: 47–63

Aronowitz, S (2000) *The Knowledge Factory: Dismantling the Corporate University and Creating True Higher Learning*, Boston: Beacon Press

Bailey, M and Freedman, D (eds) (2011) *The Assault on Universities: A Manifesto for Resistance*, London: Pluto Press

Ball, S (2012) *Global Education Inc.: New Policy Networks and the Neoliberal Imaginary*, London: Routledge

Barrett, R (2000) 'Market arguments and autonomy', *Journal of Philosophy of Education*, 34: 327–41

Blass, E (2005) 'The rise and rise of the corporate university', *Journal of European Industrial Training*, 29: 58–74

Blewitt, J (2010) 'Deschooling society? A lifelong learning network for sustainable communities, urban regeneration and environmental technologies', *Sustainability*, 2: 3465–78

Boher, K (2010) 'Universities becoming the outsourcing solution', *American Journal of Business Education*, 3: 1–6

Bowers, C (2008) *University Reform in an Era of Global Warming*, Eugene, OR: Eco-Justice Press

Bunting, M (2005) *Willing Slaves: How the Overwork Culture is Ruling Our Lives*, London: HarperCollins

Cato, M (2012) *Free Universities!*, Weymouth: Green House Publications. Available at: www.greenhousethinktank.org/files/greenhouse/home/Universities_inside.pdf (accessed 12 March 2012)

Collini, S (2011) 'From Robbins to McKinsey', *London Review of Books*, 33: 9–14

Corbyn, Z (2008) 'Did poor teaching lead to crash?', *Times Higher Education*, 25 September. Available at: www.timeshighereducation.co.uk/story.asp?storycode= 403696 (accessed 15 March 2012)

Corcoran, P B and Wals, A E J (eds) (2004) *Higher Education and the Challenge of Sustainability: Problematics, Promise, and Practice*, Dordrecht: Kluwer Academic Publishers

Coté, M, Day, R and de Peuter, G (2007a) *Utopian Pedagogy: Radical Experiments Against Neoliberal Globalization*, Toronto: University of Toronto Press

Coté, M, Day, R and de Peuter, G (2007b) 'Utopian pedagogy: creating radical alternatives in the neoliberal age', *The Review of Education, Pedagogy and Cultural Studies*, 29: 317–36

Couldry, N (2009) 'Rethinking the politics of voice', *Continuum*, 23: 579–82

Couldry, N (2010) *Why Voice Matters: Culture and Politics After Neoliberalism*, London: Sage

Crouch, C (2011) *The Strange Non Death of NeoLiberalism*, London: Polity Press

Dark Mountain (2009) *Uncivilisation: The Dark Mountain Manifesto*. Available at: www.dark-mountain.net/wordpress/dark-mountain.net/wordpress/wp-content/uploads//uncivilisation-dark-mountain-manifesto.pdf (accessed 11 January 2012)

Delbanco, A (2007) 'Academic Business', *New York Times Magazine*, 30 September. Available at: www.nytimes.com/2007/09/30/magazine/30wwln-lede-t.html (accessed 11 January 2012)

Day, R (2004) 'From hegemony to affinity: the political logic of the newest social movements', *Cultural Studies*, 18: 716–48

Edu-factory Collective (2009) *Towards a Global Autonomous University*, New York: Autonomedia. Available at: www.edu-factory.org/edu15/images/stories/gu.pdf (accessed 12 March 2012)

Foster, J (1994) *The Vulnerable Planet: A Short Economic History of the Environment*, New York: Cornerstone Books

Foster, J, Clark, B and York, R (2010) *The Ecological Rift: Capitalism's War on the Earth*, New York: Monthly Review Press

Foulkes, A (2006) 'Capitalism and democracy', *The Freeman*, 56 (9). Available at: http://www.fee.org/pdf/the-freeman/0611foulkes.pdf (accessed 12 March 2012)

Freeman, R (2005) 'What really Ails Europe (and America): the doubling of the global labor force', *The Globalist*. Available at: www.theglobalist.com/StoryId.aspx?StoryId=4542 (accessed 12 January 2012)

Freire, P (1996) *Pedagogy of the Oppressed*, Penguin: London

Giroux, H (2011) 'Beyond the swindle of the corporate university: higher education in the service of democracy', *Truthout,* 18 January. Available at: www.truth-out.org/beyond-swindle-corporate-university-higher-education-service-democracy66945 (accessed 11 January 2012)

Graeber, D (2002) 'The new anarchists', *New Left Review*, 13: 61–73

Graeber, D (2011) *Debt: The First 5000 Years*, New York: Melville House Publishing

Green New Deal Group (2008) *A Green New Deal*, London: new economics foundation. Available at: www.neweconomics.org/sites/neweconomics.org/files/A_Green_New_Deal_1.pdf (accessed 11 January 2012)

Hall, R (2011) 'Occupation: a place to deliberate the socio-history of re-production', *Roundhouse Journal: Reimagining the University*, 54–63. Available at: http://roundhousejournal.org/events/reimagining-the-university-journal/ (accessed 7 April 2012)

Handy, C (1995) *The Empty Raincoat*, New York: Random House Business

Harvey, D (1998) 'Marxism, metaphors and ecological politics', *Monthly Review*, 49: 17–31

Harvey, D (2011) *The Enigma of Capitalism and the Crises of Capitalism*. London: Profile Books

Hawken, P, Lovins, A and Lovins, L (1999) *Natural Capitalism: The Next Industrial Revolution*, London: Earthscan

Holmwood, J (ed.) (2011) *A Manifesto for the Public University*, London: Bloomsbury

Hotson, H (2011) 'Don't look to the Ivy League', *London Review of Books*, 33: 20–22

Illich, I (1973) *Deschooling Society*, Penguin: London

Jones, P, Selby, D and Sterling, S (eds) (2010) *Sustainability Education: Perspectives and Practice Across Higher Education*, London: Earthscan

Langley, C (2006) 'Science, engineering and technology: it is a military affair', *Physics Education Special Feature: War and Peace*, 41: 508–13

Langley, C (2009) 'Commercialised universities: the influence of the military', in Satterthwaite, J, Piper, H and Sikes, P (eds) (2009) *Power in the Academy,* Stoke-on-Trent: Trentham

McMurtry, J (1991) 'Education and the market model', *Journal of Philosophy of Education*, 25: 209–18

McMurtry, J (1998) *Unequal Freedoms: The Global Market System as an Ethical System*, Connecticut: Kumarian Press

Marcuse, H (1969) 'Repressive tolerance', in Wolff, R, Moore Jr, B and Marcuse, H (1996) *Critique of Pure Tolerance*, Boston: Beacon Press, pp. 95–137

Marcuse, H (1972) *One Dimensional Man,* London: Abacus

Meiksins Wood, E (2002) *The Origin of Capitalism: A Longer View*, London: Verso

Meister, J (1998) *Corporate Universities: Lessons in Building a World-class Workforce,* New York: McGraw-Hill

Mills, C (1956) *The Power Elite*, New York: Oxford University Press

Ollman, B (2003) *Dance of the Dialectic: Steps in Marx's Method*, Chicago: University of Illinois Press

Pine, J and Gilmore, J (1999) *The Experience Economy*, Boston: Harvard Business School Press

Porritt, J (2005) *Capitalism as if the World Matters*, London: Earthscan

Posner, R (2009a) *A Failure of Capitalism: The Crisis of '08 and the Descent into Depression*, Boston: Harvard University Press

Posner, R (2009b) *The Crisis of Capitalist Democracy*, Boston: Harvard University Press

Posner, R A (2010) *The Crisis of Capitalist Democracy*, Boston: Harvard University Press

Posner, R A (2011) *A Failure of Capitalism: The Crisis of '08 and the Descent into Depression*, Boston: Harvard University Press.

Poulantzas, N (1979) *Fascism and Dictatorship*, London: Verso

Prince, C and Beaver, G (2001) 'The rise and rise of the corporate university: the emerging corporate learning agenda', *International Journal of Management Education*, 1: 17–26

Provisional University (2010) *Better Questions on the University.* Available at: http://provisionaluniversity.tumblr.com/writing (accessed 15 March 2012)

Reich, R (2006) 'China: capitalism doesn't require democracy', *Common Dreams*, 10 January. Available at: www.commondreams.org/views06/0110-42.htm (accessed 11 January 2012)

Rhoades, G and Slaughter, S (1997) 'Academic capitalism, managed professionals and supply-side higher education', *Social Text*, 51, 9–38

Schweickart, D (2002) *After Capitalism*, New York: Rowman and Littlefield

Sennett, R (1998) *The Corrosion of Character: The Personal Consequences of Work in the New Capitalism*, London: W W Norton

Shore, C and Wright, S (2000) 'Coercive accountability: the rise of audit culture in higher education', in Strathern, M (ed.), *Audit Cultures: Anthropological Studies in Accountability, Ethics, and the Academy*, New York: Routledge, pp. 57–88

Slaughter, S and Leslie, L (1997) *Academic Capitalism: Politics, Policies, and the Entrepreneurial University*, Baltimore: John Hopkins University Press

Slaughter, S and Rhoades, G (2004) *Academic Capitalism and the New Economy: Markets, State, and Higher Education*, Baltimore: John Hopkins University Press

Stavrianakis, A (2009) 'In arms' way: arms company and military involvement in education in the UK', *ACME: An International E-Journal for Critical Geographies*, 8: 505–20

Sterling, S (2001) *Sustainable Education*, Totnes: Green Books

Thompson, E (ed.) (1970) *Warwick University Ltd: Industry, Management and the Universities*, Penguin: Harmondsworth

Welsh, J, Ross, E and Vinson, K (2010) 'To discipline and enforce: surveillance and spectacle in state reform of higher education', *New Proposals: Journal of Marxism and Interdisciplinary Inquiry*, 3: 25–39

Wilkinson, R and Pickett, K (2010) *The Spirit Level: Why Equality Is Better for Everyone,* London: Penguin

Žižek, S (2009a) *First As Tragedy, Then As Farce*, London: Verso

Žižek, S (2009b) 'In 1968, structures walked the streets: will they do it again?', in Satterthwaite, J, Piper, H and Sikes, P (eds) (2009) *Power in the Academy,* Stoke-on-Trent: Trentham

## Documentary

*Inside Job* (2010) directed by Charles Ferguson, USA, Sony Pictures

# 3    Another world is desirable

## A global rebooting of higher education for sustainable development

*Daniella Tilbury*

Higher education is widely understood to be key to improving the future prospects of many across the globe. It is perceived as a critical determinant of the life chances of individuals who take up its learning opportunities, and important to regional communities and countries which benefit from the influence of education at this level. The international community also has high expectations: without exception, the UN international agreements and reports of the last twenty years call for universities and colleges to respond to the major social, economic, environmental and cultural challenges of our day (GUNI, 2012). This follows an acknowledgement that universities have a record of leading economic development and social change through scientific breakthroughs as well as through the education of opinion-leaders, decision-makers and future-makers (Cortese, 2003; Elton, 2003; Lozano, 2006). The latter is well documented; studies confirm that 2 per cent of the world population attend higher education but more than 80 per cent of the decision-makers in industry, community and politics are graduates of universities (Scott *et al.,* 2012).

Academic debate is beginning increasingly to echo social dialogues calling for an alternative, more desirable world that is underpinned by a sustainable future (Alvares and Faruqi, 2012). Once again, the engagement of higher education in the construction of this global vision and pathway is seen as critical. However, this expectation is to be starkly contrasted with the growing recognition that universities continue to be responsible for the unsustainable futures we currently face. David Orr's poignant reminder that 'those who contribute to exploiting poor communities and the earth's ecosystems are those who have BAs, MBAs, MScs and PhDs and not the "ignorant" poor from the South' (2004: 7) resonates deeply with many commentators across cultural divides (Mochizuki and Fadeeva, 2008; Sanusi and Khelgat-Doost, 2008; Sterling, 1996). Universities, it is argued, have been mirroring social models that promote economic development without regard for the health and well-being of people and planet.

In their book, *Decolonising the University: The Emerging Quest for Non-Eurocentric Paradigms*, Alvares and Faruqi (2012 ) are equally critical of existing higher education systems and practices. The book calls for a rejection of the Eurocentric frameworks and assumptions that have been modelled across the globe and demands more authentic learning pathways that celebrate other cultures, languages and ways of life. The authors challenge the imperialist worldview and argue that 'another world is desirable', citing the Declaration signed at the Universiti Sains Malaysia (2011). Building on the Declaration, the book draws attention to the need to restore 'the organic connections between our universities, our communities and our cultures' and presents a vision of a university where the students 'do not experience learning as a burden but as a force that liberates and uplifts societies' (2012: 1).

These readings are persuasive in their argument that if higher education is to play a transformative role in our societies and assist with the construction of a more desirable world, it must itself be the subject of transformation. This is the discourse that paved the road to the higher education element of the 2012 UN Conference on Sustainable Development (Rio+20), as well as being the underpinning message of this book. Rio+20 mobilised the international academic community in ways which have not been seen before. For the first time, universities and colleges are becoming a more forceful presence, stepping up their engagement and acknowledging responsibility for this agenda. The *Higher Education Declaration*, a UN-led initiative for Rio+20, is a case in point. The Declaration captures the commitments and actions of chancellors, presidents, rectors, deans and leaders of higher education institutions to improve their performance and influence in this area (The Rio+20 Directory of Committed Deans and Chancellors, 2012). The Declaration is an explicit statement of responsibility, acknowledging the role of higher education in the international pursuit of sustainable development.

The *Rio+20 People's Sustainability Treaty for Higher Education* (The Higher Education Treaty Circle, 2012) also suggests that the sector is stepping up its commitments and action in this area. The development of the Treaty, facilitated by the Copernicus Alliance, brought together over 35 international and national higher education organisations, agencies and associations rooted in different parts of the world. It is supported by over 80 signatories who commit to transforming the higher education system (see Box 3.1) and builds upon previous international declarations and agreements in this area (see Table 3.1), recognising that the latter, while important symbolic milestones, have had limited direct effect to date in changing policy and practice (Bekessy *et al.,* 2007). They have mostly contributed to building a context and receptiveness to sustainability. The new Treaty upscales the commitment and expects the organisational and sector agencies signing it to work together to effect the change.

**Box 3.1:** *The People's Sustainability Treaty for Higher Education*

The Treaty recognises the following:

1) *Transformation is complex and a long-term ambition.* The higher education system, it argues, must recognise that the attainment of sustainable development goals demands a paradigm shift in education. This requires a review of institutional responsibilities as well as the reorientation of curriculum and pedagogy to better serve the needs of current and future generations. The Treaty recognises that this will take time.

2) *Transformation must be guided by vision and clarity of purpose.* A rebooting of higher education is required. The Treaty argues that universities and colleges must reassess their purpose and social visions and clarify how these align with sustainable development. This should underpin changes to structures and governance arrangements. The vision-building, it argues, 'must occur at a localised level and be explicit of the inner values as well as intangible guiding principles which define higher education institutions and their practice'.

3) *Transformation of knowledge structures is required.* The Treaty calls for more holistic and integrative approaches to knowledge generation and transfer. It advocates approaches which cut across the traditional knowledge disciplines, theoretical and methodological mainstays and engage with communities of practice.

4) *Transformation requires fostering respect for and understanding of different cultures and embraces contributions from them.* The criticality of intercultural understanding and cross-cultural dialogue in the attainment of sustainable development is highlighted here. The Treaty recognises that higher education policy and practice must be inclusive of indigenous knowledge, lifestyles, wisdom and values, as well as promote intercultural respect and dialogue across the communities it serves.

5) *Transformation of lifestyles as well as professional competencies is required.* Higher education's core business needs revisiting. The Treaty argues that 'Universities and colleges must promote less carbon intensive, less consumerist, healthier lifestyles across campuses and through external partnerships'. This needs professional

development of staff as the first step in a complex process which will lead to the integration of professional competences and graduate attributes on sustainable development within the curriculum.

6) *Transformation requires innovation.* Institutional culture and curriculum should promote qualities such as curiosity, humility, collective action, boldness and innovation as 'these are fundamental to triggering the transformation'.

7) *Transformation requires effective leadership.* Leadership support and development in this area is lacking. The Treaty recognises that there is a need for leaders who are able to adopt inclusive approaches which engage staff, students and stakeholders in the transformation process and which can translate vision and purpose into higher education reality.

8) *Transformation strategies need information and decision-making tools.* Real time information systems that can capture progress and inform strategic investment and management decisions are needed. The Treaty recognises that these tools are important in building bridges between strategic policy, academic development and operational issues.

(Adapted from Rio+20 People's Treaty on Higher Education for Sustainable Development, 2012)

*Table 3.1* International, regional and national declarations and commitments

| Year | Declaration/ Charter | Partners involved | Scope | Short description/Keywords |
|------|---------------------|-------------------|-------|---------------------------|
| 1990 | Talloires Declaration | University Leaders for a Sustainable Future (ULSF) | Global | First declaration specifically targeted to the higher education sector. *Keywords:* leadership for sustainability; support; mobilisation of resources. |
| 1991 | Halifax Declaration | Consortium of Canadian Institutions, International Association of Universities (IAU), United Nations University (UNU) | Global | The ethical and moral obligation of universities in addressing sustainability was recognised. *Keywords:* ethical obligation; shape present and future; leadership; development of policies and practices. |

| Year | Declaration/ Charter | Partners involved | Scope | Short description/Keywords |
|------|----------------------|-------------------|-------|----------------------------|
| 1993 | Kyoto Declaration on Sustainable Development | IAU | Global | Closely tied to Agenda 21 and the United Nations Commission on Environment and Development (UNCED) Conference in Rio de Janeiro 1992. It called for specific sustainability plans. *Keywords:* sustainability action plans; ethical obligation; sustainability imperative; environmental education; sustainable physical operations. |
| 1993 | Swansea Declaration | Association of Australian Government Universities | Global | The declaration stressed the commitments outlined in previous documents. *Keywords:* review of physical operations; environmental literacy and curriculum; ethical obligations; research and public service. |
| 1994 | COPERNICUS University Charter for Sustainable Development | Association of European Universities (Copernicus Alliance) | Regional (Europe) | It called for a paradigm shift in European universities. *Keywords:* core social mission; new frame of mind; whole-institutional commitment; environmental ethics and attitudes; education of university employees; programmes in environmental education; interdisciplinarity; dissemination of knowledge; cooperation and networking; partnerships; continuing education programmes; technology transfer. |
| 2001 | Lüneburg Declaration | Global Higher Education for Sustainability Partnership (GHESP) | Global | In preparation for the 2002 World Summit on Sustainable Development (WSSD) in Johannesburg. *Keywords:* key role of universities; catalyst for social change; globalisation, poverty alleviation, social justice, democracy, human rights, |

*Table 3.1* continued

| Year | Declaration/ Charter | Partners involved | Scope | Short description/Keywords |
|------|------|------|------|------|
| | | | | peace and environmental protection; generation of new knowledge; training of future trainers; curriculum re-orientation; lifelong learning. |
| 2002 | Ubuntu Declaration | UNU, UNESCO, IAU, Third World Academy of Science, African Academy of Sciences and the Science Council of Asia, COPERNICUS-Campus, GHESP, ULSF | Global | Called for the development of a global learning environment for learning for sustainability. It suggested the creation of networks and Regional Centres of Expertise (RCEs). *Keywords:* review of programmes and curricula; attract future trainers; meet the Millennium Development Goals (MDG); knowledge transfer; development of an action-oriented tool kit for universities; development of sustainability strategies for reform; development of an inventory of best practice and case studies. |
| 2003 | April 2003 – 2nd Sub regional conference for ESD in Bishkek, Statement of the conference participants. | CAREC and five Central Asian countries: Kazakhstan, Kyrgyzstan, Tajikistan, Turkmenistan and Uzbekistan, in cooperation with partners and donors | Sub-regional | These initiatives became Central Asia's commitment to the UN Decade for ESD. *Keywords:* sustainable development literacy, collaboration UN DESD. |
| 2004 | Memorandum: 'Re-Thinking Academia. Reorientation on the horizon of sustainability' | Group 2004 (interdisciplinary group of German professors, all being committed to higher education policy) | National (Germany) | Contribution to the German debate on the higher education reform. It called for a paradigm shift in German universities. *Keywords:* sustainability science, innovation of higher education, sustainable universities, inter- and transdisciplinarity. |

| Year | Declaration/ Charter | Partners involved | Scope | Short description/Keywords |
|---|---|---|---|---|
| 2005 | Graz Declaration on Committing Universities to Sustainable Development | COPERNICUS CAMPUS, Karl-Franzens University Graz, Technical University Graz, Oikos International, UNESCO | Global | Stressed the key opportunities, which the Bologna Process creates for embedding sustainability across higher education.<br><br>*Keywords:* give status to sustainability in universities' strategies and activities; sustainability as a framework for the enhancement of the social dimension of European higher education. |
| 2005 | Bergen Communiqué | European Union (EU) education ministers, European Commission and other consultative members | Regional (Europe) | EU universities should build upon sustainability principles. For the first time since 1999, made a strong reference to the Bologna Process as a key mechanism to establishing a European Higher Education Area by 2010 and promoting the European system of higher education worldwide. The process should be based on the principle of sustainability.<br><br>*Keywords:* university reform supporting education for sustainability; interdisciplinarity; innovation to address social challenges; sustainability skills and learning objectives; employability. |
| 2008 | Declaration of the Regional Conference on Higher Education in Latin America and the Caribbean (CRES) | UNESCO | Regional (Caribbean and Latin American) | CRES was intended to be a contribution to identifying the major issues of Latin America and the Caribbean, looking toward the UNESCO World Conference on Higher Education in 2009.<br><br>*Keywords*: sustainability for social progress; cultural identities; social cohesion; poverty; climate change; energy crisis; culture of peace; democratic relations and tolerance; solidarity and cooperation; critical and rigorous intellectual ability. |

*Table 3.1* continued

| Year | Declaration/ Charter | Partners involved | Scope | Short description/Keywords |
|------|----------------------|-------------------|-------|----------------------------|
| 2008 | G8 University Summit Sapporo Sustainability Declaration | G8 University Network | Global | The aim was to develop common recognition of the need for global sustainability, to discuss responsibility of universities and provide messages to G8 leaders and societies. |
| | | | | *Keywords:* universities working closely with policy-makers, leadership for sustainability; re-orientation of education and curriculum; dissemination of information; training leaders; interdisciplinary perspective. |
| 2008 | Climate Change Statement of Action | University and College Presidents | National (Canada) | Aimed to make campuses more sustainable and address global warming by bringing together institutional commitments to reduce and neutralise greenhouse gas emissions on campus. |
| | | | | *Keywords:* creation of emissions inventory; set a date for universities becoming 'climate neutral' within two years, development of action plans, inventory and progress reports made publicly available. |
| 2008 | Promotion of Sustainability in Postgraduate Education and Research Network (ProSPER.Net) Charter | UNU-IAS | Regional (Asia-Pacific) | An alliance of several leading higher education institutions in Asia and the Pacific Region that committed to working together to integrate Sustainable development into postgraduate courses and curricula. |
| | | | | *Keywords:* networking, work together to reorient curricula and research towards sustainable development. |

| Year | Declaration/ Charter | Partners involved | Scope | Short description/Keywords |
|------|----------------------|-------------------|-------|-----------------------------|
| 2009 | World Conference on Higher Education | UNESCO | Global | Called on governments to increase investment in higher education, encourage diversity and strengthen regional cooperation to serve societal needs. |
| | | | | *Keywords:* advancement of understanding of multifaceted issues and our ability to respond; interdisciplinary focus; critical thinking; active citizenship; peace, well-being, human rights; education for ethical citizens. |
| 2009 | Turin Declaration on Education and Research for Sustainable and Responsible Development | G8 University Network | Global | The aim was to acknowledge the pivotal role that higher education institutions and scientific research organisations should play in supporting sustainability at global and local levels. |
| | | | | *Keywords*: new models of social and economic development consistent with sustainability principles; ethical approaches to sustainability; new approaches to energy policy; focus on sustainable ecosystems. |
| 2009 | Living Sustainably: the Australian Government's National Action Plan for Education for Sustainability | Australian Government | National (Australia) | Sets out a framework for national action that adopts the following four strategies to respond to the needs and priorities of education for sustainability. |
| | | | | *Keywords:* leadership for sustainability; reorienting education towards sustainability; fostering sustainability in business and industry; harnessing community spirit to act. |

*Table 3.1* continued

| Year | Declaration/ Charter | Partners involved | Scope | Short description/Keywords |
|------|---------------------|-------------------|-------|----------------------------|
| 2010 | Learning for Change: Scotland's Action Plan for the Second Half of the UN Decade of Education for Sustainable Development | Scottish Government | National (Scotland) | Action plan as a response to the UNDESD. In the higher education sector, the Scottish government has set up specific recommendations for universities on how to advance the sustainability agenda. *Keywords*: monitor progress; sustainability skills; embedding sustainability in teaching and learning strategies; campus sustainability learning; student-led activities; strengthening sustainability in teaching standards; interdisciplinary work; climate change action plans. |
| 2010 | Declaration: 'Universities for Sustainable Development' | German Rectors' Conference, German Commission for UNESCO | National (Germany) | Contribution to the UN Decade of Education for Sustainable Development. *Keywords*: higher education for sustainable development, sustainable universities, UN DESD. |
| 2010 | UNICA Green Academic Footprint Pledge | UNICA Network | Regional (capitals of Europe) | Emphasised the unique position of universities in the different capitals of Europe. *Keywords:* role and purpose of universities; develop campuses as living laboratories in the area of sustainability. |
| 2011 | Declaración de las Américas: 'Por la sustentabilidad de y desde la universidad' | Inter-American Organization for Higher Education IOHE /OUI | Regional (Inter-American) | Commitment of Universities of the OUI to assume institutional responsibility to the global environmental crisis and encourage other social actors to do the same. *Keywords*: new models of social and economic development consistent with sustainability principles; whole-institutional commitment; cooperation and networking. |

| Year | Declaration/ Charter | Partners involved | Scope | Short description/Keywords |
|------|----------------------|-------------------|-------|----------------------------|
| 2012 | Science for Sustainability: The Need for a Successful Breakthrough | German Commission for UNESCO | National (Germany) | Memorandum on reorienting the German science system. *Keywords*: science and research for sustainability, inter- and transdisciplinarity. |
| 2012 | Higher Education Declaration for Rio+20 | UNESCO, UNEP, PRIME, UNU, Academic Impact | Global | Declaration supported in the lead-up to Rio+20. *Keywords*: research, education, campuses, global footprint, international frameworks. |
| 2012 | The People's Treaty on Sustainability for Higher Education | Copernicus Alliance and 35 HE agencies, associations and organisations | Global | Treaty developed to influence international negotiations and dialogues. It is a formal voluntary commitment of Rio+20. *Keywords*: transformation of systems and structures; four-stage action plan; education for sustainable development; partnerships. |

(Adapted from Tilbury, 2011)

Higher education faces multifaceted and interconnected challenges that must be navigated if it is to contribute to the transformation of socio-economic cultures and improve the future prospects of people and the planet. In short, a global rebooting of higher education is required.

Post Rio+20, it has become clear that the last twenty years have: provided clarity on the scale and the urgency of the changes required; generated a better understanding of the complexity underpinning pathways towards sustainability; provided authentic examples of student-driven changes in campuses and curricula; seen higher education funding agencies and research councils allocating resources for sustainability studies and actions; as well as witnessed glimpses of good institutional practice (GUNI, 2011; see Box 3.1).

In parallel however, higher education has seen funding cuts and increasing regulation from governments. Accompanying the new policy directives have been changing student expectations. Students are seeking to see the immediate relevance of degrees to their employability prospects, value for money, convenient access to programmes and 'just in time' support. In this context, universities and colleges are wrestling to balance growth with quality and reputation. It is thus, not surprising perhaps, that the senior leaders and key agencies responsible for higher education have struggled to prioritise the reorientation of higher education towards sustainability.

**Box 3.2: Movements towards sustainability**

*Higher Education's Commitment to Sustainability: From Understanding to Action* (GUNI, 2012) brought together a series of research studies commissioned by the Global Universities Network for Innovation (GUNI). The contributors acknowledge that changes for sustainability in higher education have not been deep or systemic but capture stories of progress from various regions and countries across the globe.

Key movements identified in the study included the following:

- Initiatives from Africa, South East Asia as well as Latin America that lowered the carbon footprints of universities and colleges. These included reducing waste and energy consumption; protecting biodiversity and natural spaces; sourcing sustainable goods and services and modelling sustainability to influence the decisions and actions of staff, students and local communities. In countries such as the US and the UK, recent government funding is providing the support needed to scale-up carbon reduction activities.
- On the research front, research council and funding agencies have also played their part in Australia, Canada and the European Union. They are supporting new conceptual and practical spaces for research and encouraging academics to go beyond their discipline boundaries as they recognise that interdisciplinary research remains a fringe activity. Perhaps more significant is the increasing interest of academics in the co-generation of knowledge and participatory research methodologies which align with models of 'research as social change'.
- The report documents a growth in the number of specialist undergraduate and postgraduate courses in sustainability. It cites examples from India, the Philippines, Saudi Arabia and South Africa, where students are being supported to pursue careers in this area. However, the teachers, architects, business leaders, doctors, event managers, engineers and journalists are still being schooled in unsustainable social assumptions and practices.
- Worthy of attention are the United Nations University (UNU) accredited Regional Centres of Expertise, otherwise known as RCEs. The last eight years have seen the establishment of 68 RCEs in Africa, Australia, Asia Pacific, Europe, the Middle East, South America, the Caribbean and North and Central America. RCEs seek to expand the span of local partnership work in sustainability as well as create learning platforms for people across the global region.

(Tilbury, 2011)

So will the next twenty years see a global rebooting of higher education for sustainable development? Several chapters in this book argue that the building of leadership capability in this area is required to turnaround the current scenario. As Scott *et al.* (2012) remind us, 'Change does not just happen but must be led, and deftly'.

Leading change for sustainability within, and across, the sector requires more than knowledge of sustainability or a commitment to transforming higher education. It requires a facility for navigating organisations through a change journey that is complex, uncertain, slow and political. This was the conclusion of an international study that analysed the experiences of 188 higher education leaders seeking change for sustainability (see Box 3.3). The 'Turnaround Leadership' study concluded that building leadership capability is key to systemic and deep change across our universities and colleges.

Higher education has been engaged in a snakes-and-ladders relationship with sustainable development. As universities and colleges advance their commitments to this agenda, changing political moods provide slippery slides,

---

**Box 3.3: Turnaround leadership for sustainability in higher education**

This study has sought to define the capabilities that characterise an effective leader in sustainability across universities and colleges in Australasia, North America, the UK and Europe. It found that effective leaders have the ability to 'listen, link and lead' – in that order. This is underpinned by a number of key capabilities summarised below. Effective higher education leaders for sustainability:

a) understand that change is not an event but a complex learning (and unlearning) process for all concerned;
b) are capable of negotiating the complex processes of change across all institutional levels;
c) have well developed emotional intelligence and the contingent way of thinking necessary to lead and engage a wide diversity of staff in deep change across disciplinary 'silos' and 'tribes';
d) are effective in assisting their staff to learn how to make a desired change work in practice;
e) understand that leadership is most tested when things go wrong and exercise judgement that is aligned with the principles of fairness, long-term vision and inclusiveness;
f) are decisive and committed to halting unsustainable practice and structures; are transparent, accountable and lead by example.

(Scott *et al.* 2012)

the economic crisis in the West present distractions; while student and social movements for sustainability gain strength and accelerate changes in this area.

If we are to transform higher education, we need a game change – replacing the throw of a dice, and the uncertainty that it brings, with a scenario where there is greater strategy and empowerment within the sector and beyond. Leadership is key to moving forward: although, as in the game of chess, every piece, every stakeholder, plays a part in contributing to a new scenario. Transforming higher education structures, systems and cultures is necessary so that they can respond to the key challenge of our century. Only then, will we advance towards a more desirable future.

## References

Alvares, C and Shad S Faruqi, S (2012) (eds) *Decolonising the University: The Emerging Quest for Non-Eurocentric Paradigms*, Pulau Pinang: Penerbit Universiti Sains Malaysia

Bekessy S, Clarkson, R, and Sampson, K (2007) 'The failure of non-binding declarations to achieve university sustainability: A need for accountability', *International Journal of Sustainability in Higher Education*, 8 (3): 301–16

Cortese, A (2003) 'The critical role of higher education in creating a sustainable future', *Planning for Higher Education*, 31 (3): 15–22

Elton, L (2003) 'Dissemination of innovations in higher education: a change theory approach', *Tertiary Education and Management*, 9, 199–214

Global Universities Network for Innovation (GUNI) (2012) *Higher Education's Commitment to Sustainability From Understanding to Action*, World in Higher Education Series No. 4, Barcelona, Spain: GUNI

Higher Education Treaty Circle (2012) *The People's Treaty on Higher Education for Sustainable Development*. Available at: www.copernicus-alliance.org (accessed 14 May 2012)

Lozano, R (2006) 'Incorporation and institutionalisation of sustainable development into universities: breaking through barriers to change', *Journal of Cleaner Production* 14 (9–11), 787–96

Mochizuki, Y and Fadeeva, Z (2008) 'Regional Centres of Expertise in sustainable Development (RCEs): An overview', *International Journal of Sustainability in Higher Education*, 9 (4): 371–79

Orr, D (2004) *Earth in Mind: On Education, Environment and the Human Prospect* (10th anniversary edition). Washington DC: Island Press

Rio+20 Directory of Committed Deans and Chancellors (2012) *Higher Education Sustainability Initative for Rio+20*. Available at: http://rio20.eurome d-management. com/wp-content/uploads/2012/03/Higher-Education-Sustainability-Initative-for-Rio-The-directory-of-Deans-Chancellors-committed.pdf (accessed 16 June 2012)

Sanusi, Z and Khelgat-Doost, H (2008) 'Regional Centres of Expertise as transformational platforms for sustainability: a case study of University Sains Malaysia, Penang', *InternationalJournal of Sustainability in Higher Education*, 9 (4): 487–97

Scott, G, Tilbury, D, Deane, E and Sharp, L (2012 in press) *Turnaround Leadership for Sustainability in Higher Education*. Sydney: Australian Office of Learning and Teaching

Sterling, S (1996) 'Education in change', in Huckle, J and Sterling, S (eds) *Education for Sustainability,* London: Earthscan, pp. 18–39

Tilbury, D (2011) 'Higher education for sustainability: a global overview of commitment and progress', in Global Universities Network for Innovation (GUNI) (2011) *Higher Education's Commitment to Sustainability: From Understanding to Action,* World in Higher Education Series No. 4., Barcelona, Spain: GUNI

Universiti Sains Malaysia Declaration (2011) 'Another world is desirable', Delegates of the International Conference on Decolonising Our Universities, 27–29 June 2011, Penang, Malaysia

# Part II

# Aspects

# 4 Promoting sustainable communities locally and globally

## The role of Regional Centres of Expertise (RCEs)

*Ros Wade*

## Introduction

Learning to live sustainably has never been more urgent. The challenges of climate change, peak oil and acute biodiversity loss are converging at a time of global economic disorder and injustice: short-term, national solutions are not an option. Local and global issues are ever more closely interconnected – the global pervades the local, just as the local pervades the global. The global economy, however, is still structured around a model that was devised at the start of the industrial revolution and huge tensions exist between stakeholders at the local and global levels. These are difficult times to try to balance the economic, social and environmental elements, which make up the process of sustainable development (SD). Policy-makers need new tools and new ways of thinking to address these challenges. If we are unable to do this, then we endanger the very fabric of our world that makes human life possible: we are already living way beyond the carrying capacity of the planet and if we do not start now to change our unsustainable ways of living, then we may not have another opportunity.

In order to address these immense challenges, new forms of learning are needed that break through current hegemony and navigate the barriers between disciplines, sectors and cultures. More of the same sort of education, which has led us to this unsustainable impasse, is not appropriate if we seriously want to develop sustainable communities, locally and globally. Education for Sustainable Development (ESD) can provide the framework and strategy for this imperative, and universities clearly have an important role to play here in promoting the development of sustainable communities, both locally and globally. However, as outlined by Sterling and Blewitt (this volume), there is often a disjuncture between key elements of ESD and those of higher education institutions (HEIs).

ESD is a complex and contested term, but UNESCO (2007) has provided a framework for international consensus on environment and development around a number of key elements. ESD:

- is based on the principles of intergenerational equity, social justice, fair distribution of resources, and community participation, which underlie SD;
- is based on local needs, perceptions and conditions, but acknowledges that fulfilling local needs often has international effects and consequences;
- engages formal, non-formal and informal education;
- builds civil capacity for community-based decision-making, social tolerance, environmental stewardship, adaptable workforce and quality of life; and,
- is cross-disciplinary.

These are challenges for universities, as elitist and technocratic models of universities that do not promote these characteristics still abound. Such universities do not tend to prioritise social justice and community participation and, frequently, the main focus of senior managers is on the financial bottom line. Indeed, to many senior managers, the idea of the sustainable university merely refers to the financial sustainability of the organisation! Elitist universities tend to be characterised by a focus on established paths of disciplinary knowledge and a reluctance to embrace new synergies of interdisciplinary thinking, while technocratic universities tend to have a top-down managerial structure, discouraging innovation and focusing on increasing student numbers and cutting costs. Many traditional university structures and alliances are outdated and, rather than provide opportunities for new ideas and innovation between the social and natural sciences, the rigid parameters around many of the disciplines actually restrict opportunities for new ways of learning and thinking. Added to this, the increasingly managerial models of many universities around the world can impede radical change and tend to be self-perpetuating in terms of hierarchies of knowledge. How many universities are willing to put into practice the principles of ESD (other than possibly write some into their mission statement)? With an increasing emphasis on the science, technology, engineering and mathematics (STEM) subjects, how many universities engage with, recognise and value indigenous knowledge or non-formal and informal community education and learning – something that is deemed essential for ESD and for our future needs in living on this planet (UNESCO, 2007)?

Non-formal educators, like Paulo Freire in the 1970s and, more recently, NGOs like the Asia South Pacific Association for Basic and Adult Education (ASPBAE) in the Philippines (Nagata, 2009; Asia-South Pacific Education Watch, 2007 and 2009), have much to tell us about valuing local and indigenous knowledge and about how people learn through interaction that is relevant and meaningful to them. Western scientific expert knowledge engaging with local cultural knowledge can be mutually beneficial and meaningful. However, without the mutual respect and regard for non-formal learning and local, indigenous knowledge, there is a danger that only Western scientific knowledge is valued and that this then becomes a commodity to be bought and sold. Vandana Shiva's 1997 work on intellectual property rights among the rural communities of India has highlighted the dangers of Western knowledge perspectives with respect to biodiversity, cultural and social loss. Knowledge and education need to be seen as social goods that are shared in order to enable us to live more sustainable lives – not a commodity to be bought and sold. Education and learning for sustainability are both a personal journey and a social process, which need participation with wide-open minds.

This chapter will argue that universities have the potential to play a leading role in creating dynamic new communities of practice (COPs), which can address some of the challenges of learning for sustainability. COPs are characterised by mutual engagement and negotiated joint enterprise 'within a larger context – historical, social, cultural, institutional' (Wenger, 1998: 79). In particular, the chapter will examine the potential of Regional Centres of Expertise in Education for Sustainable Development[1] (RCEs) to develop local/global communities of practice for transformative learning for sustainability.

Ever since the Earth Summit of 1992 identified ESD as a key element in societal transformation towards SD, there has been growing attention to, and adoption of, ESD by governments across the globe. However, most commentators would agree that very limited progress has been made and that the prime goal of reorienting education systems towards sustainability has not yet been achieved. Had this been one of the Millennium Development Goals (MDGs), then there would certainly be grave cause for concern internationally about the very poor progress. Universities have been particularly slow to take on this challenge of reorienting teaching and research towards sustainability, although a number have made good progress in terms of energy, waste and procurement practices. However, with regard to their core business of the curriculum, it is hard to escape the conclusion that both governments and HEIs have been hampered by their inability to break down the structural and conceptual barriers of past tradition and hierarchies.

The different interpretations and understanding of ESD have also hampered efforts to embed it in education policy and practice. As Mochizuki and Fadeeva point out:

> the impetus behind the RCE initiative came from the perceived need to focus on action rather than theory. The RCE initiative was proposed in order to overcome 'inertia' created by the difficulty of reaching international consensus on the nature and scope of ESD.
>
> (Mochizuki and Fadeeva, 2008: 371)

It was seen as a measure to address deficiencies in education systems, which had been very slow to respond to the commitments of Agenda 21.[2]

Despite these challenges, this chapter will try to demonstrate that the RCE initiative presents an opportunity to circumvent some of these obstacles and constraints and to mobilise ESD networks in ways that will challenge unsustainable university practices and may help to promote the development of a sustainable university.

## RCEs: An innovative initiative

'Education for Sustainable Development' (E*f*SD) means what it says: it is not just environmental education or even sustainable development education, but 'education for sustainable development'. Only when we are successful in pooling all available people and resources can we do an appropriate job. We must 'walk the talk' in order to transform all education and transcend all existing divisions to achieve our ultimate goal of a better future for all.

To develop the curricula and courseware needed – and regularly update these – and to inform teacher training and retraining in effective ways, we must aim at an inclusive and flexible process, mobilising all who have something to contribute. Specific attention should be given to online learning and the contributions of the media.

The initiative that might mobilise many and serve to give focus to their contributions could be: to create jointly a 'global learning space for sustainability', based on 'regional centres of expertise'. These centres are firmly based on the principles of respect, self-organisation and participation. We should not get confused by the extremely large regions, as these have been defined by the UN and UNESCO. At that scale, it will be impossible to be sufficiently place-specific in E*f*SD. Regions are seen here – as in common language – as parts of countries like Bretagne, Tohoku or Catalunya.

The regional centres should include institutions of primary, secondary and tertiary education, research institutions, (science) museums, non-formal education, zoos/parks, etc. The establishment and development of RCEs lies, crucially, with local/regional actors and institutions, preferably sustainable universities that are prepared to contribute and provide leadership. The framework for such leadership has been provided by the Ubuntu Alliance Group of Peers and the RCE Global Service Centre based at the UNU – Institute of Advanced Studies in Yokohama (UNU-IAS).

The 'regional centres of expertise' are identified in a comparable way to the monuments on the cultural heritage list by a peer group established by the Ubuntu Alliance. This has the distinct advantage of local/regional conditions being taken fully into account. This approach also has a tremendous mobilising potential. Characteristically, the successful RCEs would run a large number (a 'portfolio') of highly attractive and effective E*f*SD projects, each of these run by two or more member institutions coming from different sectors of society. As it is important to mobilise many people and institutions, initially, prizes could be awarded for innovative, joint projects of two or more institutions from different sectors.

*Professor Dr Hans Van Ginkel is a former rector of the United Nations University.*

# The role of networks and communities of practice

Proponents of ESD are a very diverse group, as the many debates and discussions on definitions of this term clearly testify. They come from a very wide range of ideological and philosophical perspectives; although, of course, they do all share a belief in a need to change the unsustainable way we are living. They cover a wide spectrum from those who believe that ESD can be mainstreamed within the present capitalist system to those who believe that a totally new world paradigm is needed. Arguing from a world systems perspective, Ginsburg *et al.* (1991) emphasise that educational change cannot be separated from wider social change, as the two are closely interconnected. In their analysis of educational reform in a global context, they argue that educators who seek to affect radical social change need to be active at both ends of the spectrum – that is, within the mainstream and outside the mainstream. COPs provide opportunities to do just this, as they 'build and strengthen alliances – locally, nationally, and globally – with other groups and social movements in order to intervene successfully in "educational reform" movements' (1991: 29). They point out that powerful, dominant interest groups make very effective use of such alliances (such as the oil industry's role in campaigns to challenge climate science) and the higher education (HE) ESD community may need to learn some lessons from this. For example, ESD networks could become more politically astute, develop more advocacy and lobbying skills, and start working more closely with key NGO coalitions. Social marketing is being used more widely by environmental and development NGOs. If we want to make changes, then it is not sufficient anymore (if it ever was) just to talk to each other – we need to reach wider audiences and those in positions of power. RCE communities of practice offer one way to take this forward.

COPs are not new to universities and HE generally. But, in the past, many of these have developed in relation to subject-specific areas, through, for example, the subject bodies like the Political Studies Association and the Geographical Association. It should also be noted that, in the UK, the Higher Education Academy (HEA) and the Environmental Association for Universities and Colleges (EAUC) have established influential and supportive networks for ESD. Informal networks have also emerged through shared interest (particular projects, social networks). However, a network is not necessarily the same thing as a COP, as Wenger makes clear, although sometimes they may share some of the same characteristics. COPs look at social networks more from the perspective of action and learning, rather than merely information sharing through interpersonal relationships. While some networks may remain at the level of information sharing, others can develop into COPs.

# Learning communities of practice and the role of universities in RCEs

The role of universities in developing RCEs is crucial (although not all RCEs are coordinated by HEIs). More importantly, the RCE model as a COP provides an opportunity for HEIs to develop many of the characteristics of a sustainable university, through engagement with curriculum, campus and community:

> The concept of COPs has been adopted most readily by people in business because of the recognition that knowledge is a critical asset that needs to be managed strategically. Initial efforts at managing knowledge had focused on information systems with disappointing results. COPs provided a new approach, which focused on people and on the social structures that enable them to learn with and from each other.
>
> (Wenger, 2006)

Many organisations in the business sector have taken up this idea in order to innovate and to encourage teams to make the links between learning and performance. The goal of ESD is much more complex and multi-level than any organisational goal, of course. ESD focuses on developing relationships in order to transform practice and has a responsibility to both present and future generations. Wenger also notes, however, that 'the very characteristics that make COPs a good fit for stewarding knowledge—autonomy, practitioner-orientation, informality, crossing boundaries—are also characteristics that make them a challenge for traditional hierarchical organizations' (2006: 74). These characteristics chime well with the development of new learning and knowledge for ESD.

Universities are traditionally a complex mix of the hierarchical and of the anarchic, which notions of academic freedom allow. While the latter may be able to provide the creative space for developing new ways of learning, academic freedom is frequently constrained by pressures from the drive towards more central control, government regulation and increased marketisation of education. Neoliberal capitalist agendas still dominate policy-making (despite their economic failures) and Blewitt (this volume) has outlined some of the tensions between this model and that of the sustainable university. For example, increasing competition in the HE sector is leading to the growth of private sector global providers, who are, in some cases, edging out local providers. At its best, these new providers can offer space for innovation and new thinking, but with their profit-centred ethos it is possible that many ESD activities and courses that do not relate to current market thinking may disappear. The potential for conflicts of interest means that local accountability may be lost as the main allegiance of global providers is to their shareholders.

The tension between increased marketisation and SD is already being played out in many universities, especially those that are trying to move towards a concept of the sustainable university. The RCE model is a vehicle for working through some of these challenges and for taking a key role in developing the new forms of learning that are needed and which break down barriers between disciplines and sectors:

> An RCE is not a physical centre but an institutional mechanism to facilitate capacity development for SD. An RCE is a network of existing local–

regional institutions mobilised to jointly promote all types of learning for a sustainable future. RCEs, both individually and collectively, aspire to achieve the goals of DESD.

(Mochizuki and Fadeeva, 2008: 370)

The RCE initiative of the UN Decade of ESD (2005–14) aims to develop a global knowledge network for transformative education to promote sustainable communities. RCEs have largely developed organically in response to regional contexts and needs, while at the same time being part of a wider global network. While RCEs have been endorsed by the UN University at a global level, most begin with individual universities and build upon existing networks, as well as creating new ones, which draw in a range of disciplines and partners with shared goals in promoting ESD. In this sense, they could be said to share Wenger's key dimensions of practice, which underpin the concept of the COP: 'Mutual engagement; A joint enterprise; A shared repertoire' (Wenger, 1998: 73). They usually have a loose, democratic structure of governance, with a steering group of members of key stakeholder groups who coordinate activities and the work of key projects. The project groups form the core ESD COPs set within the wider regional COP, which, in turn, is set within the global RCE community of practice.

RCEs should have four elements: governance, collaboration, research and development, and transformative education ('contributing to the transformation of the current education and training systems to satisfy ambitions of the region regarding sustainable living and livelihood' UNU-IAS, 2012). Their goals are to:

- re-orient education towards SD … tailored to address issues and local context of the community in which they operate;
- increase access to quality education that is most needed in the regional context;
- deliver trainers' training programmes; [and,]
- lead advocacy and awareness raising efforts to raise public awareness about the importance of educators and the essential role of ESD in achieving a sustainable future.

(UNU-IAS, 2012)

By May 2012, there were 101 RCEs across the globe, supported by a secretariat at the United Nations University (UNU) Institute of Advanced Studies, with a bi-annual conference for sharing ideas and developing collaborative synergy. The RCE network is also linked in with the global UN milestones and events towards SD. Mochizuki and Fadeeva began talking about RCEs in terms of COPs in 2008 and showed how the RCE concept developed from that of a knowledge hub, with knowledge transfer at its heart, to that of a COP, with social learning at its core. Most RCEs are on a spectrum somewhere between the two concepts.

## The role of RCE communities of practice in ESD: opportunities and challenges for universities

*Creating space for dialogue and development*

Sterling (this volume), Blewitt (this volume), Cullingford (2004) and others have highlighted some of the structural and conceptual challenges we face in developing sustainable universities. The RCE model as a COP can help to break down some of these barriers and, in the best traditions of the subversive educator, can also circumvent them. This is not to say that this is an easy task – far from it. RCEs challenge the managerial, technocratic model that pervades many universities and this can bring them into conflict with university systems and structures. However, as they do not draw their legitimacy from within the university but from the global UN University and network, they have a degree of autonomy that can enable them to open up a creative space for dialogue and development. This can be compared to the 'creative autonomy' that English local government developed in response to increasing government interference and control in the 1990s, identified by Atkinson and Wilks-Heeg (2000). Creative autonomy allowed a space to engage in innovative responses to sustainability and local needs, and resulted, for example, in dynamic and effective Local Agenda 21 programmes. Because RCEs often do not fit into the preferred economic and managerial model, they may find it difficult to gain resources and support from within the university; yet, at the same time this frees them from university bureaucracy and control.

COPs, according to Wenger (2006), are not limited by formal structures: they create connections among people across organisational and geographic boundaries. This is, in itself, a liberating concept, as it can free us from constricting structures and subject silos. However, the concept is inherently apolitical, and, without ethical commitments, the danger is that it could be perceived narrowly in terms of improving business efficiency, or, in authoritarian countries, could possibly even be used to develop communities that practise intolerance and repression.

What would be distinctive about a COP for sustainability would be its commitments, not merely to the goals of sustainability, but to ESD values through the practice of 'the principles of intergenerational equity, social justice, fair distribution of resources and community participation, that underlie SD' (UNESCO, 2007). RCEs offer a framework to develop COPs for sustainability because they are cross-sectoral and involve educators at all levels of formal and non-formal learning. Their aims and ethos have their lineage in the UNESCO principles above and in the framework for ESD, with its emphasis on interdisciplinarity, lifelong learning, participation, formal, non-formal and informal education (UNESCO, 2007).

Interestingly though, in their first iteration, the notion of RCEs

> equated 'knowledge' mainly with 'scientific knowledge' possessed by experts in research institutions, leaving educators at the receiving end of

'knowledge' and largely overlooking the potential of NGOs, civil society organisations and other non formal actors in contributing to 'knowledge' required for addressing diverse sustainability challenges.

(Mochizuki and Fadeeva, 2008: 372)

This conceptualisation of the RCE comes from an elitist notion of the university and has much more in common with the banking model of education (characterised and criticised by Freire), with HEIs as 'experts' and communities as passive recipients of knowledge.

This view of RCEs might also be said to leave the market model largely unchallenged, with educational expertise is seen as a key commodity that can be bought and sold. However, this is perhaps a bit unfair, as one of the market model's critics, Cullingford (2004), charges universities with no longer engaging with the big issues, which the RCE initiative provides, from the start. This RCE model encouraged multidisciplinary learning and engagement with the global and local context for the challenges of SD. In fact, one of its great strengths was its flexibility, which enabled it to develop, through engagement and social learning, into a local/regional/global COP for sustainability. Through mutual engagement and shared enterprise, RCEs can contribute as COPs for ESD in a number of ways. This engagement has the potential to act as a catalyst and enabler in promoting sustainable communities, locally and globally.

### *Addressing local and regional needs within a global context*

It is important to note that all RCEs, although they share common aims and aspirations, have different histories and different features. Most of them have started from a university base, but by no means all of them. It is also too early for any in-depth evaluation of their effectiveness and impact, although frameworks are now being developed for this purpose (Mochizuki and Fadeeva, 2008; Mader, 2011).

Most RCEs have the potential to provide knowledge hubs, as well as to link learning with action through COPs as a conduit to respond to local and regional needs. They also provide a focal point to bring together different COPs and to develop synergy between them. For example, RCE Greater Nairobi presents an example of an RCE that was very strongly founded on the principles of community development for sustainability. First developed through an NGO, the Kenya Organisation for Environmental Education, only later was it then fully adopted by Kenyatta University. In addition, one of London RCE's initiatives was developed in response to the needs of local communities around the Olympic Park. This enabled a number of local groups to come together, to make links with the local universities of East London and Greenwich to consider how to ensure a positive sustainability legacy from the Olympic development. In a sense, the RCE performed the role of broker in bringing these groups together to make common purpose, while leaving them autonomous to decide on future plans. This enabled several different groups to join together from the formal sector (school, universities) and the non-formal (community groups, NGOs).

All RCEs are committed to addressing local and regional needs and priorities, which they are asked to identify in their application for RCE status. Most have representatives from local government as well as educational and civil society organisations. RCE/EAST (Toronto) was originally an initiative of the city council, led by Toronto Zoo. However, when Toronto University became part of the RCE, this helped to promote 'the university's objective of engaging in outreach and helping to impact upon the development of public policy, through interdisciplinary engagement in environmental concerns' (Stefanovic, 2008: 423). RCE Penang, coordinated by Universiti Sains Malaysia, is also closely involved with local communities and sees its role as threefold:

- helping students to be aware of the world in which they live and to gain an understanding of 'the interactions between multifaceted economic, social and environmental problems (including the contribution of individuals to these processes) and a familiarity with perspectives on these issues from other societies and cultures
- helping societies to find through its network, social and technical solutions through academic research and professional experience
- developing partnerships between policy makers, decision makers, NGOs and key individuals who are involved in SD related activities at local, regional and international levels.

(Sanusi and Khelghat-Doost, 2008: 493)

Stakeholders of RCE Penang include a range of local, national and international organisations and individuals, such as Taiping Peace Institute, WWF, Water Watch Penang, Socio-economic and Environmental Research Institute, SE Asia Ministry of Education Regional Centre for Science and Mathematics, as well as individuals from the local recycling group. By the very nature of being part of the RCE initiative, these local and regional activities are set within the global context of SD. In addition, as part of the global RCE network, there is a ready-made forum for the sharing of ideas, issues and practices. Mutual exchanges and links occur on a regular basis, through emailing, visits and via articles in the RCE Bulletin.

### *Practising social responsibility*

One of the reasons that universities were regarded by UNU as providers of leadership for RCEs was their corporate and moral responsibilities to the local and wider communities. Universities were seen as:

institutions that have a social responsibility and moral obligation (derived from academic freedom and autonomy they enjoy), to address sustainability challenges and institutions with stable human and financial resources. It is precisely the combination of these normative expectations for IHEs [Institutions of Higher Education] and the perceived institutional capacity and stability of IHEs that makes them a backbone of the 'local knowledge base' in the RCE concept.

(Mochizuki and Fadeeva, 2008: 373)

RCEs can enable universities to fulfil some of their social commitments while also working within the tensions and constraints of a market model of survival in an increasingly competitive HE terrain. The RCE is both a part of the university but also set apart, in that it derives its legitimacy from a stakeholder group from the regional community, which is much broader. RCE endorsement from the UNU enables a university to subscribe to ethical commitments at the vision/ mission level and to support the ethical practices of the RCE in a slightly hands-off way while continuing to operate in a very competitive business environment. However, at the same time, the members and practices of the RCE have opportunities to influence the organisation and practices of the university as a whole. Some might say that this is potentially invidious by allowing the university to have its cake and to eat it – however, others would argue that this provides an iterative process with the potential for positive feedback loops from the relationships, which offer a pathway towards the ESD values and principles inherent in a sustainable university.

## Linking theory and practice

At the heart of ESD and RCEs is the commitment to the transformation of society and the reorienting of education systems towards sustainability. The links between theory and practice, therefore, are core to this endeavour. With the trend to the mainstreaming of goals of SD, this focus on praxis can be situated within current HEI concerns about economic development and employability. The RCE COP offers opportunities to apply expert knowledge, as well as to develop new knowledge in response to local and regional problems and concerns. For example, a worm-composting project at RCE Penang brought together scientists and members of the local community to use technology from University of Sains Malaysia, which enabled the local community to use waste from paddy, cow dung and general village waste to increase their income by 100 per cent (Sanusi and Khelghat-Doost, 2008: 493). RCE Penang's 'Going Bananas' project is another example of linking theory and practice through interdisciplinarity in a project that bridged the university and the community. It is designed as an income-generating project as well as an environmental conservation initiative through the production of paper from the trunk of banana trees. In order to develop the technology for this, the university needed to involve a number of different faculties with expertise in industrial technology, art, humanities and social sciences (Sanusi and Khelghat-Doost, 2008: 492).

In some ways, the theory/practice approach fits quite well into the market model of higher education, as it can, on one level, be seen as responding to the needs of the market. However, in this case, financial profit is not necessarily an end result: something perhaps much more valuable can be developed. RCE COPs can help to shape and strengthen social capital by providing opportunities to develop and test new ideas, build confidence, and thus help to create an enabling environment for sustainable communities to develop and grow.

In the case of RCE Tongyeong in Korea, the main goal is 'to raise public understanding and increase support for SD'. The local government has redirected policy significantly since the launch of the RCE programme.

> One of the most drastic redirections has been the shifting of a major city planning project towards creating sustainable urban space-making instead of widening and building roads. Many public discussions have been held by both the government and local NGOs thereby stimulating debate on SD and ESD, and demonstrating how awareness among policy makers can drive social progress towards sustainability.
>
> (UNU RCE website[3])

### RCE Tongyeong, South Korea

RCE Tongyeong seeks to promote the sustainable development of Tongyeong city through education. By encouraging sustainable practices in each area of the society, RCE spreads the vision of a sustainable society. RCE Tongyeong fosters ecotourism and edu-tourism to contribute to the development of the local economy in a sustainable manner. As a part of a global learning network, the city of Tonyeong has the following aims:

- cooperating with RCEs around the world;
- participating with various stakeholders;
- establishing a network of formal and informal education;
- providing opportunities to learn about global citizenship, cultural diversity and the environment;
- increasing awareness of the relationship between ecology and human activities;
- making a Korean model of sustainable development.

#### *Our education programmes include:*

*Education for Sustainable Development (ESD) for Citizens Working Group*

Thirty-six institutes are providing citizen education. By participating in developing the educational programmes together with the RCE, these institutes themselves are stimulated to reorient their existing programs towards ESD.

*Training Working Group*

RCE Tongyeong provides regular training programmes for government officers, corporations and local opinion leaders. By learning the philosophy of sustainable development, a sustainable way of corporate management, education, and governmental policy, they can lead Tongyeong city into a city of sustainable development.

*ESD at School Working Group*

This group promotes ESD within the formal education system.

*Youth Working Group*

ESD programmes for youth aim to promote youth club activities and provide opportunities to learn the spirit of sustainable development.

*International Relations Working Group*

This group promotes cooperation among groups worldwide on common efforts for ESD and a sustainable future.

**The future vision: a global learning space**

RCE Tongyeong hopes to transform Tongyeong city to become part of the 'Global Learning Space'. Each corner of this traditional city can be a place that exemplifies the spirit of sustainable development. In order to achieve this, we will make use of all available resources, including but not limited to: unique and important cultural heritage, the legacy of great artists, the history of General Lee Sunsin, the fisheries industry, the clean sea, and the 192 beautiful islands in and around Tongyeong.

*Kim Seung Woo is from RCE Tongyeong.*

## Interdisciplinarity

Parker points out that, while not ignoring the importance of specialist disciplinary knowledge,

> our assessment of positive ways forward and strategies for adaptation, mitigation and restoration life systems (including human social systems) must be based on more joined up forms of knowledge. Hence though interdisciplinarity may be difficult, finding ways to deal with it productively

could be considered to be one of the main intellectual tasks of the SD research programme and HE should be at the forefront of this effort.

(Parker, 2010: 327)

While some RCEs have a particular focus, as RCE Ireland does on teacher education, interdisciplinarity is integral to all the work. This is not to say that interdisciplinarity is easy, but RCEs provide an opportunity to work through some of the issues and problems. Most RCEs bring together, in mutual engagement over shared agendas, members from a range of academic disciplines and from different sectors of the community. Traditional subject silos and HE hierarchies can then be broken through and a more left-field approach can be adopted:

> When opportunities present themselves to universities to link their research activities with hands-on problems facing communities, the demands for interdisciplinary approaches are real. Problem based research requires multiple perspectives, as no one discipline will suffice to capture social, cultural, regulatory, technological, scientific, economic and ecological dimensions of lived experience.
>
> (Stefanovic, 2008: 424)

RCE Penang's citizenship programme is an example of community engagement in training high school students to identify problems and issues in their communities, and using an interdisciplinary approach to try to solve them (Sanusi and Khelghat-Doost, 2008: 492).

One of the major hurdles to interdisciplinarity is that of shared understanding and language and

> many proponents of the importance of ecological systems do not fully recognise that the study of human social systems has *also* already involved theories about the ways in which different elements of economic and social organisation interact with and affect each other. Equally, the social sciences still do not fully engage with the ecological contexts of societies.
>
> (Parker, 2008: 327)

Most RCEs have been addressing some of these issues, for example, RCE/EAST and the University of Toronto are collaborating with public policy-makers, decision-makers, broader civil society environmental organisations, academics and students (Stefanovic, 2008: 422). RCE Penang has also set up a number of initiatives with the Universiti of Sains Malaysia, such as the establishment of a research cluster that requires all project proposals to include faculties from diverse backgrounds, and, as much as possible, to integrate the art and science dimensions of the research to be carried out (Sanusi and Khelghat-Doost, 2008: 492).

The challenge is both for academics from different disciplines to gain some understanding of each other's specialisms, and for them to make these more accessible to the wider community and public. However, '[d]ifferent expectations of methodological rigour may have impeded collaboration, as may differing epistemological interpretations of what constitutes real knowledge or ethical interpretations of what constitutes the "right" thing to do' (Stefanovic, 2008: 424). Even for proponents of systems thinking, this is still a major challenge, but often breakthroughs can occur simply through mutual and shared endeavour and engagement, overcoming misunderstanding and mistrust. This, in turn, can lead to developing shared meanings and a shared repertoire, which is an important characteristic of a learning COP.

London RCE, coordinated by London South Bank University (LSBU), has also been working through some of these issues through their HE ESD Curriculum project. This comprises academics from the Faculties of Arts and Human Science, Engineering and Business. ESD became part of the curriculum policy commitments of the university in 2009, but progress in putting this into practice has been very slow, with no real drivers from the top. However, a shared agenda and commitment to sustainability have enhanced mutual understanding and respect and brought over 60 colleagues to work together on an interdisciplinary curriculum conference in March 2012.

### Enabling links between formal, non-formal and informal education

ESD has recognised the importance of making links between formal and non-formal educators, and the need to breakdown some of the hierarchies of knowledge that transcribe this. RCE Yogyakarta in Japan has been actively promoting SD in this way as part of its mission. In 2011, it conducted a workshop on the implementation of ESD into HE curricula through community-based topics. This was designed to explore ideas and methods on how to integrate the concept of community-based SD into university subjects or courses as a part of an accelerated process of implementing ESD in society. Nine faculties from Universitas Gadjah Mada were involved, and the topics of proposals included conservation and forest management, community-based herbal plantations, sustainable laboratory practice, house waste management, green philosophy, and sustainable housing management (UNU-IAS, 2011).

Educational expertise in responding to community needs has generally resided with community and adult educators rather than with educators from the formal sector, so these kinds of initiatives can bring expert scientific knowledge into a mutually beneficial relationship with local community knowledge and expertise. Although non-formal education has often lacked government support and validation, the formal sector has much to learn from its experience and expertise. The work of Latin American educators like Paulo Freire has demonstrated methodologies for transformative learning and these are crucial for ESD as has been shown by, for example, the work of NGOs, like ASPBAE (Wade, 2010: 103).

In many cases, this type of work is undertaken by NGOs or by alliances of NGOs with universities. RCE networks can provide the framework for this to happen, as in RCE Greater Nairobi, which was originally set up by the NGO Kenya Organisation for Environmental Education, and brings together expertise from NGO community educators, from government organisations like the National Environmental Management Authority (NEMA) and from the Universities of Nairobi and Kenyatta.

The London Teacher Education RCE project also brings together teacher educators with NGOs working in ESD and is developing collaborations around place-based learning. However, there are still many constraints to such collaborations, as the agenda is usually set by formal sector educators because of the parameters of national curricula and government regulation in which they have to work. In England and Wales, teacher education is governed by a set of 'standards' and requirements, which generally require great imagination and commitment to tune to ESD. In addition, the emphasis on targets does not encourage mutual learning and collaboration between the formal and non-formal sectors, and it takes exceptionally creative, confident and committed teachers to find ways to develop this.

### *Flexibility: opportunities for cross-sectoral partnerships for ESD research and development*

Another of the strengths of the RCE network is that it has had the flexibility to develop since its early beginnings (Mochizuki and Fadeeva, 2008: 372), probably owing in no small part to its loose, informal structure. While new RCEs do have to go through quite a rigorous application procedure, they are allowed a great deal of flexibility in the model they develop. This has enabled them to respond to local, regional needs with regard to ESD, and has been a catalyst for a wide range of groups and networks to come together. RCEs encourage multi-sector and cross-sectoral engagement, which is illustrated by a look at the composition of their steering groups and stakeholders (as with RCE Penang and RCE Toronto). In the case of RCE Toronto, this engagement was between the city council and the university and allowed for real world participation of the university as part of the local community. RCE Munich, similarly, was started from the office of the mayor, and stakeholders include the university, local businesses and a bank, as well as youth organisations and local education providers. This kind of learning community opens up exciting new possibilities for mutual learning and for addressing local needs with regard to sustainability:

> What an RCE is primarily meant to attain is the provision of a platform for multi-stakeholder dialogue to share information and experience and seek ways to promote inter-disciplinary and multi-sectoral collaboration for ESD at the regional/local level. RCE can be interpreted as a mobilization mechanism to achieve much-coveted 'locally-relevant and culturally appropriate' ESD and a concrete manifestation of the 'partnership approach' emphasized in DESD International Implementation Scheme (UNESCO, 2005).
>
> (Mochizuki and Fadeeva, 2008: 372)

The UNU brand has also provided a recognition and an impetus for universities to become engaged with the RCE agenda, despite the complete lack of any funding attached – this, in itself, is a major achievement in the current economic climate! The partnership approach has also led to more formal partnerships developing between different RCE COPs, for example, a Comenius project on science education between European RCE partners.

However, flexibility and informality can be a weakness as well as a strength, as there are inevitable tensions with the formal HE hierarchical structure and this means that RCEs can be ignored or sidelined by senior management in universities. In turn, this can lead to resourcing problems as many universities fail adequately to fund or staff RCEs. Most RCEs survive on volunteerism, which is increasingly hard to manage as budgets and staff time get cut back. A number of RCEs (such as RCE London and RCE Skane) have no official funding or staff time allocated to them, making it difficult to maximise the potential they offer. In addition, their loose management structures can make accountability and evaluation quite difficult, leaving them open to be captured by particular interest groups. However, this is mitigated by their strong equality and social justice ethos, and their democratic, participatory forms of governance and active stakeholder involvement. This flexibility and informality also strengthens shared endeavour and their role as learning COPs, which are essential for the sustainable university.

## RCEs in Europe

RCEs in Europe build a community. Most European RCE members know each other pretty well. They partially share the same European history and they have a good understanding of each other's cultures in this little part of the world.

The first RCEs to be acknowledged (in 2005) were the Dutch RCE Rhine-Meuse and the RCE Cataluña in Barcelona, Spain. Nowadays, the European RCE community consists of 25 partner-organisations, unequally spread over Europe. Most partners are to be found in the northwestern part of the continent, with an emphasis on Germany and the UK. RCEs are under-represented in southern and eastern Europe. An active campaign is in preparation to stimulate new partnerships, especially in these latter regions.

Like elsewhere in the world, the majority of RCEs in Europe have originated from universities. But, in general, regional partnerships around these innovating universities have grown rapidly. Because RCEs in Europe are situated relatively close to each other, they have an opportunity to meet and cooperate (inter)regionally quite easily. The main activities of RCEs are in the field of education, training and ecology, and they cover subjects which focus primarily on the themes of Sustainable Consumption and Production, Well-being, and Sustainable Livelihoods.

The main challenges for RCEs in Europe are considered to be:

*   how to support the European Education for Sustainable Development (ESD) policy of the European Union in its European Strategy 2013–20;
*   how to build partnerships for (inter)regional projects in new fields like social entrepreneurship and open educational learning in ESD. Many RCEs have their individual experiences in these fields but we still lack the community touch;
*   how to inspire and better communicate RCEs as partners of preference for projects on ESD.

*Jos Hermans is an advisor to the European RCEs.*

## Threats and constraints

RCEs, by their very nature, can develop organically in response to the needs and context of each region. This allows them space for creativity and synergy, both intellectually in knowledge creation, and culturally in cutting across sectors and bringing together a wide range of interests for mutually shared benefits. However, at the same time, this perhaps gives them a fragility and a lack of formality, which means that they can be ignored or sidelined. Some can be criticised for a lack of strategy and many have difficulties in embedding their practices in the university or community. For effective, long-lasting change, engagement needs to be top-down, as well as bottom-up. It is too early for an in-depth evaluation of their successes, but this chapter has pinpointed the many opportunities that they offer for developing the new thinking and new learning that is needed in order to meet the immense challenges of the next 50 years. It is important to note that the RCE initiative is but one strand of the multi-layered and complex social fabric that is needed to effect substantial change towards a more sustainable planet. The same can be said of universities. Bringing all the necessary elements together is a delicate process, with many threats and constraints to this major project. RCEs, clearly, can have a strong contribution to make towards the development of the sustainable university in relation to campus, curriculum and community, by engaging with and trying to address real-world problems. At the same time, RCEs also seem to have the potential to act as a mobilising force for sustainable communities, locally and globally and, in themselves, can be seen as a social movement of solidarity, where the whole is larger than the sum of its parts.

## How to become an RCE

The Ubuntu Committee of Peers for RCEs was created in 2006 by the Ubuntu Alliance, an alliance of 14 of the world's foremost educational and scientific/technological institutions that, together, signed the Ubuntu

Declaration during the World Summit on Sustainable Development in Johannesburg in 2002. Members of the Ubuntu Committee of Peers for RCEs meet once a year to discuss ways to promote RCEs, review applications and provide recommendations to the UNU on the acknowledgement of new RCEs.

Those who are interested in establishing an RCE are requested to submit an application, in which they describe regional challenges to sustainable development and specify the expected role of interested parties and their responsibilities in the RCE. They should also identify a shared vision and objectives, as well as clear strategies to achieve the goals and objectives of the RCE. Applications should be developed in a way that takes into consideration the four core elements of an RCE: governance, collaboration, research and development, and transformative education. For further information see: www.rce-network.org/elgg/#

## Conclusion

The sustainable community can be conceptualised in many different ways, of course. Bell and Lane (2009: 648), for example, have argued that 'the notion of human beings living in a community but not compromising social or environmental limitations is a key element of a sustainable city'. Yet this begs many questions, such as, what is meant by social and environmental limitations? It also ignores the fact that most of us are members of a large number of different communities, based around fields of work, particular disciplines, personal interests and hobbies, family concerns as well as geography and place. The RCE COP allows for this complexity with the potential for the development and engagement of communities for the purpose of learning and mobilisation for sustainability. Universities have a key role to play in developing these COPs and in contributing as an active and responsible member of the community. In terms of policy, they need to create an enabling environment that allows and supports the development of relevant COPs for sustainability across disciplines, engaging with local and indigenous knowledge, embracing and valuing non-formal and community education, and engaging with local and regional concerns and issues.

Universities have an opportunity to play an integral role in sustainability solutions – but to do this they need to stop being part of the problem. They have an opportunity to create an enabling environment, with the campus as a learning site and the curriculum guiding the students on their own journeys towards sustainability. As active participants in both the local and global community, they have a key responsibility and duty of care to present and future generations in the development of sustainable communities and ways of living.

The RCE provides an excellent vehicle for this, as it frames the discourse on the sustainable community within the wider global imperatives of sustainability. At the same time, it supports an enabling culture and ethos where learning communities of ESD practice can develop and flourish at both the local and global scales.

## Notes

1   For more information about the global RCE network, please see: www.rce-network. org/elgg/#
2   Agenda 21 refers to the commitments that world governments signed up to at the 1992 UN Earth Summit. For details, please see: www.un.org/esa/dsd/agenda21/
3   See note 1, above.

## References

Asia-South Pacific Education Watch (2007) *Solomon Islands: Summary Report. Educational Experience Survey*, Mumbai: Asia South Pacific Association for Basic and Adult Education

Asia-South Pacific Education Watch (2009) *Papua New Guinea: Summary Report. Survey of Education Experience*, Mumbai: Asia South Pacific Association for Basic and Adult Education

Atkinson, H and Wilks-Heeg, S (2000) *Local Government: From Thatcher to Blair: The Politics of Creative Autonomy*, Cambridge: Polity Press

Bell, S and Lane, A (2009) 'Creating sustainable communities: a means to enhance social mobility?', *Local Economy*, 24 (8): 646–57

Cullingford, C (2004) 'Sustainability and higher education', in Blewitt, J and Cullingford, C (eds) *The Sustainability Curriculum: The Challenge for Higher Education*, London: Earthscan

Ginsburg, M, Cooper, S, Raghu, R and Zegarra, H (1991) 'Educational reform: social struggle, the state and the world economic system', in Ginsburg, M (ed.) *Understanding Educational Reform in a Global Context: Economy, Ideology and the State*, New York: Garland Publishing

Mader, C (2011) *RCE Community: Progress, Challenges, and Aspirations.* (PowerPoint presentation at International RCE conference.) Available at: http://www.ias.unu.edu/ resource_centre/Session%202%20Opening_1.ppt (accessed 20 June 2012)

Mochizuki, Y and Fadeeva, Z (2008) 'Regional Centre of Expertise on Education for Sustainable Development (RCEs): an overview', *International Journal of Sustainability in Higher Education*, 9 (4): 369–81

Nagata, Y (ed.) (2009) *Tales of HOPE II: Innovative Grassroots Approaches to ESD in Asia and the Pacific*, Tokyo: ACCU

Parker, J (2008) 'Situating education for sustainability: a framework approach', in Parker, J and Wade, R (eds) *Journeys around Education for Sustainability,* London: LSBU

Parker, J (2010) 'Competencies for interdisciplinarity in higher education', *International Journal of Sustainability in Higher Education*, 11 (4): 325–38

Sanusi, Z and Khelghat-Doost, H (2008) 'Regional Centre of Expertise as transformational platform for sustainability: a case study of Universiti Sains Malaysia, Penang', *International Journal of Sustainability in Higher Education*, 9 (4): 487–97

Shiva, V (1997) *Biopiracy: The Plunder of Nature and Knowledge*, Cambridge, MA: South End Press

Stefanovic, I (2008) 'Educational alliance for a sustainable Toronto: the University of Toronto and the City's United Nations University (UNU) Regional Centre of Expertise', *International Journal of Sustainability in HE*, 9 (4): 416–27

UNESCO (2005) *Report by the Director-General on the United Nations Decade of Education for Sustainable Development: International Implementation Scheme and UNESCO's Contribution to the Implantation of the Decade*. Available at: http://unesdoc.unesco.org/images/0014/001403/140372e.pdf (accessed 6 June 2012)

UNESCO (2007) *Introductory Note on ESD: DESD Monitoring and Evaluation Framework*, Paris: UNESCO

United Nations University Institute of Advanced Studies (UN-IAS) (2012) *Regional Centres of Expertise*. Available at: www.ias.unu.edu/sub_page.aspx?catID=108&ddlID=183 (accessed 6 June 2012)

UNU-IAS (2011) *RCE Bulletin, Issue 15.* Available at: www.ias.unu.edu/resource_centre/RCE%20Bulletin%2015%20FINAL.pdf (accessed 6 June 2012)

Wade, R (2010) 'EFA-ESD synergy: Redrawing the educational map: Journeys to sustainability', in *ESD Journey of HOPE: Final Report of the Asia-Pacific Forum for ESD Educators and Facilitators*. Available from: www.accu.or.jp/litdbase/pub/pdf02/004.pdf (accessed 6 June 2012)

Wenger, E (1998) *Communities of Practice: Learning, Meaning and Identity*, New York: Cambridge University Press

Wenger, E (2006) 'Communities of practice in 21st century organizations'. Foreword to *Guide de mise en place et d'animation de communautés de pratique intentionnelles*, Québec: CEFRIO

# 5   Leadership

*Chris Shiel*

Universities are uniquely placed to play a leading role in the pursuit of sustainable development ... [The challenge] for HE senior management is to lead the academic community in its engagement in this process and to facilitate this wider impact.

(Pearce *et al.*, 2008: 47)

Although universities are 'uniquely placed' (something consistently reinforced since Agenda 21[1]), the rhetoric is rarely matched by reality. Most universities are failing to address this leadership role; few universities are tackling the agenda in a systemic and holistic way. Examples of transformative change are rare. Evidence suggests that progress in many institutions has been driven bottom up, rather than through strategic, top-down intervention and conventional leadership approaches. Why is this, given that the Higher Education Funding Council for England (HEFCE, 2005, 2008) has strategically sought to raise the profile of sustainable development to ensure that it is 'mainstream'; and that university leaders around the world have signed international declarations of support for sustainable development?

Mention 'leadership' and most people automatically think of those roles at the top of a hierarchy – within universities, the role of the vice chancellor and his or her immediate team. Yet, while the leadership responsibilities of the university's senior team are critical, particularly their role in endorsing strategy for sustainable development, distributed leadership is equally important. As Marshall *et al.* suggest, 'we doubt if change for sustainability can often be brought about by directed intentional action, deliberately followed through' (2011: 9). Should more bottom-up progress be promoted as the single enabling factor? Or will effective leadership at senior management level, as an additional driver to the work of champions, speed up progress?

This chapter will focus on leadership for sustainable development (SD) and, specifically, the leadership of SD within universities. A number of questions will be explored: What can we learn from theories of effective leadership? What would effective leadership for SD involve? Does the university setting require different leadership? What practices would be involved and what type of behaviours might be anticipated? The implications of the findings will be explored, followed by two of my own experiences. Finally, suggestions will be offered as to how senior managers might create an enabling environment, and facilitate a culture where everyone feels empowered to contribute to SD.

## A context that calls for 'directed intentional action'

The role that higher education needs to play in contributing to SD is well documented and reinforced by other chapters within this book. Essentially, universities need to be at the forefront of social change in the transition towards a more secure and sustainable future (Bowers, 2001). In discharging this significant leadership role, universities need to ensure that the educational experience for all learners contributes to the development of a generation of leaders and decision-makers who have the knowledge and skills to address sustainability (Copeland, 2008; Jones *et al.*, 2010). Research needs to contribute by addressing the challenges posed by *un*sustainable development, and university/campus operations need to exemplify social responsibility and best practice in environmental management (Pearce *et al.*, 2008). Universities also have a key role to play within their communities to enhance economic, social and environmental sustainability.

Corcoran (2010) reinforces the challenge and the need for transformation, suggesting that 'changes are needed in curricula, pedagogy, policy and institutional structures' (2010: xiii). Sterling (2004b) argues that system-wide and transformative change of policy and practice are necessary to address the complexity and magnitude of what is required for SD; however, institutional inertia impedes progress (Corcoran, 2010).

This is particularly evident in the UK higher education sector. The lack of coherent leadership for SD is frequently cited, at the same time as acknowledgements are made that champions lead innovations, working bottom up (Copeland, 2008; Jones *et al.*, 2010). This lack is disheartening, given that sustainability is hardly novel, as Wals and Blewitt (2010) remind us. They do go on to explain that, within universities, SD is often 'just another course or research project, expendable if it does not pay its way' (2010: 70), or limited to 'campus greening'. In regard to the latter, environmental managers have exhibited passionate leadership in greening estates, but, in most universities, this has not been matched by curriculum change (Sterling and Scott, 2008); examples of campus greening activities influencing the core business are few (Tilbury, 2011). Environmental managers, in many cases, have operated in a silo (not necessarily through choice). The transformative ways of thinking and working necessary for SD have been largely absent because few higher education leaders have grasped the wider implications of the agenda, beyond getting someone in the organisation to deal with it, usually an environment manager.

The UK's Universities that Count (2009: 37, 11) indicated areas of integration that scored low in their benchmarking survey results, which included:

- integration of corporate responsibility and environmental management into strategic decision making, and;
- building corporate responsibility and environmental management into the development of senior managers.

The report highlights that the higher education sector average is considerably below the business sector average in these areas, reinforcing the idea that SD is too often a low-priority agenda item for university executive teams; it rarely appears on the agenda of boards or councils in the governance process.

As referred to above, HEFCE has sought to increase engagement, making university responsibilities explicit. HEFCE's carbon reduction strategy (2010) underscores SD as a strategic issue, which 'extends beyond the traditional estates function' to other activities 'including teaching, research and public communications' (2010: 8), and which requires 'behavioural change and new ways of working' (2010: 15) led from the top. Further, 'it is a crucial area for governors who should be informed and involved in decision making on the institution's approach to reducing its emissions' (2010: 17).

The reference to governance is important: the university board has to be the main advocate for sustainability if employee engagement programmes are to succeed (Brighter Planet, 2010). Perhaps a lack of advocacy from those with university governance responsibility contributes to the overall lack of strategic engagement. The institutionalisation of sustainability, which requires culture change and transformational ways of working (Doppelt, 2010), must be strategically coordinated – but coordination also requires the support of board members (HEFCE, 2008) who hold the executive team to account.

Undeniably, some universities are leading the way, but they are the exception rather than the norm. Even within institutions where there has been substantial progress, only 'one or two people' (HEFCE, 2008: xi) championed the initiatives (Copeland, 2008; Jones *et al.*, 2010). Few senior leaders have seen the potential of an integrative approach to SD; champions leading change within particular domains of activity rarely have the power (without senior management backing) to challenge silo mentality and transform the system.

Tilbury (2011) also highlights the need for coordination and leadership from senior managers within global higher education. She notes that, although most university leaders have signed international declarations for sustainability, visibly espousing their commitment, change requires more than a commitment to principles. She concludes that senior management teams hold the key to transforming higher education to contribute towards a sustainable future. Indeed, even higher education leadership development programmes have been of limited value in supporting leadership for sustainability.

'Directed intentional action' may not bring about SD, but the lack of action, on the part of university senior teams, is currently blocking progress. Unless those at the top are fully engaged, universities will fail to contribute to the large-scale transformation necessary for a more just and sustainable future.

Senior leadership development for sustainability is one piece of the jigsaw, so the obvious question is: what would effective leadership for SD involve? We will explore what is required to be an effective leader (the assumed content of leadership development programmes), through the lens of SD.

## A brief tour of leadership theory, applied to SD

A plethora of research studies on leadership exists – much of it contradictory, quite a lot of it criticised (particularly in terms of methodological flaws). Far less research explores leadership for SD, there is little that is empirical and virtually nothing suggests what it would look like within higher education.

Throughout history, we have been fascinated with what it takes to be a successful leader. What are the essential qualities that enable some leaders to achieve the impossible, inspiring loyalty and sacrifice, while others achieve very little? If this were better understood, then it might be easier to articulate what is required to effectively lead on SD. Speculation on the essence of leadership spans centuries; however, scientific research on this issue did not really begin until the twentieth century, from which point studies took so many different directions that some writers actually question whether it is a useful scientific construct (for example, Alvesson and Sveningsson, 2003).

Stogdill wrote: 'There are almost as many different definitions of leadership as there are persons who have attempted to define the concept' (1974: 259). Yes, very little has changed since Bennis commented that, '[T]he concept of leadership eludes us or turns up again with its slipperiness and complexity. So we have invented an endless proliferation of terms to deal with it … and still the concept is not sufficiently defined' (1959: 259).

This criticism suggests a similar landscape to that of SD. SD means different things to different people, with criticisms that the concept is too broad/complex, too ambiguous, being interpreted in a variety of ways (Leal Filho, 2000) and giving rise to 'a veritable industry of deciphering and advocating what [SD] really means' (Kates *et al.*, 2005). Leadership, too, has attracted similar press. No surprise, then, that little has been published on the intersection between leadership and sustainability (Brown, 2011).

The leadership literature is vast and more confusing, in terms of the variety of frameworks and contradictions, than the literature on sustainability. Leadership books that offer 'ten tips' hold great allure. Unfortunately, just as there is no guarantee that prescriptions for effective leadership will secure effective organisations, a toolkit for SD will not save the world. The leadership literature does, however, offer a useful touchstone for reflection.

### *What the key theories say*

Key theories of leadership have sought to explore the characteristics of leaders, the characteristics of followers, and the characteristics of the situation. Studies emphasise different variables at the expense of others, and definitions take on particular emphasis, depending on the variables of study. Are some theories applicable to leading SD?

Early leadership studies failed to identify universal leadership traits (Stogdill, 1974); however, the possession of particular traits (for example, self-confidence, intelligence) and skills (technical skills, conceptual skills, inter-personal skills) may enhance success. When identifying the universal traits of effective leadership failed, research moved on to correlate leadership behaviour with leadership effectiveness, before considering the influence of contextual variables on the leadership process (situational leadership). A categorisation of important leadership behaviours includes: task accomplishment behaviours, relations-related behaviours, and behaviours necessary for leading change. In a nutshell, it is suggested that effective leaders show a high concern for task, people, and managing change, adapting their behaviours to situational contexts. If the task were to achieve SD, then an effective leader would simply focus behaviours on task, people and change, adapting to context. Unfortunately, although logical, this represents a huge simplification: the enormity of the task, the complexity of the challenges and the need for deep, transformational change should not be underestimated (Sterling, 2004a; Jones *et al.*, 2010).

### *Learning from transformational leadership theories*

Transformational change necessitates transformational leadership, so transformational leadership theories may be more helpful to the task at hand. These theories emerged from the need to explain the exceptional influence of some leaders, the emotional reactions in followers, and to distinguish transformational leadership from transactional leadership (although the two are not mutually exclusive).

According to Bass (1985, 1996), a transformational leader inspires such loyalty and respect that followers have an enhanced perception of the importance of the task, are more likely to transcend their own self-interest and will activate higher-order needs. As a transformative behaviour, 'inspirational motivation' involves the leader communicating an appealing vision, using symbols to focus attention and effort, and then modelling appropriate behaviours (Bass and Avolio, 1997).

Intuitively, this sounds particularly appropriate for leading a journey towards SD: if the right attention is given to communicating the vision, with leaders inspiring motivation and role-modelling SD, the task seems relatively straightforward. So why is more of it not going on? Unfortunately, transformational change is not particularly easy within hierarchical structures; if leaders are more familiar with transactional approaches, deploying technical/rational methods for compliance, then transformation requiring inspiration may be alien and emotionally demanding.

### Some contemporary leadership theories

More recently, leadership competencies have been identified that seem pertinent to leading SD. These include:

- emotional intelligence (Goleman, 1995), which helps leaders solve complex problems, manage crises and enhance decision-making;
- systems thinking (Senge, 1990), (the need to think systemically is a necessary quality for SD); and;
- the ability to learn (Argyris, 1991), which is fundamental if new, more sustainable ways of working are to be developed, particularly as SD necessitates a 'learning as we go' approach (Parkin, 2010).

However, Argyris also highlights the nature of the challenge: the 'smartest people' are not always predisposed to acknowledge that they have anything to learn (and perhaps, within this, is an important reminder for those seeking to engage university leaders).

Contemporary conceptions of effective leadership are not far off from what intuitively might be required to lead SD effectively. Yukl offers a good example. He considers leadership as both a specialised role but also an influencing process, proffering a broad definition: 'Leadership is the process of influencing others to understand and agree about what needs to be done and how to do it, and the process of facilitating individual and collective efforts to accomplish shared objectives' (2006: 8). He suggests that leadership through influence (which may be rational or emotive), with effective facilitation, enables a group or organisation to meet 'future challenges'.

It is in this broader conception of leadership that it is possible to see how leadership theory might be extended to embrace leadership for SD, where influencing strategies and mobilising collective efforts focus on the biggest future challenge imaginable: a sustainable future. If leaders would agree that the sustainability of the planet must be the 'shared objective', then effective leaders would unite collective efforts, on a journey forward. Again, with the right attention given to task, relationships, etc., the work seems straightforward.

Yukl encapsulates, in ten functions, the essence of effective leadership. Effective leaders:

1 help interpret the meaning of events…
2 create alignment on objectives and strategies…
3 build task commitment and optimism…
4 build mutual trust and cooperation…
5 strengthen collective identity…
6 organize and coordinate activities…
7 encourage and facilitate collective learning…
8 obtain necessary resources and support…

9    develop and empower people ... and
10   promote social justice and morality.

(Yukl, 2006: 456)

Although Yukl was not writing with reference to SD, it is easy to see how these functions, with some elaboration, are not inconsistent with what is required to lead SD (the tenth point in particular). Yukl stresses the importance of leaders interpreting complexity and ambiguity, and facilitating sense-making, both being critical to SD. He also emphasises that collective learning is essential, with effective leaders increasing commitment and confidence to move forward through their own enthusiasm – again relevant and applicable to leading SD.

But does this all seem too managerial? Without unpicking the management/ leadership debate here, suffice to say it is largely a question of interpretation. Bennis (1959) suggests that management (which he interprets as administration) is about control, doing things right, and the bottom line; leadership, on the other hand, involves innovation, challenging the status quo, inspiration, and keeping an eye on the horizon (external challenges). He reinforces the need for leaders to be engaged in learning to address problems that are ill defined (Bennis, 1998), something which clearly resonates with sustainability. He suggests that the indispensible qualities of leadership are: guiding vision, passion, integrity, trust, curiosity and daring – qualities which champions of sustainability demonstrate in abundance. Although Bennis, like Yukl, was not writing with explicit reference to SD, it is easy to see the connections: 'vision' must embrace a sustainable future; passion and integrity must be directed towards, and embrace, living within environmental limits and social justice. Following this logic, leadership theory is not inconsistent with what is required to lead SD.

### *Theories on leadership for sustainable development*

Literature that explicitly explores traits in relation to SD is sparse. But, Martin and Jucker, in questioning the failure of universities to develop the right kind of leaders, suggest qualities needed to enable sustainability, which include 'humility, respect for all forms of life and future generations, precaution and wisdom, [and] the capacity to think systemically and challenge unethical actions' (2005: 21). Such qualities, although not empirically tested, suggest an opportunity for a new dimension in mainstream leadership studies, which have sought to identify a magic formula for success (in bottom-line terms), but not in relation to SD.

Egri and Herman (2000) provide one of the few empirical studies, exploring traits in relation to successful environmental initiatives. Their results suggest that environmental leaders appear to be more eco-centric, with higher levels of openness to change. They are also strong in 'self-transcendence', with values related to benevolence, universalism and motivation to promote nature, and the welfare of others. They identified several leadership approaches that were shared among their interviewees, including interdependent inclusiveness with stakeholders and a preference for egalitarian, decentralised and participative decision-making.

Quinn and Dalton (2009), in an exploratory study of senior executives, conclude that the leadership practices required for leading sustainability are similar to those required of 'effective' leaders – but with the need for some 'additional capacity'. They suggest that leaders of SD have a different mindset as to the purpose of the organisation, build capacity in their systems and culture to address sustainability, and include a wider group of stakeholders.

Ballard confirms that there are points of commonality suggesting that mainstream leadership theory is extremely helpful in elaborating the general characteristics of leadership and not entirely inconsistent with what is required to lead SD:

> A leader for sustainable development will share many of the qualities that distinguish great leaders in other fields. For instance, she will demonstrate integrity (interpreting it appropriately for work in the field of [SD]) and courage and will both earn and demonstrate trust.
>
> (Ballard, 2005: 17)

Ballard proposes that, as well as general qualities, leadership for SD must address the contextual barriers to change, which may be subjective (ways of seeing the world, at both a personal and organisational level) or objective (for example, where technology or systems lock in the continuance of *un*sustainable practice). Leaders need to recognise contextual barriers and address three conditions for change, in parallel: awareness, agency and association.

Developing awareness is a critical task. Few people recognise the scale, urgency and relevance of the issues; some may actively choose to avoid thinking about them. In building awareness, attention needs to be drawn to the urgency, complexity and structure of the issues and the systemic nature of the challenge. Finally, enhancing awareness also includes developing a sophisticated understanding of how mental models block progress, limiting not only perceptions of the problem, but also development of solutions.

Ballard highlights the importance of 'agency' (and the role of the leader in empowering action), and 'association' (working with others), which is 'at the heart of work for [SD]' (2005: 6). Unless all three conditions are addressed in parallel, work will fail: without agency, association and awareness will result in a talking shop; if awareness has not been fully developed, but there is association and agency, actions may be either trivial or pointless; and, if association is missing, the result may be the isolated activist who is ignored, or possibly stressed (a familiar scenario to those who have sought to work bottom-up).

Ballard's work provides an insightful reminder of the enormity of the 'awareness' agenda. He also reinforces, as other authors do (for example, Dunphy *et al.*, 2007), the centrality of learning for SD, particularly the importance of double-loop learning.[2] Further, in drawing attention to how mental models promote unsustainable practices and how awareness will be insufficient without agency and association, Ballard offers some explanation for why progressing the agenda in universities has been slow.

Few senior university leaders exhibit the depth or breadth of awareness of SD, which Ballard suggests is critical; they are generally unaware of the limitations of mental models, and they demonstrate little awareness of urgency. Insufficient attention may also been given to developing awareness, agency and association, in parallel, within universities. Further, it might be suggested that, although universities are sites of learning, insufficient attention is paid to double-loop and continual learning in an organisational learning sense, something which is a 'building block for long-term success on the path toward sustainability' (Doppelt, 2010: 215).

Doppelt (2010) sheds further light on the nature of the problem, suggesting that organisations are often hierarchical and locked into traditional ways of working (something said of universities); building capacity for SD in a way which synergises collective effort may be impossible in such structures. He observes that discussions on technological solutions and policy development have dominated the dialogue on sustainability, with insufficient attention given to the enormous shift that is required in terms of changing engrained thinking, cultures and practices. He suggests that patriarchal governance models (based on hierarchical authority structures, and reactive, command/control leadership) impede progress; holistic approaches towards SD will fail where the system blocks information flows, encourages silo mentality, disempowers employees, and where vertical lines of authority hinder collaboration.

Doppelt outlines a guide for change, based on a review of how leaders of both public and private organisations have either succeeded or failed with the implementation of sustainability programmes. On the basis of identifying seven blunders that contribute to failure, Doppelt (2010: 107) proposes solutions in a 'wheel of change'. He suggests that vision and leadership are key steps towards sustainability and that the first lever (or entry point for change) has to be to challenge and shift the dominant business-as-usual *mindset* (a tough call given that participants may feel insecure operating with non-traditional ways of working in a patriarchal governance system). The second lever for change is to rearrange the *parts* of the system to eliminate silo mentality. The third lever (drawing on the work of Meadows, 1997) is to alter the *goals* of the system so that SD is explicit. The fourth lever (again identified by Meadows) is to restructure the *rules of engagement* (which includes developing new operating modes and governance strategies). The fifth is to address *information flows*, which involves ensuring communication in relation to sustainability permeates the organisation. The sixth is to address the *feedback loops* to enhance learning. The seventh, finally, is to adjust the *parameters* that influence stakeholder behaviour (charts, job descriptions, etc.).

Doppelt offers a stimulating model for planning change, which may appeal to those whose worldview has been influenced by management, with the caveat about the allure of anything that proffers the 'ten best': beware the dangers of reductionism. Meadows (2002) warns of the dangers of prescription, suggesting that for anyone raised in the industrialised world, it is easy to be 'blinded by prediction and control'. She, like others, reinforces the idea that entrenched worldviews manifest significant barriers to change; she warns of the dangers of adopting a simplistic systems perspective with notions of control reminding us that: 'We can't control systems or figure them out.'

Marshall *et al.* (2011) express similar criticism, while not completely disagreeing with Doppelt's suggestions. They comment that his advice is 'quite prescriptive and procedural' (2011: 9), referring to Meadows' work on systems thinking and indicating a preference for an approach to leadership for SD that involves a spirit of inquiry, and leadership as a relational practice.

Ferdig (2007) also emphasises the complexity and the challenges of progressing SD, when views of how the world works and views on leadership are deeply entrenched. She suggests that we need to challenge commonly held assumptions of phenomena based on mechanistic models (and too largely influenced by scientific management) and look to complexity science, which 'reveals a radically different worldview that challenges Newtonian assumptions of empirical truth, reductionism, stability, certainty, predictability, and control' (2007: 27). She cautions that the only certainty is uncertainty: a leader who stands ' "above" the people and situations he or she is leading' will not suffice.

She suggests that traditional approaches to leadership, where the leader is 'in control' and is deferred to, contributing to a 'learned helplessness', may not address the decline of natural systems, or deliver the changes needed for a sustainable future. In a similar vein to Marshall *et al.* (2011), who suggest that leadership exists at various levels in various guises, Ferdig proposes that '[a] sustainability leader is anyone who chooses to engage in the process of creating transformative change with others aimed towards a more sustainable future' (2009: 1).

Ferdig asserts that an unprecedented approach to leadership that increases the total capacity for change within the system is necessary. The capabilities associated with traditional leadership, such as strategic thinking, planning and mobilising effort, are more critical than ever, but a radically expanded view of leadership, shifting leadership to 'with' rather than 'over', is vital. This approach acknowledges that anyone can be a leader; a sustainable future will only arise with the collective engagement of all human beings. Those in formal leadership roles must see themselves in relation to others and through authentic relationships, synergizing human action. They need to think holistically (building capacity for holistic thinking) and engage in a process of enquiry and learning.

Ferdig's proposals are far less prescriptive than Doppelt's, although, to be fair, Doppelt (2010) does suggest that any of his seven levers offer an entry point for change, anyone can begin the process, and that 'building support from the grass-roots upwards is often the way change begins' (2010: 120). Ferdig's ideas (2007, 2009), however, may have greater appeal for those who have been championing change for SD bottom up, particularly with the emphasis on leadership as a relational activity, and the need to create spaces for constructive conversations.

## A note on the unique higher education setting

Some take exception to the use of 'toolkits' for change, which emphasise managerial, rather than emancipatory, approaches. However, provided caution is exercised against over-prescription (in terms of what counts as SD, see Scott and Gough [2004] – and also in terms of the best approach to change), then there is no reason not to use them.

Marshall (2007) notes that universities are increasingly recognising the significance of effective leadership and applying lessons learned from the private sector; the 'art' of leading change is high on the institutional agenda. Further, there is an increasing recognition of the importance of distributed leadership within universities, and realisation that this has to be counter-balanced by leadership from the top. Senior teams need to communicate the vision for change, developing the enabling structures and consolidating the work of champions.

Unfortunately, hitherto, most institutional change initiatives (and leadership development), with the exception of Green Academy (see Luna and Maxey, this volume), have focused on issues related to competitive advantage (and financial sustainability) rather than SD. There is, however, no reason why SD should not be the focus of leadership and change – all that is required is for the senior team to acknowledge that SD is a priority. Therein lies the challenge! But, as Terry Williams (2012), Dean of Hull Business School, writes:

> [I]t is not easy to be an advocate of change. Systems and processes tend to endure, regardless of changing circumstances. Although it is important to stay true to key goals and objectives, urgency often displaces importance, a re-occurring concern of senior leaders. This sin is often compounded by senior leaders who are over-scheduled and unaware of systems thinking, which can provide an invaluable holistic understanding on multiple levels allowing for better decision making.

Yet champions working bottom up can only take the agenda so far. The enabling of cross-institutional working and institutional change sits with the senior team. If sustainable thinking and behaviour is to permeate universities and become 'the central organising governance principle of the future' (Doppelt, 2010: 271), then senior leaders must take up the torch.

### Perspectives from environment managers

As environment managers, the starting point for our own leadership is our own passion and enthusiasm; we are working in this field because we are committed to making a difference.

Building relationships is central to making progress. It is necessary to identify allies and seek union, to engage and cooperate with potential champions across the organisation by looking for interconnections with other areas of work, but also engaging with those who are potential blockers. We must be prepared to challenge but also know when influencing strategy requires other less challenging forms of advocacy to avoid polarisation of positions.

Making the case, securing the resources, and being able to translate ideas into action are central to effective dispersed leadership. But perseverance (particularly when it feels like you're pushing a large boulder up a very steep hill) is also crucial. When you are trying to affect change over a sustained period, you need to consider how you will look after yourself and your team so that you are able to keep going. But it is also important to make the most of it when 'the wind is blowing in the right direction', highlighting achievements in order to maintain momentum (one of the benefits of pursuing awards and accreditations).

The need to track and respond to constant change, both within the organisation and externally, means we must trial different approaches, learn from the failures and change styles as required.

Factors that inhibit leadership at this level may include a lack of clear leadership from above, the need to mitigate the impact of competing strategies and highlight the interdependencies between them, and lack of either resources (financial/human) or institutional buy-in.

Leaders at all levels can help by providing clarity of direction, considering sustainability in every decision that they make, taking the opportunity to champion sustainability at the different forums that they have access to, and considering the interdependencies between competing strategies and policies.

*Amanda Williams is the Environment and Energy Manager at Bournemouth University. Pat Pica is the Energy and Environment Manager at University of Sussex.*

## Two personal experiences

### *Bournemouth University*

My own experience of championing change confirms the importance of creating learning spaces, collaborating across boundaries (discipline boundaries and academic/professional services boundaries) and initiating challenging conversations (the most uncomfortable encounters have yielded the best fruit).

Anderson and Bateman (2000) show that sustainability champions are more likely to be successful in gaining attention and stimulating commitment for action when they frame the issues using business language and a logical approach; this has been particularly important in building capacity and gaining support – dramatic or emotional appeals have been less successful. However, critically important in leading change has been the development of a repertoire of persuasive rationales for engagement, tailored to appeal to different audiences and different contexts. Approaches have to resonate with organisational cultures and participants' worldviews.

Ideas from the leadership of change, in particular 'participative evolution' (Dunphy and Stace, 1993) and Quinn's (1980) suggestions for creating 'pockets of commitment' through integrating processes and interests, have proven valuable for capacity building and developing a holistic approach within a higher education setting (see Shiel, 2007).

Experience working with colleagues and other champions across the sector suggests that a leader for SD (at any level) shares many of the traits and qualities of an effective leader, but with higher levels of openness, integrity, humility, ability to think creatively, and with a greater respect for the planet. In focusing on the task, effective leaders channel their energy directly towards the sustainability challenges, with particular attention to inclusive behaviours that enhance collective engagement. As intimated by transformational leadership theory, they inspire motivation and establish a shared vision of a sustainable future. Champions indicate that a sustainability leader:

- exemplifies commitment to learning for SD (building personal awareness, 'mindfulness') and facilitating learning with, and for, others;
- explores own/others' worldviews and challenges own/others' limitations;
- stimulates task commitment and optimism for action;
- demonstrates responsibility for the environment in the personal, professional and community spheres;
- exemplifies passion;
- displays creativity in planning for the future (visioning);
- respects the complexity of systems and the interconnected nature of the challenges;
- encourages multiple perspectives, learning from conflict;
- identifies new ways of working and opportunities for learning;
- develops alliances, to build commitment and momentum;
- seeks to implement a holistic approach, uniting everyone in collective responsibility and action;
- assesses all actions and decisions in relation to SD (more futures-oriented decision-making);
- inspires hope;
- proactively seeks positive solutions and displays daring in challenging the status quo;
- endures in adversity.

Senior leaders in universities need to consider all of the above, but particularly their strategic role in enabling holistic approaches to flourish within their institutions; decisions about strategy, culture and infrastructure have to come from the top.

## The role of leadership in cultivating organisational change towards sustainability

Plymouth University has adopted an integrated approach to sustainability, one that blends both the passion and practical wisdom needed to make both big and small changes. A key part of this cultural change is building an environment based on shared values, recognising the importance of self-leadership within an organisational accountability framework underpinning collaborative leadership.

At Plymouth, number one in the People and Planet league table of 'green' universities 2010 and top performer since 2007, we are embedding sustainability at the heart of our university strategy. For us, it is about the way we do things around here – the way we support and enable student learning, our global research activities addressing the grand challenges of our day, and our role in driving social inclusion and economic growth in the communities we serve.

Our journey started with a small group of enthusiasts made up of academics and specialists who gave us a pure vision about what was possible; they set out a dream. As senior leaders, we saw the merits and benefits ahead for the institution and our people, in terms of both reputational gain and financial sustainability. But perhaps most important to turning our ambition into action was the fact we felt we had a responsibility to do the right thing. Our approach now involves a large group of people, across the whole university: people in estates, procurement, research, the energy team, policy, teaching, and, in particular, our students.

Ultimately, sustainability involves a clear partnership between an institution's leadership, staff and students. For us, this is collaborative leadership in action. It is clear that resilience – personal, professional and institutional – together with strategic agility will be key to securing institutional sustainability in the future. This approach helps to build and sustain an enabled and flexible university culture that is better able to withstand recessionary pressures and succeed in situations characterised by high levels of ambiguity.

Collaborative leadership (Ibarra and Hansen, 2011) is important as our sector reshapes itself for the future. Collaborative leaders think globally and embrace diverse talent to produce results – simultaneously innovative and efficient, agile and scalable – an exciting and new model of leadership, fundamental to building a successful sustainable university.

*Professor Wendy Purcell is Vice-Chancellor of Plymouth University. Caroline Chipperfield is Policy Adviser to the Vice-Chancellor at Plymouth University.*

### Reference

Ibarra, H and Hansen, M (2011) 'Are you a collaborative leader?', *Harvard Business Review*, 89 (7/8): 69–74

## HEFCE LGM project

In relation to leadership development for senior teams, my experience suggests that some senior leaders are very willing to engage, providing they can set aside the time to listen and learn. The HEFCE Leadership Governance and Management Fund (LGMF) project represents a small-scale project that aims to address a shortfall in senior leadership development and engagement. The project has engaged senior leaders and board/council members at four universities, in exploring their roles in relation to leadership behaviours to embed SD, and actions to achieve challenging carbon reduction targets (Shiel, 2012). The project, facilitated by myself, two environment managers and an external ESD expert, has deployed a social learning approach (Bandura, 1977) and undertaken action learning (Revans, 1982) to build capacity for top-down leadership for SD. Experience in delivery has reinforced the importance of speaking the language of leadership with participants, critical as a prelude to engagement, but also ongoing, before challenging worldviews on leadership and SD.

The challenges of engaging with this stakeholder group should not be underestimated. Critical to implementing the project has been the time invested in managing three parallel strands of activity – the content agenda, the control agenda and the process agenda – coupled with substantial amounts of 'backstage activity', all of which are imperative in change projects (Buchanan and Boddy, 1992). The first hurdle is gaining commitment to engage; the second (and bigger hurdle) is translating commitment into time for workshop learning within the limits of a full governance calendar (and where gatekeepers either facilitate or limit opportunities). The barriers are not insurmountable but such projects test tenacity and emotional resilience.

The workshops have provided space for senior participants to engage in collaborative learning for SD. The depth and breadth of awareness suggested by Ballard (2005) is being developed, agency is being addressed through action planning, and learning and reflection is enhancing capacity. Conceptions of leadership behaviours for SD are being explored; worldviews are being challenged.

Discussion around leadership behaviours for SD appears to need to be grounded in the theories that inform worldviews on leadership. Participants understood less about strategic leadership behaviours than anticipated; some had a narrow understanding of SD, while some remained cynical. Participants found it relatively easy to articulate blocks to SD, but more challenging to identify which behaviours accelerate momentum. They also understood relatively little about the influence of worldviews. Developing actions plans represented more familiar territory. However, participants did not always appreciate that, in action learning, *they* own the actions.

Although the project is still underway, outputs are already resulting in organisational change and culture change, where the language is shifting and, increasingly, references SD. However, university structures and processes are tough to challenge; they may impede radical change. Unless university chairs and vice chancellors serve as passionate champions of change, ensuring that senior teams consider what it takes to lead the profound systemic and cultural change that SD requires, progress will falter.

Leadership development programmes are undoubtedly needed to build capacity for SD, but will only work if they are seen as part of a journey; participants have to be willing to engage in participatory approaches.

Participants in the project suggest that the big questions for senior teams are the following:

- How can sustainability be driven into the core business of higher education?
- How can an enabling environment be created, whereby all participants feel empowered to contribute to SD?
- What institutional structures and processes currently inhibit progress? What alternatives would enhance collaboration, inspire engagement and synergise action?

## A perspective from the chair of a university board

It is critically important that the chair of the university board is visibly committed to ensuring that the university fully engages with sustainable development (SD). The challenges that we face in terms of global sustainability are complex, requiring leadership, at all levels within institutions, and participative approaches for developing new ways of thinking and behaving. Universities are uniquely placed to respond to the challenge through research and education, and in the way they discharge their corporate responsibility both as a business and as a stakeholder in the community.

Leadership has to come from the top. It is the board or council that holds the university's senior management team to account; it is the board that needs to exemplify commitment and ensure that SD is not only addressed in all decisions but also championed throughout the organisation.

My role as chair is important and my passion for supporting change for SD is critical to progress. However, understanding what we need to do is challenging, requires learning, and those who are sceptical need to be convinced.

At Bournemouth University, the board and senior management team have had the opportunity to engage in a HEFCE Leadership, Governance and Management project, which has sought to build capacity in terms of leadership for SD. The objectives of the project were to:

- provide participants with the opportunity to increase their knowledge of SD;
- explore the breadth of concerns (in relation to sustainability);
- identify their potential role in supporting culture change;
- develop approaches to securing commitment to carbon reduction and SD.

The project has provided participants with the space to reflect on SD and to consider what leadership action and leadership behaviours are necessary for change. I have been delighted to play an important role in supporting this project and to following actions through, although, at this point, we are in the early stage of a journey.

One significant action a chair can take at an early stage is to ensure that all papers that go through committees, and come to governance structures for decisions, have considered SD. The chair can also ensure that the governance body recruits champions of SD.

At a personal level, exemplifying passion for the agenda, providing support for those who are taking the agenda forward, and seeking to lead by example are imperative. If we wish to build broader support for SD, then those at the top have to visibly lead by example and ensure that their behaviour is congruent with the message.

*Susan Sutherland OBE is Chair of Bournemouth University Board and former Chief Executive of the Poole Hospital NHS Foundation Trust.*

## A proposal for taking practical actions

In terms of some practical actions, senior leaders need to:

- articulate a compelling vision that embraces SD;
- develop approaches and communication strategies that mobilise the engagement of the entire university population;
- ensure champions exist at all levels (including on the board);
- set bold SD performance objectives (SD embedded in the curriculum, zero waste, etc.);
- ensure SD is considered in every decision;
- exemplify passionate, visible leadership, working towards SD.

Further, senior leaders need to reconsider accepted views of leadership (and ways of working); reflect on what might be more radical and inspiring; appreciate the importance of dispersed leadership; and move forward in a spirit of enquiry.

Champions at all levels highlight the need for passion and commitment, coupled with high levels of resilience. Collaboration is also critical: forging alliances generates impetus for change, sustains momentum and nourishes champions. Through collaboration, encounters with remarkable people occur, further opportunities and new ways of seeing are developed, and a world of possibilities emerges – 'predictable miracles' in the sense used by Jaworski (1998).

Further actions which contribute to the momentum, which all staff can take, include:

- challenging the status quo and dispersing leadership;
- giving feedback to various leaders – from line managers, (lecturers in the case of students) to the vice-chancellor;
- including students in this thinking, especially as the UK is going down the path of 'student choice', 'the student experience' and the 'student as partner/ customer'.

If students increasingly ask for SD leadership, they'll increasingly get it.

So, if leadership is about 'learning how to shape the future' (Senge in Jaworski 1998: 3) and enhancing capacity to participate in the world, then university leaders, at all levels, need to ensure that their institutions lead and shape a future that is sustainable. Espousing commitment must be followed by action and an unprecedented approach to leadership.

## In conclusion

This chapter has considered leadership for SD in the context of higher education where much of the leadership, to date, has been exemplified by champions, and has very rarely been matched by leadership at the top. In response to calls for effective leadership that transforms universities, a brief overview of leadership theory has been considered, to sketch the terrain for those who are less familiar, but also to look for clues as to whether what is required to lead SD is very far removed from the general leadership literature. The literature offers a touchstone for considering what constitutes leadership effectiveness (attention to task, relations, etc.). Transformational leadership theories highlight the importance of inspirational motivation, developing appealing visions and uniting collective effort. Research that borrows from the strategic change literature offers ideas, which may be more or less useful in different situational contexts.

A leader for SD needs to understand what constitutes effective leadership while also developing a heightened awareness of the complexity and urgency of the SD agenda. Learning SD is critically important for senior university leaders; learning leadership is equally important for those championing change at any level.

Although this chapter started with a consideration of the lack of leadership for SD by senior university leaders, it has been proposed that leadership is a relational activity and a collective responsibility. As Ferdig states: 'Together, we can begin to explore how collaborative, self-organising leadership, can generate innovative and sustainable solutions, and wiser investments, for a more sustainable world' (2007: 34).

Drawing on my own experience of leading SD projects, some behaviours of leaders for SD have been suggested. The list is not exhaustive; each leader will develop their own ideas of what works in their context. Key are:

- learn and lead;
- inspire;

- facilitate;
- envision a better future, exemplify the change you want to see in the world, and endure.

Leading the necessary breadth, depth and pace of change within the complexity of higher education organisations is a tough challenge; moving forward from the initial victories of champions towards profound systemic and cultural change will only happen with the support of the senior team. Some practical considerations have been suggested.

Critical at this juncture is that the senior team are not just signed up in principle, but that they take radical action for a journey forward. Collaborative learning must be a central feature of that journey, at the heart of the change, and prominent in leadership development programmes to build capacity. Leaders at all levels must be as deeply committed to their own learning as they are to facilitating learning for sustainability; challenging the constraints of individual and institutional mental models is essential. The challenge is considerable but within it lies the possibility of mapping a route to a more just and sustainable future.

## Notes

1    Agenda 21 was the action plan agreed at the UN Conference on Environment and Development in Rio de Janerio, Brazil in 1992. See UNCED, 1992.
2    '*Double-loop* learning occurs when error is detected and corrected in ways that involve the modification of an organization's underlying norms, policies and objectives' (Argyris and Schön, 1978: 2–3). This is opposed to single loop learning, where error results in a correction, but not a questioning of the underlying frameworks and values.

## References

Alvesson, T and Sveningsson, S (2003) 'The great disappearing act: difficulties in doing "leadership"', *Leadership Quarterly*, 14, 359–81
Anderson, L and Bateman, T (2000) 'Individual environmental initiative: championing natural environmental issues in U.S. business organizations', *Academy of Management Journal*, 43 (4): 548–70
Argyris, C (1991) 'Teaching smart people how to learn', *Harvard Business Review*, 69 (3): 99
Argyris, C and Schön, D (1978) *Organizational Learning: A Theory of Action Perspective,* Reading: Addison Wesley
Ballard, D (2005) 'Using learning processes to promote change for sustainable development', *Action Research*, 3 (2): 135–56
Bandura, A (1977) *Social Learning Theory,* New York: General Learning Press
Bass, B (1985) *Leadership and Performance Beyond Expectations*, New York: Free Press
Bass, B (1996) *A New Paradigm of Leadership: An Inquiry into Transformational Leadership*, Alexandria: US Army Research Institute for Behavioural Sciences

Bass, B and Avolio, B (1997) *Full Range Leadership Development Manual for the Multifactor Leadership Questionnaire*, Palo Alto: Mindgarden

Bennis, W (1959) 'Leadership theory and administrative behaviour: the problem of authority', *Administrative Science Quarterly*, 4: 259–60

Bennis, W (1998) *On Becoming a Leader*, London: Arrow Books

Bowers, C (2001) *Educating for Eco-justice and Community*, Athens: University of Georgia Press

Brighter Planet (2010) *Employee Engagement Survey Report.* Available at: www.brighterplanet.com/research (accessed 31 October 2011)

Brown, B (2011) 'Conscious leadership for sustainability: how leaders with late-stage action logistics design and engage in sustainability initiatives', *Human and Organizational Systems*, Santa Barbara: Fielding Graduate University

Buchanan, D and Boddy, D (1992) *The Expertise of the Change Agent: Public Performance and Backstage Activity*, New York: Prentice Hall

Copeland, G (2008) 'Sustainable development: strategic considerations for senior managers', in Shiel, C and McKenzie, A (eds) *The Global University: The Role of Senior Managers*, London: BU/DEA

Corcoran, P (2010) 'Foreword', in Jones, P, Selby, D and Sterling, S (eds) (2010) *Sustainability Education: Perspectives and Practice Across Higher Education*, London: Earthscan

Doppelt, B (2010) *Leading Change Towards Sustainability*, Sheffield: Greenleaf Publishing

Dunphy, D and Stace, D (1993) 'The strategic management of corporate change', *Human Relations*, 46 (8): 905–20

Dunphy, D, Griffiths, A and Benn, S (2007) *Organizational Change for Corporate Sustainability*, London: Routledge

Egri, C and Herman, S (2000) 'Leadership in the North American environmental sector: values, leadership styles, and contexts of environmental leaders and their organizations', *Academy of Management Journal*, 43 (4): 571–604

Ferdig, M (2007) 'Sustainability leadership: co-creating a sustainable future', *Journal of Change Management*, 7 (2): 25–35

Ferdig, M (2009) *Sustainability Leadership.* Available at: www.sustainabilityleadership institute.org/atomic.php (accessed 30 September 2011)

Goleman, D (1995) *Emotional Intelligence*, New York: Bantam Books

HEFCE (2005) *Sustainable Development in Higher Education.* Available at: www.hefce.ac.uk/pubs/hefce/2005/05_28/ (accessed 31 January 2012)

HEFCE (2008) *Sustainable Development in Higher Education: Update to Strategic Statement and Action Plan.* Available at: www.hefce.ac.uk/pubs/hefce/2009/09_03/ (accessed 3 January 2012)

HEFCE (2010) *Carbon Reduction Target and Strategy for the Sector*, Bristol: Higher Education Funding Council for England

Jaworski, J (1998) *Synchronicity: The Inner Path of Leadership*, San Francisco: Berret-Koeler Publishers

Jones, P, Selby, D and Sterling, S (eds) (2010) *Sustainability Education: Perspectives and Practice Across Higher Education*, London: Earthscan

Kates, R, Parris, T and Leiserowitz, A (2005) 'What is sustainable development? Goals, indicators, values, and practice', *Issue of Environment: Science and Policy for Sustainable Development Report*, 47 (3): 8–21

Leal Filho, W (2000) 'Dealing with misconceptions on the concept of sustainability', *International Journal of Sustainability in Higher Education*, 1 (1): 9–19

Marshall, J, Coleman, G and Reason, P (eds) (2011) *Leadership for Sustainability: An Action Research Approach*, Sheffield: Greenleaf Publishing

Marshall, S (ed.) (2007) *Strategic Leadership of Change in Higher Education: What's New?*, London and New York: Routledge

Martin, S and Jucker, R (2005) 'Educating Earth-literate leaders', *Journal of Geography in Higher Education*, 29 (1): 19–29

Meadows, D (1997) 'Places to intervene in a system (in increasing order of effectiveness)', *Whole Earth*, Winter 1997, 78–85

Meadows, D (2002) *Dancing with Systems*. Available at: www.sustainer.org/pubs/ Dancing.html (accessed 15 February 2012)

Parkin, S (2010) *The Positive Deviant: Sustainability Leadership in a Perverse World*, London: Earthscan

Pearce, S, Brown, E, and Walker, J (2008) 'The role of universities in sustainable development', in Shiel, C and McKenzie, A (eds) *The Global University: The Role of Senior Managers*, London: BU/DEA

Quinn, J (1980) 'Managing strategic change', *Sloan Management Review*, 21 (4): 67–86

Quinn, L and Dalton, M (2009) 'Leading for sustainability: implementing the tasks of leadership', *Corporate Governance*, 9 (1): 21–38

Revans, R (1982) 'What Is action learning?', *Journal of Management Development*, 1 (3): 64–75

Scott, W and Gough, S (2004) 'Education and sustainable development in UK universities: a critical exploration post-Rio', in Blaze-Corcoran, P and Wals, A (eds) *Higher Education and the Challenges of Sustainability: Problematics, Practice and Promise*, Dordrecht: Kluwer Academic Publishers

Senge, P (1990) *The Fifth Discipline: The Art and Practice of the Learning Organisation*, New York: Doubleday/Currency

Shiel, C (2007) 'Developing and embedding global perspectives across the university', in Marshall, S (ed.) *Strategic Leadership of Change in Higher Education*, London and New York: Routledge

Shiel, C (2012) 'Enabling university leaders to serve as role models for sustainable development', in Leal, W (ed.) *Sustainable Development at Universities: New Horizons*, Frankfurt: Peter Lang Scientific Publishers

Sterling, S (2004a) 'An analysis of the development of sustainability education internationally: evolution, interpretation and transformative potential', in Blewitt, J and Cullingford, C (eds) *The Sustainability Curriculum*, London: Earthscan

Sterling, S (2004b) 'Higher education, sustainability, and the role of systemic learning', in Corcoran, P and Wals, A (eds) *Higher Education and the Challenge of Sustainability: Problematics, Promise and Practice*, Dordrecht: Kluwer Academic Publishers

Sterling, S and Scott, W (2008) 'Higher education and ESD in England: a critical commentary on recent initiatives', *Environmental Education Research*, 14 (4): 386–98

Stogdill, R (1974) *Handbook of Leadership: A Survey of the Literature*, New York: Free Press

Tilbury, D (2011) 'Higher education for sustainability: a global overview of commitment and progress', in GUNI (ed.) *Higher Education in the World 4. Higher Education's Commitment to Sustainability: From Understanding to Action*, Barcelona: GUNI

UNCED (1992) *Agenda 21: Programme of Action for Sustainable Development*, New York: UN Department of Public Information

Universities that Count (2009) *Annual Report*. Available at: www.beta.yudu.com/item/details/107038/Universities-that-Count-Annual-Report-2009 (accessed 30 September 2011)

Wals, A and Blewitt, J (2010) 'Third-wave sustainability in higher education: some (inter) national trends and developments', in Jones, P, Selby, D and Sterling, S (eds) (2010) *Sustainability Education: Perspectives and Practice Across Higher Education*, London: Earthscan

Williams, T (2012) *The Role of Higher Education in Creating Sustainable Leaders*. Available at: www.guardian.co.uk/sustainable-business/role-education-sustainable-leaders?CMP=twt_gu (accessed 5 June 2012)

Yukl, G (2006) *Leadership in Organisations*, Upper Saddle River: Pearson/Prentice Hall

# 6 The journey towards sustainability via community

## Lessons from two UK universities

*Rehema M White and Marie K Harder*

## Introduction

Universities are leaders in thinking and action, with a moral imperative to lead their regions in sustainability thinking and action (Orr, 2004). So why is it that our funds are not invested ethically? Why is all our printing not completed double sided, using environmentally-friendly inks and recycled paper? Why do we persist in teaching that making money is the ultimate in self-satisfaction? In this chapter, we reflect on stories from two UK universities, the University of Brighton and the University of St Andrews, in order to demonstrate how we have found that nurturing a community of values around sustainability leads to sustainability action, and vice versa.

The aims of this chapter are to explore what community means within a university, and how it relates to sustainability action. We distinguish community within a university from the usual sense of 'university community engagement'. The latter denotes the relationship and interaction between a university and its surrounding community (town as opposed to gown), whereas we intend the internal community created among university members: academic and support staff, estates and operations staff, senior management and students. These aims are pursued within the context of individuals adopting pro-environmental behaviour, groups of people collectively focusing on sustainability values and shifting social norms, and the institution as an organisation undergoing change through effective governance for sustainability.

'Community' is a contested topic, holding different meanings in different contexts and literatures. We acknowledge the existence of longstanding debates on community of place (McMillan and Chavis, 1986), communities of interest (Gusfield, 1975), communities of practice (Wenger and Snyder, 2000; Hart and Wolff, 2006) and more. However, in this chapter, we define community in our university as being a 'community of values'. We intend, first, to unfold the narratives of the journeys towards sustainability in our respective institutions, and, second, to analyse our experiences. Our premise is that the creation of community (values) in an institution enhances sustainability action (behaviour), which in turn facilitates the development of community again.

We offer our institutional experiences as rich case studies, describing characteristics of the university, historical context and the journey towards sustainability. Examples illustrate our reflections. These narratives are developed from personal and deep engagement in our institutions, and reflect our individual perspectives.

## The University of Brighton

The University of Brighton evolved from a significant polytechnic, which has since merged with a nursing and midwifery institute and set up a medical school jointly with another university. Most of our courses have strong professional credentials, and our research is applied. We have over 21,000 students and 2,600 staff based on five sites, which are very different, ranging from inner city to rural, with the recent addition of a 'hub' in Hastings, which offers a portfolio of courses to a district with little higher education provision. Most sites are themselves split, for example, by main roads, and some subject disciplines are split over three sites in different towns.

### The seeding of the 'community'

Before 2005, the University of Brighton had very few identifiable formal sustainable development (SD) elements. One was a rudimentary environmental policy, and another an Environmental Action Group, set up and chaired by a pro vice-chancellor (PVC) with strong SD interests, who acted as a facilitating type of SD leader throughout Brighton's journey. The group was open to all, rotated between the various campuses, but attracted different participants each time, resulting in no increase in momentum.

The PVC SD leader tasked an assistant to develop an SD policy, involving one-to-one consultations across the university. These consultations exposed some staff to SD concepts for the first time, precipitated limited discussion, and identified some enthusiasts (and opponents!). At this point, a few staff had been loosely 'connected' with others who had an interest in SD; they were not a 'community', but at least they knew they were not alone.

### The first circle of the 'community' and its ripples

With the resultant SD policy discussed and approved by the governors, the PVC shifted the remit of the previous Environmental Policy Group to one appropriate for the Sustainable Development Policy Group (SDPG), and widened the membership from operational managers, such as estates and catering, to include faculty in closely related disciplines, such as Environmental Sciences, and members from cross-university thematic centres, for example, Economic and Social Engagement, the Learning and Teaching Centre, and the Community–University Partnership. The SDPG had no direct management authority itself, but would be an advisory group to the University Management Group.

Discussions at SDPG meetings were enriched by inviting staff who were individually active in various aspects of SD to present overviews. Membership was then offered to allow more 'SD enthusiasts', and their involvement helped stimulate managers to identify and envision SD in their own arenas. Within a year, a good overview of SD elements in most aspects of university work across sites and academic disciplines was given through informal presentations, audits and reports.

A 'community of values' was beginning to form, as SD topics were discussed, different practices shared, and the group began to form and articulate a university aspiration. There was encouragement from other enthusiasts, and from observing implicit senior management approval for new SD actions. A sense of SD 'community' was growing at these regular SDPG meetings, giving participants confidence to emanate SD ideas and take small actions in their own areas. For example, the Purchasing Officer was more aware of the importance of her participation in a regional forum for sustainable purchasing that she had stopped going to, but now rejoined. Some SDPG faculty members felt they could raise, briefly, the topic of SD in academic meetings. The new corporate plan, which was full of references to SD, became a live document to SDPG members who could use it as ballast for introducing SD concepts in working groups.

Resulting actions were numerous:

- Estates began to report on, for example, sustainability in new buildings, a recycling contract, solar heating for the swimming pool, studies on space use, and plans for energy metering.
- Estates prioritised the need for a replacement Energy Manager and upgraded the role from part-time to full-time.
- A professor was asked to identify researchers across the university working on any SD themes, and set up an internal conference.
- A key member of staff was asked to 'audit' the SD content and processes in the courses of one university school.
- The Centre for Learning and Teaching began SD discussion groups each term, open to all academic staff.
- The Business Services Office reported on all of its projects that had SD elements.

### The second circle of community and its ripples

The SD-sympathetic PVC then initiated three modest new mechanisms for implementing the SD policy. One was to request faculties to consider how they might incorporate the SD policy into their academic work, triggering some to set up working groups or assign staff to this. The second was the appointment of an SD Facilitator (0.4 FTE), with the remit to support and facilitate SD implementation across all areas, from estates to research. The role entailed very active troubleshooting and encouragement – not direct implementation. The third mechanism was the relaunching of the Environmental Action Group – but as a network, with a meeting at each of the five sites, each term.

These actions brought SD to the attention of more staff (and students) at the university. Faculty discussions were not easy, but several groups eventually developed their own approach to SD and produced authentic and original faculty policies. In so doing, they became groups of staff with 'shared values' in SD; they knew they could refer to each other for moral support, felt a sense of community, and became linked to the wider community of values of the SDPG members, which carried with it the endorsement of a PVC chair. Some individuals discussed new ideas with working colleagues, stimulating localised SD awareness-raising.

The Facilitator then sought any previously unnoticed staff with SD interests, introducing them to each other, and exploiting opportunities to involve them in activities with each other or in coalescing groups. The community of values had clearly grown beyond the 20 to 30 SDPG members; another 50 to 60 staff were involved in active SD groups, mostly based around academic subjects or sites.

The relaunching of the localised, multiple Environmental Action Network (EAN) groups generated little activity until, on one site, they were held in 'Open Space' format, where participants could suggest addenda items on the day, and the group then collectively decided how to proceed, usually breaking up into 'action-based' sub-groups. This approach generated participant ownership and proved successful in project outcomes. Other sites followed to various extents, beginning to instigate small, self-driven action groups, usually for short-term projects, such as investigating leaking taps on site, or setting up a composter for a tea room. Momentum built up slowly, assisted by the SD Facilitator being available to remove logistical barriers (e.g. Who in estates has to be asked to site the composter? Who do we ask about ethical investment?). A second key factor was that the minutes of these EAN groups were publicly guaranteed to be taken forward to a local senior manager, and summaries went to the SDPG for discussion, giving participants confidence that their input was being heard. Participants did not all attend regularly, but rather came when they had an 'action' they wanted to put forward, or for which they needed to seek collaboration.

Many actions occurred as a result of the EAN meetings:

- Hastings EAN had an SD Away Day visiting Pestalozzi Village.
- Grand Parade EAN organised an Education in SD Day, partly to consider how to minimise waste from student art materials.
- Cockcroft/Watts EAN designed and ran a water awareness stall for Sustainable Development Awareness Week.
- The Aldrich Library Green Group asked and received funding for a site to set up two composters.
- Individuals from Hastings are requesting permission to create a biodiversity area.
- The Cockroft/Watts EAN asked about action around travel plans, prompting a presentation from estates.

- Staff developed interest in the Cockroft building retrofit measures started by estates.
- Calls for clarification on the Environmental Purchasing Policy involved Purchasing in a debate on procurement.
- A new external website, with additional internal pages, summarises SD action across the university and provides all the relevant links.

At this stage, several small groups of staff and students had met and developed shared SD values, knew where to find each other, and were inspiring each other to take action (around 150 active individuals in total).

Despite this level of activity, awareness of broad SD aspects was poor among the majority of staff. Many felt that SD was of no relevance to their core work, especially in the curricula. Where this lack of knowledge was overcome, usually via small group discussions, newly informed staff began to investigate SD relevance to their work, and discuss this with colleagues. Although this 'ripple' effect was deep, effective and encouraging, its pace was considered very slow by SD enthusiasts.

Being part of one or more of these various SD groups slowly gave staff confidence to take SD action, and middle managers became encouraged to moot SD concepts to their departments. Communication of SD actions was crucial at this time, and taken on first by the SD Facilitator and then, increasingly, by the Marketing and Communications Department, stimulating further SD action. As a rough indication of overall progress, the university moved twenty places up in the People and Planet League,[1] on the basis of such informal actions.

### *The interlinking circles stage*

As staff who felt members of one group began to hear about the actions of others, the strengthening links of 'a wider SD community' and changing social norms began to make it significantly easier to cross traditional domain boundaries in the university. For example, estates asked if EAN fora could be used to help develop the upcoming transport policy, with assistance from the SD Facilitator; previously, they would have organised their own, top-down events with restricted participation. Academics asked if estates would consider some joint built environment projects, and it was agreed. The Students' Union asked a Head of School if he would endorse student SD 'champions' to promote SD teaching in some courses: this was agreed. Business Services asked a group of academics if they would join in a bid for a regional Environmental Services programme: this was successful. The EAN groups began to make proposals regarding mainstream university policies: can we have an Ethical Investment Policy? Can bottled water be replaced by tap water and fountains? Can cyclists' safety issues near campus sites be taken to the local authorities? An EAN 'Away Day' for technical and administrative staff reported to SDPG that they could support SD more effectively if there were school/faculty policies explicitly reinforcing SD action. The institutional corporate plan was felt to be too 'far away'.

By now, informal SD action appeared to be overtaking formal policy. The PVC encouraged the academic schools (within the faculties) to consider their own SD policies, triggering working groups where there was sufficient interest. Resulting policies were brought to the SDPG then recommended to the Academic Board for consideration for formal development. The Academic Board then recommended that all faculties formally consider SD in their academic work. SD was now in the formal job domain of all academics in the university, albeit softly spoken.

Significant external and internal drivers towards embedding SD at this time included HEFCE's carbon reduction requirements,[2] the increasing impact of the People and Planet Green League, and the UK government's Carbon Reduction Commitment.[3] Previous external drivers had not been very successful, because staff did not have a strong mandate to action SD. Now, individuals in loosely connected groups across the university wanted to make changes. In addition, a new Director of Estates, determined to accelerate the university towards becoming a low-carbon user, was appointed. With the groundwork of SD awareness and interest well prepared, the university was now ready to move ahead quickly, and senior management and governors agreed on an ambitious target of cutting 50 per cent of the carbon footprint within five years, widely and repeatedly publicised, and with the required support of a new Environment Team in estates, a behaviour change programme, expert advice from the UK Carbon Trust, and a raft of actions for immediate implementation – including solar panels, insulation, new computer servers, exploration of a geothermal aquifer and co-production of energy. Dissertation and placement students were involved, and staff participated in consultations.

The SD Facilitator was thus able to focus on nurturing SD in remaining pockets of SD-poor operational activities, such as purchasing and discharges. Since most staff were aware of the importance of SD in the university, it was less difficult to create new mechanisms to allow targets, monitoring and reporting of SD progress in these areas. Old domains could now be visited, encouraging and supporting even higher-level achievements, e.g. by assisting placement students to develop detailed biodiversity action plans for each site, exploring diverse options for waste diversion and reduction, and facilitating *others* to facilitate Food Fairs and rambles and 'green maps'. In consultations for the new, revised corporate strategy, staff (double the turnout from six years previously) and students made it very clear that SD – all round SD – was a core value. A tipping point had been reached: many staff felt they belonged to a particular SD group and recognised the existence of a wide-reaching loose network of such groups; SD was significantly embedded. Not surprisingly, this state was reflected in a ranking of third place in the UK People and Planet Green League in 2012.

## The University of St Andrews

The University of St Andrews is a small and ancient institution, inhabiting clusters of old and new buildings throughout the coastal town of St Andrews in

Fife, Scotland. Its 600th birthday was celebrated in 2012. Its 7,200 students are drawn from across the UK and from overseas (one-third of the student population). It has one of the highest proportions of residential student cohorts in the UK (47 per cent), and the majority of other students live in flats in the historic town within walking distance of the university. Students socialise together based on subject studied, residence, interest or other connections, due to the proximity of their lodgings and university traditions. For example, students form close-knit relationships through their 'academic families', celebrated by an annual Raisin Monday foam fight in the ancient quadrangle. The 2,000 staff live mostly in town or in nearby rural villages or small towns, with 76 per cent of staff and students resident within the university postcode area. The university was identified as one of the most research intensive in the UK in the 2008 Research Assessment Exercise, and maintains a strong focus on research excellence as well as student experience. It consistently scores within the top five in UK university league tables and is placed on the *Times Higher Education* Top 100 Universities (2011) list. Increasingly, it is gaining a reputation as an intellectual hub integrating sustainability research, teaching and operation. The university has won acclaim through, for example, the *Times Higher Education* Award for Outstanding Contribution to Sustainable Development in 2006 and recognition in the UK Universities that Count 2010 report.[4]

A workshop with Forum for the Future and other stakeholders inspired a group of academics at the University of St Andrews to initiate an interdisciplinary module on sustainable development (SD). Staff and students then wished to take this forward, and the outline for the SD Undergraduate Programme was born in 2005, and an SD Coordinator (later renamed Director) appointed. Uniquely, our SD Programmes, offering BSc and MA Honours in SD, draw on staff from schools across the university to deliver core interdisciplinary modules in SD. This approach has been successful in engaging large numbers of staff (more than 60 staff from eleven schools have contributed) and students (our first-year class has been as large as 300 students; thus reaching approximately 25 per cent of the university's first-year cohort of students). These core SD modules are combined with student-selected partner modules to develop specialist SD pathways. The Programme is underpinned by the principles of critical enquiry, inter-disciplinarity, transformative learning, integration of theory and practice, and exploring examples and links between local and global aspects of SD. These principles have enabled us to explore institutional sustainability and undertake innovative field trips and assessments, thus explicitly linking academic enquiry with operational activities and reflective analysis. The SD Programme won the Green Gown Award for Courses in 2009 in recognition of its innovative structure and success, but its success and pace of growth outpaced the resourcing from the university centre.

Our approach has enabled us to establish SD as a topic of enquiry in its own right, to develop rigorous research programmes and contribute to academic debates. We have pioneered linkages between this area and existing disciplines across the university, with SD seen as an exciting area for study and research, fertilising other areas, rather than a government-sponsored add-on to existing courses. However, we are aware of the need to also encourage integration of education for SD into other areas of study. Attempts to encourage staff to include sustainability in the curriculum have been made, for example, through provision of resources, web links and a focus on SD issues during Green Week.

The St Andrews Sustainability Institute (SASI) was invested in for three years by the Principal's Office in order to facilitate interdisciplinary research in the area of SD and to develop a postgraduate degree in SD. SASI has been successful in developing partnerships and facilitating interactions across the university, and established an MSc in SD Programme in 2009 alongside the successful model of the undergraduate programme. SASI also played a role in the integration of different sectors across the university, nurturing research on the institution itself and facilitating workshops to explore strategic sustainability direction at the institution ('Making It Real I, II and III'), and engaging senior management structures to gain support for sustainability direction. A growing cohort of PhD students with a self-registered focus on SD reside in different schools across the university.

The SD Programmes and SASI established core groups of staff and students, across the entire academic sector, with a key focus on sustainability but with different action interests. This had three major effects. First, SD was proposed as a credible field of scholarship within an institution focused on traditional notions of academic excellence. Second, there was engagement across the university, with key individuals identified in almost every school, a number of groups forming, and with links established between estates and academia. Third, the number of active, informed students with interests in sustainability increased.

While academic activities thus catalysed the SD journey in the institution, the next phase was characterised by increased activity from the student and estates sectors. A feature of the closely-knit student community is the high number of clubs and societies. A number of these, notably the SD Society and One World Society, have explicit sustainability goals. A few key students, in discussion with academic and estates staff, initiated Transition University of St Andrews (Transition UStA), with a formal launch in 2009. Transition is a global social movement, which aims to address the challenges of peak oil, climate change and economic instability through the development of community energy descent plans and task-based working groups (Hopkins, 2008). While Transition Initiatives is coordinated by the Transition Network, Transition Universities is coordinated by People and Planet (see Higgins *et al.*, this volume; Hazan, this volume). Transition UStA increased the potential for grassroots action, serving as an interface between students and academic and estates staff. For example, in 2011–12, all university members received email or memo communication, over 750 students received face-to-face presentations, and 813 staff and students voluntarily attended events. Funding was obtained in its second year, and for the following three years, in collaboration with the town community, enabling us to employ staff to work on particular projects, such as the community garden, waste and recycling, the inter-hall energy competition, a Local Exchange

Trading Scheme, energy audits in houses and flats, engagement with schools, community events such as food festivals, carbon conversations, and developing a St Andrews-specific carbon calculator. Transition UStA has succeeded in reducing carbon emissions and establishing progress in areas of operational management but, most importantly, it has provided a platform by which to engage the internal university community. It is run through a broad Steering Group committee comprising staff and students, functions through activity-defined working groups, has an extensive website and social media presence, and runs monthly Open Fora, to which anyone can come and discuss sustainability issues. However, it does not yet effectively interface with university senior management, and thus currently has limited ability to impact directly on policy.

Key individuals in estates have had an enormous impact on the university's progress towards sustainability, in two ways. First, they have highlighted that the university energy bills are rising at 15 per cent per year. The university has many ancient, listed buildings, which are difficult to heat, several buildings from the 1970s which leak energy, and high energy use in science buildings, hence our energy bill is significant. The cost implications of this have led a member of senior management to recognise the need for investment in energy savings and to generate increased resources and support for sustainability action. Supported by the Salix fund, and with advice from the Carbon Trust, the university has embarked on an ambitious strategy to become carbon neutral by 2016 through production of renewable energy, establishing a small wind farm on farmland out of town, creating a biomass plant in an old paper mill site in a neighbouring village, and installing solar panels and other forms of technology throughout the institution where possible. All new build is now undertaken to at least excellent, but ideally outstanding, UK building standards (BREEAM).[5] Other operational concerns, such as waste management, procurement and biodiversity, are also being addressed enthusiastically. Second, estates interface very well with Transition, through key individuals. Hence, any small structural barrier to sustainability action inspired by Transition can often be overcome. For example, if a project needs a shed for its activities, a space can usually be found. During changeover in student representation on the Steering Group, estates and academic staff play a role in ensuring continuity. An annual Sustainability Intern post has been established, in which an SD graduate is employed to work in estates on sustainability projects. The SD Programme encourages dissertation projects on aspects of university sustainability and Transition. Hence, there is deliberate blurring of the boundaries between university sectors.

Other examples of cross-sectoral sustainability-related interactions are:

- estates staff teach on first- and third-year modules in the SD programme; currently Transition also presents to first-year modules;
- Honours and Masters students undertake research dissertations on aspects of the SD programmes or university operations;
- estates employs an SD graduate as a Sustainability Intern;

- estates staff are invited to talk at an energy research seminar; Transition students are invited to talk at an SD research session;
- a research project led by Social Psychology explores pro-environmental behavioural change in different schools.

Senior management approved a renewed Sustainable Development Strategy for 2010–15 (although approval was delayed for two years), and revamped an old committee to form a Sustainable Development Policy Group, which is intended to offer a platform for the different sectors to discuss sustainability and promote policy and action changes. However, currently this group has not engaged closely with either grassroots sustainability action or the SD Programmes.

Hence, sustainability at St Andrews has shifted from being driven separately from academic and operational sources, to an integrated movement in which a community of values has developed among staff and students across different sectors. Permission from senior management to pursue money-saving energy investment projects has enabled, but not driven, a large-scale programme of carbon emission reduction, together with the support of small-scale sustainability action. Small groups meet to work on particular issues. While the University of St Andrews already has high levels of social capital among students, features of this emerging community are its focus on sustainability action, reflection on values and cross-sectoral links.

## Discussion

### *A virtuous circle*

We have observed a virtuous circle in which the growth of *community* within our universities has enhanced sustainability *action*, which in turn has further contributed to community (Figure 6.1). We have found that sustainability enthusiasts[6] nurture the seed of community around sustainability values and other attributes. The circle can also be turned through formal governance and appropriate structures, such as the establishment of new means of communication or new cross-sectoral bodies established to progress sustainability (e.g. Brighton's SDPG, St Andrew's Transition).

There appears to be no single explanation in the literature for our experience. We touch below on literatures from organisational change and learning, community mobilisation, participation, social learning and social capital, and the links between values and pro-environmental behaviour to analyse our findings.

First, we provide an analysis of this virtuous circle. In the University of Brighton, there was little SD community before sustainability action was undertaken. Key individuals ('enthusiasts') were brought together in the SDPG to consider the meaning of SD, and as they developed shared understandings and values, they formed the core of the first SD community. They, and later groups, had no common professional role, although all worked within the university; it

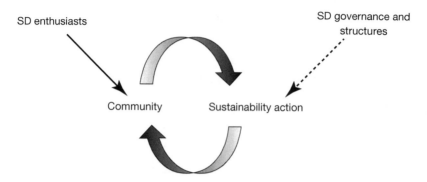

*Figure 6.1* The relationship between community and sustainability action

Development of community can lead to sustainability action, which in turn enhances the sense of community. SD enthusiasts can promote community within their organisation, and effective institutional structures can support sustainability action.

was not a community of practice in the original sense (Lave and Wenger, 1991) although they might be closer to later definitions in terms of 'mutual engagement', 'joint enterprise' and 'shared repertoire' (Wenger, 1998) or, indeed, identity (Wenger *et al.*, 2002; Wade, this volume). They were not usually on the same site; they were not a community of place (Chavis and Wandersman, 1990). They shared a broad interest in sustainability, but differed in their specific goals, hence they were not, at least initially, a community of interest. Their shared vision of sustainability and determination to act based on conviction created a sense of community, as outlined by McMillan and Chavis (1986) and later validated, with limitations, by Peterson *et al.* (2008). They had an identity related to SD, a feeling of belonging to this diffuse collective, the ability to make a difference, an identified need, and a shared emotional connection. We thus describe what we saw as a community of value: a shared aspiration in the journey towards sustainability, where 'community' provides links between people and a strengthened motivation for benevolence, localism and other aspects of SD. In this way, we follow Frazer (1999), who also interpreted community as a value.

At the University of St Andrews, the journey was different. The initiation of sustainability action occurred largely from individuals in grassroots and middle management levels. Three separate strands of action – academic scholarship, operations, and student activism – were woven together through the enthusiasm of key individuals and the emergence of platforms such as Transition UStA to facilitate cross-sectoral interaction, reflection and action. Different sustainability action groups formed, a momentum was gained, people were enthused to take further action, and more people were engaged. The nature of some projects encouraged individual reflection and action (e.g. StAnd ReUse, Carbon Conversations), but most projects facilitated collective engagement. The cross-sectoral interaction and group activity generated a sense of community of values, as described for the University of Brighton above, but via a different pathway. Our case studies focus on sustainability values, but evidence suggests that values in favour of the common good, or intrinsic values, generally enhance sustainability values (Crompton, 2010; Kasser, 2011).

### Internal community mobilisation

Without the input of the SD enthusiasts who triggered sustainability action and promoted collective debate, it is unlikely that such success would have been achieved at either institution. However, it is important to note the processes were fundamentally different at each. The University of St Andrews already recognised a sense of institutional community as being an important part of its identity, even stating in the introduction to its current prospectus that '[p]erhaps our most distinguishing feature ... is the strong sense of community reinforced by the size and location of the town and the many cultural and social activities in the [u]niversity'.[7] The identity of this community shifted to accommodate sustainability values, at least to some extent, largely due to the activities of those in estates, Transition UStA and the SD Programmes and associated groups. Increased commitment from senior management would encourage the University of St Andrews to revise mission statements, or develop statements of intent to which staff members could sign up, as at some other institutions. However, some values-based actions are diffusing into established structures, because induction processes for new staff now include awareness and discussion of topics such as walking, cycling and car-share commuting options.

In contrast, the University of Brighton focused more on mobilising small, locally rooted SD action groups, which eventually interlinked and contributed to an institutional community of values. It is expected that the distinction between SD-focused and mainstream work groups will dissolve further over time, as SD becomes fully embedded in all work streams.

An extensive literature and practical advice is available on community development and community mobilisation (Hopkins, 2008; Sarkissian *et al.*, 2009; Wilding, 2011).[7] However, while this literature offers advice, examples and lesson learned, all authors make clear that community mobilisation cannot be strategically planned by a management team, but rather is an interactive process engaging community members in decision-making. Feeling influential is a critical aspect of a sense of community (McMillan and Chavis, 1986). This process offers a challenge for university management, where token democracy often succumbs to autocracy.

A cautionary note on the mobilisation of internal community: vigorous promotion of organisational change and the use of participatory processes are required. Universities are full of competitive, driven academics, steeped in politics, embedded in tradition yet swayed by contemporary drivers, and often experiencing archaic and contemporary forms of organisational management concurrently. For example, social learning may be limited by gender and power imbalances (Buchy, 2004). Networks or social groups may exclude individuals, possibly widening the existing gender gap in academia or exacerbating the exclusion of ethnic minorities or those with disabilities or responsibilities as carers. Some people may not wish to belong to a university community, preferring to see work as a place for individual activity. There are theoretical and practical concerns that 'a denial of difference contributes to social group oppression' (Young, 1990: 10). The promotion of a single, homogenous 'community' could thus be exclusive and undesirable. Early SD activities at both Brighton and St Andrews prompted objections, accusations of propaganda and resistance from the majority of staff. At Brighton, the option to voluntarily join a sustainability action group, and at St Andrews, the encouragement of critical debate regarding the concept of SD as well as voluntary grassroots engagement, have, to some extent, mitigated against an imposed set of values.

### Universities as both communities and organisations

In organisational change terms, the shift towards sustainability is a radical change for universities to make, requiring double-loop learning, second-order[8] change and an alteration of the framework within which the organisation functions (Boyce, 2003). Successful organisational change requires cultural change together with action – both bottom-up and top-down action (Boyce, 2003; Beringer, 2007; Gudz, 2004; Jansen, 2003; Thomas, 2004). Differences between espoused theory (what the administration says) and theory in use (what people believe and how they behave) can lead to friction (Albrecht *et al.*, 2007), so it is essential that universities 'walk the talk'.

Universities are considered to be 'loosely coupled systems with diffuse decision making', and often with goals that are less explicit than those of business or other organisations (Boyce, 2003). This makes it easy for them to make small, adaptive changes, but means it is more difficult for them to achieve, in one step, the radical change required for the journey towards sustainability. In this sense, universities occupy a space between businesses, which are often tightly coupled systems with explicit goals, rarely values-driven; and communities, which are very loosely coupled, motivated by values and often with divergent goals. This may explain why universities lend themselves well to the practice of community when trying to instigate change or inspire university members to strive together for a common purpose.

In addition, universities offer a very different space in the public arena. They have a mandate to debate the direction and form of society and a moral imperative to be a leader in sustainability (Orr, 2004). They are charged with researching and developing the form of our future socio-technological relationships and with educating the leaders of the future to be responsible, critical-thinking citizens. Students are usually exposed to university at a time when they leave home and are questioning and (re)forming their values. If universities are to fulfil their role of being centres for leading thinkers and practitioners of sustainability (Orr, 2004; Rowe, 2007), then they need to actively stimulate and engage staff and students in values-based considerations through both operational decision-making and rigorous scholarship.

How, then, can universities integrate top-down and bottom-up activities for sustainability change? Unlike a community of place, which has no pre-established agenda, an institution such as a university has a purpose. However, a university is a less tightly coupled system than a business. It can thus facilitate integrated governance through structural incentives, as described above, and through participatory processes. Facilitating participation of university staff and students in decision-making can enhance social learning (Meadowcroft, 2004), create benefits from the diverse ontological, academic, ethnic, gender and age perspective experiences across campus, improve acceptance of the decisions made and establish a community with greater civic involvement. Participation, together with accountability and transparency, is also essential for organisational learning (Albrecht *et al.*, 2007). Participation can take many forms, varying from the provision of information to individuals (essentially a form of communication) to devolvement of power to the individuals concerned (see Arnstein [1969] for the theory of the ladder of citizen participation). More engaged forms of participation can be an effective way to stimulate debate across the diffuse decision-making structures of the university (Boyce, 2003), engage individuals who are not aligned with a sustainability group, build trust and relationships and strengthen identity, thus enhancing a sense of community (McMillan and Chavis, 1986). However, institutions should be clear when they invest in such processes and when they are merely consulting, in order to avoid merely imposing a corporate branding rather than nurturing a real sense of identity. An example of effective participation, with the goal of reducing campus water usage and demonstrating the novel use of community mapping to engage students in campus landscape development, is offered by Johnson and Castleden (2011). In addition to the management of participatory processes, universities should, however, be prepared to respond to significant emerging grassroots initiatives.

External drivers such as globalisation, the promotion of evidence-based practice and science in society, discipline-confined excellence, and the commodification of education have led universities to falter in their mission of value-based scholarship and enquiry. Reconvening university members around a new goal requires deep consideration of purpose, internal reflection of core values, and collective action. Our community of values concept is thus closely related to the concept of organisational culture, which is believed to be required to make radical change (Boyce, 2003). Value-based, systems thinking management approaches are considered to be more effective than target-based, audit-heavy management approaches, not only for sustainability, but also for organisational function (Seddon, 2008).

Universities are increasingly being asked to contribute directly and explicitly to society, as evidenced by the emergence of the drive for research 'impact' (Pain *et al.*, 2011), and by the push for more 'community engagement' (with the external community). We suggest that the university needs to also foster its internal community of values, and create a genuine and shared platform from which individuals can interact with external stakeholders on behalf of the university. The existence of a community of values may also facilitate the core functions of universities – research and teaching – by establishing better relationships among and between staff and students, and by facilitating reflexive, transformative and interdisciplinary scholarship.

The increased prominence of the workplace in our lives today increases the importance of pleasant working conditions. The development of strong social connections with good relationships has a major influence on individual happiness (Berscheid, 2003). Social capital, as assessed by community ties, as well as family, neighbourhood and religious parameters, supports both physical health and subjective well-being (Helliwell and Putnam, 2004). In other words, 'community' improves the health and well-being of individuals, enhancing productivity and creativity and opposing rising levels of stress and depression in universities today (Sparkes, 2007; Sayce, this volume).

### *Community influences on sustainability behaviour and actions*

Sustainability action does not just happen without individual people taking action. While the collective mode may encourage people to take action, we also need individual behavioural change. Several complex and opposing theories explain some pro-environmental behaviours (Ajzen, 1991; Jackson, 2005; Kollmuss and Agyeman, 2002; Stern, 2000). These models share acknowledgement of the links between values, attitudes and behaviour; realisation that values are influenced by social norms and culture; acceptance that, even if values and attitudes shift, pro-environmental behaviours may be impeded by structural barriers. In turn, sustainability action can enhance community through shifting social norms. The potential for bottom-up action empowers people, facilitating a sense of belonging and of agency. Through acting on belief, the university becomes more than a place of work – it becomes a place of discussion and mutual and collective action.

In universities, the analysis of the pro-environmental behaviour of staff is complicated by the ambiguity of job descriptions. Unless the entire institution has a clear sustainability mandate infused in all activities, in most cases individuals require some degree of commitment to optimise sustainability within their roles. The link between individual and institution is the group, and in our case studies, sustainability action groups facilitated individual action. In turn, group and organisational learning are linked (Albrecht *et al.*, 2007).

Community activities can lead to social learning, the acquisition of knowledge that occurs within a social group and the process by which individuals learn from the behaviour and experiences of others (Ganis and Gilfus, 2009). An outcome may be increased social capital – a term that captures the importance of bonds and networks between people and the development of social norms, as well as the benefits of building reciprocity and trust (Putnam, 2000). While the concept of social capital has been critiqued because it represents a quantitative assessment in a point of time rather than a contextualised understanding of social power relationships (Flyvbjerg, 2001), it remains a useful indicator of engagement.

# Conclusions

Can a university achieve sustainability without community? We believe it can to some extent, but that it would be a top-down, incomplete attempt. Sustainable development demands a change in mindset, a different culture, a new paradigm, a values-based society (Jackson, 2005; Kates *et al.*, 2006). In order to achieve this radical change, university staff and students need to feel they belong, that they share this vision, that they can make a difference; in other words, they need to feel a sense of community.

Our case studies have demonstrated that the emergence of a community of values appears to facilitate people joining groups in which to undertake sustainability action (such as the St Andrews community garden and the Brighton composting group). Additionally, the opportunity to join a group highlights a sense of belonging, identity, need and agency for individuals, thus enhancing the sense of community, which in turn facilitates the networking of groups within a wider community of values. The SD enthusiasts and key structural interventions are essential in turning this virtuous circle.

We have drawn from literature on organisational change and learning, community mobilisation, participation, social learning and social capital, and the links between values and pro-environmental behaviour to analyse our findings, but we have not isolated a single framework within which the concept of a community of values in a university sits. Little research specifically links these concepts, and it is a fruitful area for further studies, not only on the sustainability of universities, but also on the relevant areas of enquiry identified.

While universities are still far from being sustainable, we can look back and see how far we have come. As we travel further, we envisage an embedding of sustainability within our institutions. At this stage, we would hope to see SD actions within normal working processes. For example, the University of Brighton chose not to invest heavily in externally supported SD initiatives, but to encourage their adoption into mainstream activities. The University of St Andrews has been successful in developing student engagement and encouraging academic reflection. We cannot yet glimpse our destination, but through an emphasis on community in our institutions, we are at least on the right road to a sustainable future.

## *Acknowledgements*

The authors would like to acknowledge useful comments from two anonymous referees, insights from Katherine Ellsworth-Krebs and Winston Emmerson, and access to selected literature provided by Micheil Gordon of the University of Brighton.

## Notes

1   People and Planet is the UK's largest student campaigning network. It compiles an
    annual Green League, which is the only comprehensive and independent league table

of UK universities ranked by environmental and ethical performance. This league is published in the *Guardian* newspaper, serves as a public indication of performance on sustainability issues, and is now a strong incentive for institutions to improve. See: http://peopleandplanet.org/greenleague

2   The Higher Education Funding Council for England (HEFCE), as distinct from the Scottish Funding Council, has a strategy for carbon reduction in English higher education. This strategy comprises sector-level targets for carbon reduction and a requirement for institutions to set their own targets and develop carbon-management and reduction plans, including annual monitoring and reporting. See: www.hefce.ac.uk/whatwedo/lgm/sd/carbon/

3   The Carbon Reduction Commitment is a mandatory scheme from the UK government that is designed to improve energy efficiency and cut carbon emissions in large public and private sector organisations. The scheme features a range of reputational, behavioural and financial drivers, which aim to encourage organisations to develop energy management strategies that promote a better understanding of energy usage. High-energy-user universities have to purchase carbon credits. For further details see: www.decc.gov.uk/en/content/cms/emissions/crc_efficiency/crc_efficiency.aspx

4   Universities that Count was an earlier attempt through the Environmental Association for Universities and Colleges (EAUC) to encourage universities to report and reflect on aspects of operational sustainability, including teaching and research. It drew on previous corporate social responsibility initiatives in business and has now been superseded by a sector-specific system, Learning in Future Environments (LiFE). The nature of this assessment is likely to influence future direction in the sector in the near future. See: www.eauc.org.uk/projects

5   BREEAM is a widely used environmental assessment method and rating system for buildings, with 200,000 buildings with certified BREEAM assessment ratings and over a million registered for assessment since it was first launched in 1990. See: www.breeam.org

6   We do not call them champions, because champions are often formally appointed and our experience is that these individuals self-identify. They are not all leaders, although some possess excellent leadership qualities. Such charismatic individuals are essential for sustainability action (Thomas, 2004).

7   The 2012 Undergraduate Prospectus, available at www.st-andrews.ac.uk/media/2012Prospecus-Full%20for%20web.pdf (accessed 6 June 2012)

8   For example, see Transition (www.transitionnetwork.org/) and Fiery Spirits community of practice (http://fieryspirits.com/). Literature specifically on the development of community at universities is beginning to emerge and websites currently indicate areas of activity (e.g. www.transitionedinburghuni.org.uk/ and http://transitionuniversityofstandrews.com/).

9   First-order change (such as increasing efficiency in an administration office, adding a new course) is incremental, linear and functions within existing heuristics; it is often aligned with single-loop learning, which responds to information and enables adaptations to be made but does not question underlying assumptions. Second-order change delivers transformational and irreversible change, altering the theory of action or assumptions of the institution, in response to double-loop learning (Boyce, 2003).

## References

Ajzen, I (1991) 'The theory of planned behavior', *Organizational Behavior and Human Decision Processes*, 50: 179–211

Albrecht, P, Burandt, S and Schaltegger, S (2007) 'Do sustainability projects stimulate organizational learning in universities?', *International Journal of Sustainability in Higher Education,* 8: 403–15

Arnstein, S (1969) 'A ladder of citizen participation', *Journal of the American Institute of Planners,* 35: 216–24

Beringer, A (2007) 'The Luneburg Sustainable University Project in international comparison: an assessment against North American peers', *International Journal of Sustainability in Higher Education,* 8: 446–61

Berscheid, E (2003) 'The human's greatest strength: other humans', in Staurdinger, U (ed.) *A Psychology of Human Strengths: Fundamental Questions and Future Directions for a Positive Psychology,* Washington, DC: American Psychological Association

Boyce, M (2003) 'Organizational learning is essential to achieving and sustaining change in higher education', *Innovative Higher Education,* 28: 119–36

Buchy, M (2004) 'The challenges of 'teaching by being': the case of participatory resource management', *Journal of Geography and Higher Education,* 28 (1): 35–47

Chavis, D and Wandersman, A (1990) 'Sense of community in the urban environment: a catalyst for participation and community development', *American Journal of Community Psychology,* 18: 55–77

Crompton, T (2010) *Common Cause: The Case for Working With Our Cultural Values.* Available at: http://assets.wwf.org.uk/downloads/common_cause_report.pdf (accessed 14 June 2012)

Flyvbjerg, B (2001) *Making Social Science Matter: Why Social Enquiry Fails And How It Can Succeed Again,* Cambridge: Cambridge University Press

Frazer, E (1999) *The Problems of Communitarian Politics: Unity and Conflict,* Oxford: Oxford University Press

Ganis, F and Gilfus, S (2009) '"Social learning" buzz masks deeper dimensions', Washington, DC: Gilfus Education Group

Gudz, N (2004) 'Implementing the Sustainable Development Policy at the University of British Columbia: an analysis of the implications for organisational learning', *International Journal of Sustainability in Higher Education,* 5: 156–68

Gusfield, J (1975) *The Community: A Critical Response,* New York: Harper Colophon

Hart, A and Wolff, D (2006) 'Developing local "communities of practice" through local community-university partnerships', *Planning Practice and Research,* 21: 121–38

Helliwell, J and Putnam, R (2004) 'The social context of well-being', *Philosophical Transactions of the Royal Society of London. Series B, Biological Sciences,* 359, 1435–46

Hopkins, R (2008) *The Transition Handbook: From Oil Dependency to Local Resilience,* Totnes: Green Books Ltd

Jackson, T (2005) *Motivating Sustainable Consumption: A Review of Evidence on Consumer Behaviour and Behavioural Change,* UK: Sustainable Development Research Network, 154

Jansen, L (2003) 'The challenge of sustainable development', *Journal of Cleaner Production,* 11: 231–45

Johnson, L and Castleden, H (2011) 'Greening the campus without grass: using visual methods to understand and integrate student perspectives in campus landscape development and water sustainability planning', *Area,* 43: 353–61

Kasser, T (2011) 'Cultural values and the wellbeing of future generations: a cross-national study', *Journal of Cross-Cultural Psychology,* 42: 206–15

Kates, R, Leiserowitz, A and Parris, T (2006) *Great Transition Values: Present Attitudes, Future Changes*, in Kriegman, O and Raskin, P (eds) *Frontiers of a Great Transition* (GTI Paper Series No. 9), Boston, MA: Tellus Institute

Kollmuss, A and Agyeman, J (2002) 'Mind the gap: why do people act environmentally and what are the barriers to pro-environmental behavior?', *Environmental Education Research*, 8 (3): 21

Lave, J and Wenger, E (1991) *Situated Learning: Legitimate Peripheral Participation*, Cambridge: Cambridge University Press

McMillan, D and Chavis, D (1986) 'Sense of community: a definition and theory', *Journal of Community Psychology*, 14: 6–23

Meadowcroft, J (2004) 'Participation and sustainable development: modes of citizen, community and organisational involvement', in Lafferty, W (ed.) *Governance for Sustainable Development: The Challenge of Adapting Form to Function*, Cheltenham: Edward Elgar Publishing

Orr, D (2004) *Earth in Mind: On Education, Environment and the Human Prospect*, Washington: Island Press

Pain, R, Kesby, M and Askins, K (2011) 'Geographies of impact: power, participation and potential', *Area*, 43: 183–88

Peterson, N, Speer, P and McMillan, D (2008) 'Validation of a brief sense of community scale: confirmation of the principal theory of sense of community', *Journal of Community Psychology*, 36: 61–73

Putnam, R (2000) *Bowling Alone: The Collapse and Revival of American Community*, New York: Simon and Schuster

Rowe, D (2007) 'Education for a sustainable future', *Science*, 317: 323–24

Sarkissian, W, Hofer, N, Shore, Y, Vajda, S and Wilkinson, C (2009) *Kitchen Table Sustainability: Practical Recipes for Community Engagement with Sustainability*, London: Earthscan

Seddon, J (2008) *Systems Thinking in the Public Sector: The Failure of the Reform Regime and a Manifesto For a Better Way*, Axminster: Triarchy Press

Sparkes, A (2007) 'Embodiment, academics, and the audit culture: a story seeking consideration', *Qualitative Research*, 7: 521–50

Stern, P C (2000) 'Toward a coherent theory of environmentally significant behavior', *Journal of Social Issues*, 56 (3): 407–24

Thomas, I (2004) 'Sustainability in tertiary curricula: what is stopping it happening?', *International Journal of Sustainability in Higher Education*, 5: 33–47

Wenger, E (1998) *Communities of Practice: Learning, Meaning, and Identity*, Cambridge: Cambridge University Press

Wenger, E and Snyder, W (2000) 'Communities of practice: the organizational frontier', *Harvard Business Review*, January–February, 139–45

Wenger, E, McDermott, R, and Snyder, W (2002) *Cultivating Communities of Practice: A Guide to Managing Knowledge*, Cambridge, MA: Harvard Business School Press

Wilding, N (2011) *Exploring Community Resilience in Times of Rapid Change*, Dunfermline, UK: Fiery Spirits Community of Practice

Young, I (1990) *Justice and the Politics of Difference*, Princeton: Princeton University Press

# 7 Times of change

## Shifting pedagogy and curricula for future sustainability

*Alex Ryan and Debby Cotton*

## Introduction

Under the influence of globalisation, technological advances and economic challenges, higher education (HE) is in an era of rapid change. Its role in the transmission and transformation of culture is arguably becoming more complex, as sources of knowledge and learning multiply and as the imperative to meet the educational needs of more diverse groups of people increases. If HE is to continue to fulfil its educational function effectively, it must grasp this challenge while also responding to the full range of societal concerns around sustainability. Educational practice geared to these concerns is variously known as 'education for sustainability' (EfS), 'learning for sustainable futures' or 'education for sustainable development'. However, despite exciting developments in this field, there has been difficulty galvanising change for EfS in HE and bringing its potential to life. The international literature confirms that embedding EfS in the HE curriculum is the most difficult area of sustainability practice in which to gain traction. A study by the *Global University Network for Innovation* (GUNI) evidences this trend worldwide (Tilbury, 2011a) and reviews have demonstrated similar concerns in the UK (Policy Studies Institute, 2008; Ryan, 2009; SQW Consulting, 2009).

The performance of UK universities remains internationally impressive, as reflected in league tables, research output and overseas student recruitment. UK HE has retained its reputation for high quality education, bolstered by waves of teaching enhancement funding since the 1980s (Smith, 2005). However, the system is under enormous pressure due to changes in the national funding base, steep increases in student fees, and immigration policy regulations that affect its global competitiveness. The growing influence of prospective employers and students on curriculum priorities and quality systems necessitates greater transparency about teaching practices and graduate employment prospects – both to improve the student experience and to meet the demands of competitive recruitment markets. Some issues may be specific to the UK context, but similar pressures are in evidence worldwide. They present both risks and opportunities for EfS, which competes for attention in a changeable HE landscape with multiple influences on education.

In this chapter, we consider what twenty-first century education systems geared to sustainability could look like and assess the potential of EfS to challenge and influence mainstream HE learning. Educators who embrace EfS have varied approaches, areas of expertise and political viewpoints, but share a concern to change education systems, practices and methods. EfS revolves around shared pedagogic principles and learning processes, as documented in a review of academic literature and educational projects for UNESCO as part of the Decade of Education for Sustainable Development, 2005–14 (Tilbury, 2011b). To explore the potential for systemic change, our primary material is drawn from EfS in the UK HE sector, including policy context, national initiatives and leading practice in specific universities. We consider the value and place of EfS in changing pedagogy and learning experiences across the formal, hidden and informal curriculum. Taking a systems perspective, we reflect on the necessary further steps required to lever change at all these levels for the pedagogic evolution that EfS requires, and consider how far the transformative potential of EfS offers a chance of renewal and redirection at a time of fundamental change in higher education.

## Changing the formal curriculum: engaging communities of practice

EfS is geared to innovation in pedagogy, targeting not just the 'what' but the 'how' of education. This represents a significant challenge in terms of the formal curriculum, not just to encourage teaching 'about sustainability', but to reframe the purposes and aims of learning across entire programmes of study. Examples of curriculum innovation in EfS are easy to locate – pioneering work has emerged in a range of subjects and these path-finding endeavours are increasing and diversifying (Blewitt and Cullingford, 2004; Corcoran and Wals, 2004a; Haslett *et al.*, 2011; Jones *et al.*, 2010). The UK has produced ground-breaking programmes guided by EfS, such as the MA in Sustainable Development at the University of St Andrews and the MSc in Education for Sustainability at London South Bank University. The national Higher Education Academy (HEA) has funded small-scale projects since 2005, commissioned research and development work and supported production of the *Future Fit Framework*, a guidance tool for academic staff (Sterling, 2012).

However, EfS developments have occurred largely in the margins of the HE curriculum and it has proven extremely challenging to move to the next stage, where the work of enthusiasts begins to inform mainstream programmes, and the concerns of pioneers begin to change the shared discourse of entire subject communities. Research has established that both staff and students struggle to understand the conceptual range of the term 'sustainability', focusing first on its environmental dimensions and missing the integration of social and economic aspects (Bone and Agombar, 2011; Kagawa, 2007). Sustainability is troublesome, contested knowledge (Blewitt, 2008; Hall, 2011); to be effective as an educational tool, it must be reflected in different subject areas and by groups with differing priorities and interests. This conceptual challenge also frustrates efforts to assess educational engagement with sustainability, especially where curriculum audits are confined to descriptions of course content alone.

To introduce EfS more widely and achieve real impact on student learning experiences, change is needed at subject level. EfS must enhance existing programmes, recognising the different starting points of academic disciplines, both conceptually and pedagogically. Some disciplines have stronger connections with the language of sustainability, for example in geography, earth and environmental sciences, or the built environment, landscape and design. Existing engagement with sustainability may have been driven by research agendas and by policy directives to improve application of the knowledge base. The foundations of EfS in constructivist epistemology and critical pedagogy means that disciplines grounded in these approaches (for example, in humanities and social sciences) find easier alignment between EfS and their existing pedagogic orientation. In disciplines guided by more traditional pedagogies, there is often a struggle to grasp the challenge of EfS and other approaches may be necessary, for example by engaging with the realities that graduates will face in relation to policy engagement or their professional practice (developments in professional engineering regulations in the UK offer a valuable example).

EfS has experienced the problem common in academic innovation driven by enthusiasts: if the 'community of practice' (Lave and Wenger, 1998) is not engaged, the pioneer may move on, their priorities can be sidelined by other agendas, and their module ceases to exist. For longevity, EfS principles must guide the enhancement work of teaching teams, with the advantage that all students gain some experience of EfS, not just those taking particular (often optional) modules. Engaging the communities of practice around academic programmes takes time and careful development, but will yield more secure and credible results (see Health and well-being co-benefits of living sustainably, p. 218). Academic communities respond to EfS in ways that reflect their own trajectories and their conceptual and pedagogic roots. This means that EfS in the formal curriculum can be strongly driven by external influences from professional bodies, governmental agencies and sector level organisations. External agendas are therefore critical and should be harnessed, whether through subject communities or through mechanisms within universities for infusing EfS into teaching and learning practice.

## EfS as an enhancement priority in business management

In 2009 work began to use EfS to refresh the undergraduate Business Management programme, addressing sustainability as a cross-business issue and providing a more unified understanding of professional responsibilities and leadership for sustainability. The prompt for this change included the need to improve programme distinctiveness, recruitment and employment prospects, to address the concerns of external businesses and connect the expertise of individual team members more effectively across the programme.

The first substantial change was to develop and introduce a year-long level 1 module in which students work in project teams to explore sustainability in real organisational contexts, using the university's campus sustainability strategy as case study material. Students gain initial exposure to sustainability in relation to business strategy and present proposals to expert panels with recommendations for improvement. A graduate intern project reviewed student reflections on the learning process at level 1, finding that students needed very clear guidance from staff to get past their initial focus on environmental aspects or seeing sustainability as 'business as usual' in terms of financial viability.

Staff development sessions were organised for the teaching team, to introduce EfS principles and methods, providing space for discussion of how they could apply across the programme and where the expertise of team members could contribute and also be directed towards related publication activities. Reflection on the outcomes showed the importance of professional incentives and management recognition, and the need for best practice exemplars and organisational case studies. Building stronger external partnerships was also critical, in bringing outside perspectives to the programme and creating new workplace learning opportunities. Internship project work engaged with local businesses to understand their professional needs and practical concerns around sustainability, to inform new initiatives for students to apply sustainability skills in 'real-world' projects and placements.

The first wave of activity took place over two academic years and is reflected in a more coherent and focused undergraduate programme. Business Management was then shortlisted for a competitive national *Green Gown* award in 2011 and again in 2012. Institutional research confirmed that many student placements that year were awarded on basis of the insights and abilities of students in sustainability. The next steps are to improve the depth of engagement with EfS at levels 2 and 3 and in the MBA curriculum. Broadening the global focus remains an enhancement priority, to raise levels of dialogue around sustainability among a diverse international student cohort.

*Dr Jim Keane and Dr Alex Ryan, University of Gloucestershire*

## Reorienting curriculum development: strategic institutional change

Curriculum change in EfS must be understood not only in subject context but also in institutional context, which includes corporate strategies and routine curriculum development processes. Increasingly in HE, enhancement priorities affect most (if not all) subjects, for example to improve institutional employability credentials. In recent years, there have been moves for curriculum reform in universities across the globe, focused on learning aims such as tackling complexity, engaging in inter-disciplinary work and developing inter-cultural competence. High profile examples have emerged from institutions with strong research profiles, such as the University of Aberdeen (UK), University of Melbourne (Australia) and University of Hong Kong. Although these reforms often developed in response to strategic objectives, for example recruitment, differentiation and competitiveness, they have embraced principles aligned with EfS such as inter-disciplinarity, global citizenship, community engagement and professional responsibility (Ryan, 2012).

Often linked to these reforms, the development of institutional 'graduate attributes' encourages coherence and integration in curriculum development. As Barrie and Prosser have shown , universities use graduate attributes to 'specify an aspect of the institution's contribution to society' (2004: 43), and EfS aims are both implicit and explicit in numerous examples found worldwide. Generic attributes can reinforce teaching and learning strategies, although they can lack shared understanding, for both staff and students. There are substantial challenges in embedding graduate attributes across very different programmes, as although they can seem 'relatively innocuous and uncontentious outcomes ... they have their roots in the contested territory of questions as to the nature of knowledge and the nature of a university' (Barrie and Prosser, 2004: 244). However, these signs that universities are thinking more deeply about how they evolve their teaching and learning practices are significant for EfS and offer platforms for discussion about the aims and purposes of the curriculum, to improve its relevance as well as quality (see EfS through university quality assurance and quality enhancement systems, below).

## EfS through university quality assurance and quality enhancement systems

To support institutional development in sustainability, the UK Higher Education Council for England funded a strategic project from 2010 to 2012, *Leading Curriculum Change for Sustainability: Strategic Approaches to Quality Enhancement*, involving a consortium of five universities. Taking a 'systems' view of curriculum development practices in universities, the project explored ways to bring EfS principles into quality assurance processes and quality enhancement initiatives, and worked with the UK Quality Assurance Agency to explore links between EfS and national policy in this area.

Each of the five universities carried out a development project to connect EfS with its existing quality systems and to reflect current educational priorities and organisational strategies:

- *Aston University* – making connections with the institutional focus on industry engagement and investment in 'green ICT', to align its focus on sustainability with educational quality issues.

- *Oxford Brookes University* – using the established platform of institutional 'graduate attributes' and prior work to internationalise the curriculum, to improve awareness and practice of EfS.

- *University of Brighton* – bringing EfS more deeply into each stage of the course development process using an integrated approach, to avoid EfS being viewed as an extra layer of education policy.

- *University of Exeter* – drawing on investment in sustainability research and the lead of one academic college, to build support and extend policy development for EfS in the curriculum.

- *University of Gloucestershire* (Project Lead) – building an inclusive enhancement approach to enable diverse EfS work across faculties but with an explicit, transparent assurance mechanism.

Project outcomes included the development of policy frameworks and sector guidance materials to progress EfS as an educational quality issue, as well as the broader engagement of senior managers and staff with oversight of curriculum quality, with principles of EfS. National sector dialogue was extended through the involvement of key agencies and experts, with several outcomes, including the development of the online *'Guide to Quality and Education for Sustainability in Higher Education'*.

*Professor Daniella Tilbury and Dr Alex Ryan*

There are important synergies for EfS at the intersection between the institutional and subject levels, and a recent review shows various tactics being adopted worldwide to stimulate interest in EfS across institutions (Ryan, 2012). As has been noted in other studies and in change management literature, the approach must fit the institutional context (Brooks and Ryan, 2008; de la Harpe and Thomas, 2009). In addition, concerns raised about the imposition of educational agendas only underline the critical need for understanding the social practices that influence change. The contradictory and conflicting nature of curriculum development in universities means that there is a need to understand the process, rationale and implementation of change, as well as levels of resistance to it, to achieve strategic progress in EfS (de la Harpe and Thomas, 2009: 77–8), which is apparent in prominent examples of 'whole institution' EfS initiatives in the UK (Ryan, 2012).

Promoting EfS as a paradigm-shifting educational proposition requires that it is aligned with broader curriculum priorities. To miss this point is to risk divorcing EfS from exciting opportunities to shape and inform the emerging discourse and practice of university education. However, to work strategically to shift the formal curriculum means that not every innovation that might be labelled as EfS *will* be labelled that way. Explicit association with EfS may improve the credentials of some programmes, but at other times, connections to different priorities will better serve the needs of recruitment, professional accreditation or institutional strategy. Hopkinson *et al.* (2008) pointed to 'self-recognition' issues among academic staff in EfS, an issue also noted by Kagawa *et al.* (2010) and in a review of practice in Scotland (Ryan, 2009). For both educational and tactical reasons, not all academics connect their efforts with EfS explicitly, preferring implicit approaches to the learning process. As EfS involves dialogue around values and negotiation of contested meanings, both directness and concealment have their advantages (Blewitt, 2008; Cotton *et al.*, 2009; Wals and Jickling, 2002). This ambiguity also provides opportunities to use the invisible dimensions of learning, which are revealed powerfully in the dynamics and spaces in which learning unfolds.

## Under the radar: the hidden curriculum

It is well known that students learn significantly more than what is explicitly taught: for example, the formal curriculum may involve taking students on a field trip to South Africa, where the learning outcomes are connected to field techniques and identification of plants and animals. However, what students learn from such an experience may include awareness of cultural differences, of political context, of the challenges associated with sharing a living space with 30 other students, of the difficulties of conducting fieldwork in an unfamiliar climate, and so on. Even in more familiar contexts, the environment may impact on the student experience in powerful and significant ways: what the campus 'says' about sustainability may leave a lasting impression on students who live and work there for a sustained period, and communication about sustainability through official channels can be subverted easily through 'noise' caused by competing messages in the campus environment (Djordjevic and Cotton, 2011). These aspects of learning form part of the 'hidden curriculum' of educational settings, a term first used to describe, in schooling, the 'unpublicised features of school life' (Jackson, 1968: 17), and which is made manifest through the ethos and values of an institution (Skelton, 1997).

The hidden curriculum is a complex and ambiguous term, used in a range of different ways to describe the (sometime) disconnect between what is overtly taught in educational institutions and what students actually learn. In HE, this may include societal, institutional or individual values transmitted (normally unconsciously) to students through the campus environment or the attitudes and values of university staff. For example, lecturers' attitudes and beliefs about EfS have been shown to have a strong impact on what is taught in the formal curriculum as well as on how they interact with students in formal and informal settings (Cotton *et al.*, 2007). Lawton claims that 'Every statement that a teacher makes in a classroom is value laden, connected with ideas about the purpose of education, probably connected with more general values and beliefs, and maybe with the purpose of life' (Lawton, 1989: 3). While many lecturers believe that they should be (or even that they are) offering a balanced and unbiased view of their subject area, research indicates that this is all but impossible for many sustainability-related topics (Cotton, 2006). In HE, where lecturers have significant control over the content of the curriculum as well as the means by which it is expressed, a hidden curriculum based on their beliefs and values is potentially very significant.

Clearly, the hidden curriculum can take a negative form; early research investigated the ways in which schools reproduce the social inequalities of wider society (for example, Willis, 1978), and the hidden curriculum may act to reinforce discrimination. Yet it can also have positive impacts: recent research indicates that participation in HE may lead to an increase in environmental commitment, for example, which may be partly due to changes in social identity that occur outside the formal curriculum (Cotton and Alcock, 2012). In addition, there have been several high-profile campus development projects in sustainability suggesting that 'being a student in such an environment may influence commitment towards environmental sustainability' (Cotton and Alcock, 2012: 12). If correct, this indicates that the hidden curriculum of HE can act as a force for good in terms of EfS – but it also suggests that the risks of good work in the formal curriculum being undone by a lack of sustainability in the hidden curriculum are significant (see Harnessing informal learning about sustainability, below).

Consideration of the relationship between formal and hidden curricula in HE raises serious questions for policy-makers and practitioners in EfS: it confirms the need to prepare students to 'make sense of and respond to exposure to contradictory information, values, beliefs and practices' and to ensure that they are 'cognisant and critical', rather than 'over-determined, passive recipients of hidden curriculum messages' (Skelton, 1997: 177). The lessons the hidden curriculum teaches are often experienced daily (if embedded in the learning environment) and, given the duration of students' experience of formal education, the importance of understanding these messages cannot be underestimated. This aspect is directly connected to the sphere of informal learning, and considering links between the formal, informal and hidden curriculum is crucial to enhance all aspects of sustainability learning.

## Integrating informal learning: broadening the terrain

Much of the existing HE research and development effort has focused on EfS in the formal curriculum in various contexts. In some ways this is both inevitable and understandable since the formal curriculum is the most visible part of a university's activities. The formal curriculum is marketed to students: students engage in HE initially through their course or programme and marketing is focused on specific disciplines and subject areas as well as contact hours and pedagogic approaches. However, practitioners and researchers focused on informal learning have encountered an arena which is eminently suitable for, and already influential in, enhancing EfS opportunities for students. Informal learning offers a potential route which bypasses the disciplinary silos and sometimes negative academic attitudes which can hinder the embedding of EfS in the formal curriculum. The campus, for example, provides a subject-neutral forum through which sustainability can be experienced, discussed and critiqued regardless of the 'limitations of [disciplinary] tunnel vision' (Jucker, 2002: 13). Yet, while the informal curriculum may be more important than the formal curriculum in sustainability learning, its impact in HE is only just starting to be explored. Moreover, developing and researching informal learning can be problematic: informal learning is largely invisible, may not be recognised as such by the learners, and can be hard to describe (Eraut, 2004).

Informal learning is usually understood to mean learning from other people outside the formal educational context and in a range of different locations. According to Eraut, informal learning 'draws attention to the learning that takes place in the spaces surrounding activities and events ... and takes place in a much wider variety of settings than formal education or training' (2004: 247). In relation to sustainability, Lipscombe (2008) defines the informal curriculum as consisting of extra-curricular activities and experiences such as volunteering, internships, membership of clubs and societies and attending sustainability events. In HE, this may constitute a very important part of the learning in which students engage, since they are often living in a different area, possibly independently for the first time, and with significant social contact outside formal classes. They may be based in accommodation on campus, thereby being open to opportunities for learning from their physical environment as well as through dialogue and activities with others. Kagawa describes the campus as a potential site for learning and EfS through a 'sustainability orientated pedagogy of place' (2007: 320). In addition, recognising the potential for campus learning experiences to contribute to EfS may go some way to addressing concerns that 'the student experience at most universities typically has a fragmented connection of the values, ideals and practical aspects of living, studying or working in a sustainable way' (Hopkinson *et al.*, 2008: 439). Evidence of the potential impact of informal learning can be gathered easily through small-scale studies such as that at Plymouth University (see below).

## Harnessing informal learning about sustainability

Given the power of the informal context to influence student learning about sustainability, this is an area with which aspiring sustainable universities would do well to engage. A small-scale research project at Plymouth University (using video diaries to capture student experiences of sustainability on campus) indicated that student learning about sustainability through the campus environment was variable. Students were very conscious of issues surrounding energy and carbon, and perceived recycling as an important issue for the university, but were concerned (and sometimes confused) about the way energy was seemingly used unsustainably for lighting or computer power in parts of the institution. Although the university was seen as setting a good example in some areas (new buildings, automatic lighting), the students felt that more could be done to develop the campus as part of the learning environment (with more signage and more involvement of students in decision-making about sustainability issues).

Although the institution has an enviable external reputation for sustainability, internally this was seen as being in conflict with other university agendas, and what students learnt through their experiences outside the formal curriculum acted as a mediator of their learning about sustainability. Recent moves to address this issue include harnessing students' informal learning through a volunteering module, encouraging extra-curricular activities in sustainability through the Plymouth Award, and increasing efforts to involve students in projects and placements which link the campus with the formal curriculum – supported by the Office of Procurement and Sustainability (OPS) and the Centre for Sustainable Futures (CSF). However, there is always more which could be done: signage could be improved, and confusion about whether and why it is more sustainable to use an automatic revolving door remains a live issue. Communication with staff and students about sustainability developments is challenging, and enthusiasm for drawing on informal learning to enhance the formal curriculum is scattered. However, the structures in place (including an Institute for Sustainability Solutions Research, as well as OPS and CSF) offer opportunities for further linking informal and formal learning through the campus, curriculum and community.

*Dr Jennie Winter and Professor Debby Cotton, Plymouth University*

While informal learning often takes place without much structure, harnessing its full power may involve its integration with formal learning by encouraging reflection on everyday activities or experiences. Marsick and Watkins identify three ways in which informal learning may be enhanced:

> critical reflection to surface tacit knowledge and beliefs, stimulation of proactivity on the part of the learner to actively identify options and to learn new skills to implement those options or solutions, and creativity to encourage a wider range of options.
>
> (Marsick and Watkins, 2001: 30)

As support for campus greening expands across HE, driven in part by the need for carbon reduction plans, universities increasingly provide leadership as sustainable organisations: 'Universities can be a model for the community about how a sustainable organisation ought to operate' (Ferrer-Balas *et al.*, 2008: 296). Linking campus and operations development with student learning offers the next step towards a fully integrated EfS, which encompasses the formal, informal and hidden curricula, and provides a student experience which contributes both to sustainability and employability. By providing such a holistic experience, universities can capitalise on the potential of sustainability-related informal learning to contribute beyond the current stand-alone 'volunteering module' or HE award scheme. Explicit efforts to link the formal and informal curriculum (for example, through sustainability placements, campus-based projects or portfolios in which students reflect upon extra-curricular activities) can go some way to overcoming the view that 'the typical campus is mostly regarded as a place where learning occurs but, itself is believed to be the source of no useful learning' (Orr, 1993: 597).

## Pedagogic evolution: EfS at the heart of the HE system

This discussion of different curricular possibilities has illustrated the potential for EfS to play a significant role in the twenty-first century HE landscape. However, bringing EfS to life is no small task, given the scale and complexity of HE systems, which are influenced by various educational, political and financial agendas (Corcoran and Wals, 2004a; Wals and Jicking, 2002). Academic autonomy is enshrined at institutional and disciplinary level, which protects innovation but can also be an excuse to resist change (Bawden, 2004; Corcoran and Wals, 2004b; Cotton and Winter, 2010). If sustainability as a component of the curriculum is controversial, EfS as a pedagogic approach, touching all forms of learning and all levels of the curriculum, is even more so.

One of the challenges to be addressed is the need for clearer articulation of the ways that EfS principles relate to existing educational literature and academic practice in HE. Too often, EfS appears to present a 'special case' for pedagogic change, with a remedy that can be applied equally in all parts of the HE system. In fact, some pedagogic approaches fundamental to EfS, such as critical thinking, are part of the discourse and practice in many HE disciplines (Barnett, 1997; Moon, 2005). However, other pedagogies advocated for EfS are not in widespread use in HE and there may be practical or philosophical limitations to their use in some disciplines (Cotton *et al.*, 2009). This includes approaches such as clarifying personal values, envisioning more positive and sustainable futures, thinking systemically and exploring dialectics between tradition and innovation (Tilbury, 2011b). Universities have been built around transmissive and didactic models of learning, devised for the efficient transfer of abstract (academic) knowledge from teacher to student. Personal and societal transformation has not traditionally been an imperative, positioned as central to the purpose of educational business, yet it underpins EfS.

Many prominent commentators have insisted that EfS should have transformative intent. The concept of transformative learning originated in adult education and encompasses a range of participatory pedagogies to promote critical self-reflection, leading to transformed 'habits of mind' (Mezirow, 2000). Arguably, any university education worthy of the name should be transformative: students should see the world differently at the end of their course of study (Barnett, 2011). However, the appearance of transformative learning in HE is relatively new and its models may need refinement to support effective EfS. As Sterling has observed, transformative learning must engage the intellect alongside affective and existential domains, encouraging empowerment and action. Surface approaches – 'the mainstream emphasis on cognitive learning, with a little "values education" thrown in' – simply will not suffice (Sterling, 2011: 27). Truly transformative education involves integration and change, as outlined in Gregory Bateson's work on third-order learning (Bateson, 1972). This moves beyond first-order change ('doing things better') and second-order change (meta-learning – or 'doing better things'), to 'seeing things differently', where engagement with ethical frameworks, belief systems and interpersonal relationships is deeply implicated.

The benefits of connecting EfS and transformative learning are easy to anticipate, in the pursuit of higher-order learning that entails links with the wider community as well as the ability to deal with complexity and uncertainty (Cranton, 1996; Sterling, 2011). In many ways, transformative learning provides a model for effective education in the twenty-first century and its applied, action-oriented tenets are appearing in influential educational forums and dialogues, such as the 1996 International Commission on Education for the Twenty-First Century report to UNESCO, *Learning: the Treasure Within* (Delors, 1996). Its value for EfS is reflected in the 2011 publication of *Learning for the Future: Competences in Education for Sustainable Development,* produced by an expert group for the UNECE Steering Committee on Education for Sustainable Development, drawing on the four Delors educational principles: *learning to know, learning to do, learning to live together, learning to be.*

Given the increasing range of influences on HE, concerns have been raised about 'how far mainstream higher education is able to provide transformative learning experiences, or whether [they are] inevitably associated with innovative learning environments outside the constraints of conventional education' (Sterling, 2011: 17). It is profoundly challenging to attempt to develop more democratic and innovative approaches to pedagogy in a system which is not itself democratic and is under considerable structural pressure. In the UK, consumerist economic pressures and the persistence of managerial ideologies (Deem and Brehony, 2005) engender certain types of staff performativity and discourses around student employability, which have serious implications both for EfS and for HE in general (Blewitt, this volume; Sterling, this volume). However, there may also be some unanticipated benefits for EfS in the 'performance culture' of leagues and rankings, which has also permeated the sustainability agenda. These schemes can support EfS through informal learning, campus greening and 'whole-institution' development for sustainability, as in the UK *Learning In Future Environments* benchmarking initiative and People and Planet's *Green League*.

Rapid globalization, ICT advances and economic crisis are profoundly affecting the ways that universities design and deliver education, prompting changes in learning models and relationships. HE is being forced to reconsider traditional approaches to teaching and to refresh its curricula and pedagogies in ways that could provide strong foundations for EfS. There are indications that the current system serves particular interests: for example, the ongoing underperformance of male and ethnic minority students at all levels is a growing concern (Richardson, 2008). Yet there is evidence of a shift to more democratised modes of learning, as cohorts become more diverse and hierarchies between lecturers and students begin to dissolve. Policy trajectories that support the commodification of learning tend to encourage passive forms of education and inhibit the creative curriculum development needed to address the issues students will face in their future lives and workplaces. Yet there are significant drivers for EfS, not least from students and employers concerned about the need for literacy and capability to deal with global sustainability challenges. The twenty-first century HE landscape offers opportunities as well as risks for EfS, and there is growing awareness of the need for learning cultures that foster enquiry, challenge, flexibility, connectivity and responsibility (Barnett, 1997; Boyer, 1990). As this need increases, EfS can help the HE system to achieve the transformation it requires. Taking this systems view, 'sustainability is not just another issue to be added to an overcrowded curriculum, but a gateway to a different view of curriculum, of pedagogy, of organisational change, of policy and particularly of ethos' (Sterling, 2004: 50).

## Conclusion

Arguably, HE is approaching a tipping point, but despite the many constraints upon the system, we remain optimistic about the potential for EfS to thrive within it. The HE system is in flux and if there was ever a time for evolutionary change, perhaps it is now. In this chapter we have attempted to draw attention to the 'feasible utopias' (Barnett, 2011: 4) that can already be glimpsed in our universities, offering great promise and possibility for a different kind of education. Piecemeal approaches may not survive the larger shifts affecting HE, but an infusion of EfS thinking and practices could help to protect the best of our educational traditions and recent learning innovation. For those who think progress in EfS may have stalled, that the agenda has been corrupted, or that nothing will ever shift the HE curriculum, a historical view might suggest that the sustainability agenda has rapidly achieved a substantial presence in public life and educational practice. As Kuhn's original model of scientific 'crisis' and 'wars' showed, significant paradigm shifts do not take place quickly (Kuhn, 1962). EfS is informed by several fields of thought and practice, so to expect revolution without strategic effort and systemic change would be unrealistic.

Two key messages are important if this movement is to continue to evolve. First, EfS needs to be more effectively positioned in relation to the broader pedagogic development literature and strategic approaches to curriculum change. Engagement with entire communities of practice and scholarship will be essential for future EfS, within teaching teams and institutional settings, with external organisations and international subject associations. The emphasis on reframing individual practices must not be to the detriment of efforts to shift education paradigms in terms of the systems themselves (Sterling, 2004; 2011; Tilbury, 2007; 2011b). This means increasing the level of dialogue and engagement with other movements for educational change, to avoid EfS being seen as a special political petition and becoming marginalised from the core educational concerns and practices of HE.

The second key issue is the need to consider all dimensions of curriculum and pedagogy. Universities are places of inspiration and creativity, and if EfS is to become part of the mainstream, it must engage all parts of the system. This means finding its place in contributing to academic innovation, institutional change and improved learning environments. It requires pedagogic approaches that enable EfS to flourish in the formal curriculum and through informal learning. Proactive support and continued dialogue will be needed, within universities and across the sector, to ensure that conceptual changes are translated into action on the ground. All of the key UK sector agencies and funding councils have shown willingness to join the vanguard and support change in this area, through funding schemes, formal declarations and practical support – a sign of their trust in the importance of the issue.

This chapter has sought to unpack issues that merit further attention in order to progress understanding of EfS and ensure its place in the future of HE. Its role in this book is as part of a holistic system, in which each component contributes to the growth of effective EfS, which in essence should mean effective learning in general. Connecting informal and formal learning, integrating learning across different parts of HE institutions and shifting boundaries between universities and their surrounding communities, using the full range of EfS pedagogies and understanding the change processes involved, are all crucial to this endeavour. Conceived in this way, EfS offers a vision of education that would serve the global community well. To move in this direction will be neither swift nor easy, but it will be a satisfying journey.

## References

Barnett, R (1997) *Higher Education: A Critical Business,* Milton Keynes: SRHE/OUP
Barnett, R (2011) *Being a University,* London: Routledge
Barrie, S and Prosser, M (2004) 'Generic graduate attributes: citizens for an uncertain future', *Higher Education Research and Development,* 23 (3): August 2004, 243–6
Bateson, G (1972) *Steps To An Ecology of Mind,* San Francisco: Chandler
Bawden, R (2004) 'Sustainability as emergence: the need for engaged discourse', in Corcoran, P B and Wals, A E J (eds) *Higher Education And The Challenge of*

*Sustainability: Problematics, Promise, and Practice*, Dordrecht: Kluwer Academic Publishers, pp. 21–32

Blewitt, J (2008) *Understanding Sustainable Development*, London: Earthscan

Blewitt, J and Cullingford, C (eds) (2004) *The Sustainability Curriculum: The Challenge for Higher Education*, London: Earthscan

Bone, E and Agombar, J (2011) *First-year Attitudes Towards and Skills In, Sustainable Development*, York: Higher Education Academy. Available at: www.heacademy.ac.uk/assets/documents/sustainability/FirstYearAttitiudes_FinalReport.pdf (accessed 20 April 2012)

Boyer, E (1990) *Scholarship Reconsidered: Priorities of the Professoriate*, The Carnegie Foundation for the Advancement of Teaching, New York: John Wiley and Sons

Brooks, C and Ryan, A (2008) *Education for Sustainable Development: Strategic Consultations Among English HEIs*, York: Higher Education Academy

Cotton, D (2006) 'Teaching controversial environmental issues: neutrality and balance in the reality of the classroom', *Educational Research*, 48 (2): 223–41

Cotton, D and Winter, J (2010) 'It's not just bits of paper and light bulbs: a review of sustainability pedagogies and their potential for use in higher education', in Jones, P, Selby, D and Sterling, S (eds) *Sustainability Education: Perspectives and Practice Across Higher Education*, London: Earthscan

Cotton, D and Alcock, I (2012) 'Commitment to environmental sustainability in the UK student population', *Studies in Higher Education*. Available at: www.tandfonline.com/doi/abs/10.1080/03075079.2011.627423

Cotton, D, Warren, M, Maiboroda, O and Bailey, I (2007) 'Sustainable development, higher education and pedagogy: a study of lecturers' beliefs and attitudes', *Environmental Education Research*, 13 (5): 579–97

Cotton, D, Bailey, I, Warren, M and Bissell, S (2009) 'Revolutions and second-best solutions: education for sustainable development in higher education', *Studies in Higher Education*, 34 (7): 719–33

Corcoran, P and Wals, A (eds) (2004a) *Higher Education and the Challenge of Sustainability: Problematics, Promise, and Practice*, Dordrecht: Kluwer Academic Publishers

Corcoran, P and Wals, A (2004b) 'The problematics of sustainability in higher education: a synthesis', in Corcoran, P B and Wals, A E J (eds) *Higher Education and the Challenge of Sustainability: Problematics, Promise, and Practice*, Dordrecht: Kluwer Academic Publishers, pp. 87–90

Cranton, P (1996) 'Types of group learning', *New Directions for Adult and Continuing Education*, 71: 25–32

De La Harpe, B and Thomas, I (2009) 'Curriculum change in universities: conditions that facilitate education for sustainable development', *Journal of Education for Sustainable Development*, 3 (1): 75–85

Deem, R and Brehony, K (2005) 'Management as ideology: the case of "new managerialism" in higher education', *Oxford Review of Education*, 31 (2): 217–35

Delors, J (1996) *Learning, The Treasure Within: Report to UNESCO of the International Commission on Education for the Twenty-first Century*, Paris: UNESCO

Djordjevic, A and Cotton, D (2011) 'Communicating the sustainability message in higher education institutions', *International Journal of Sustainability in Higher Education*, 12 (4): 381–94

Eraut, M (2004) 'Informal learning in the workplace', *Studies in Continuing Education,* 26 (2): 247–73

Ferrer-Balas, D, Adachi, S, Banas, C, Davidson, A, Hoshikoshi, A, Mishra, Y, Onga, M and Otswals, M (2008) 'An international comparative analysis of sustainability transformation across seven universities', *International Journal of Sustainability in Higher Education,* 9 (3): 295–316

Hall, B (2011) 'Threshold concepts and troublesome knowledge: towards a "pedagogy of climate change"?', in Haslett, S, France, D and Gedye, S (eds) *Pedagogy of Climate Change,* York: The Higher Education Academy

Haslett, S, France, D and Gedye, S (eds) (2011) *Pedagogy of Climate Change,* York: The Higher Education Academy

Hopkinson, P, Hughes, P and Layer, G (2008) 'Sustainable graduates: linking formal, informal and campus curricula to embed education for sustainable development in the student learning experience', *Environmental Education Research,* 14 (4): 435–54

Jackson, P (1968) *Life in Classrooms,* Chicago: Chicago Teaching College

Jones, P, Selby, D and Sterling, S (eds) (2010) *Sustainability Education: Perspectives and Practice Across Higher Education,* London: Earthscan

Jucker, R (2002) *Our Common Illiteracy: Education As If Earth and People Mattered,* Frankfurt: Peter Lang

Kagawa, F (2007) 'Dissonance in students' perceptions of sustainable development and sustainability', *International Journal of Higher Education,* 8 (3): 317–38

Kagawa, F, Blake, J and Jones, P (2010) 'Sustainability in the University of Plymouth curricula: as perceived by heads of school', University of Plymouth: Centre for Sustainable Futures

Kuhn, T (1962) *The Structure of Scientific Revolutions,* Chicago: University of Chicago Press

Lave, J, and Wenger, E (1998) *Communities of Practice: Learning, Meaning, and Identity,* Cambridge: Cambridge University Press

Lawton, D (1989) *Education, Culture and the National Curriculum,* London: Hodder and Stoughton

Lipscombe, B (2008) 'Exploring the role of the extra-curricular sphere in higher education for sustainable development in the United Kingdom', *Environmental Education Research,* 14 (4): 455–68

Marsick, V and Watkins, K (2001) 'Informal and incidental learning', *New Directions for Adult and Continuing Education,* 89: 25–34

Mezirow, J (2000) *Learning as Transformation: Critical Perspectives on a Theory in Progress,* San Francisco: Jossey Bass

Moon, J (2005) *We Seek it Here … A New Perspective on the Elusive Activity of Critical Thinking: A Theoretical and Practical Approach,* York: Higher Education Academy/ ESCalate Subject Centre for Education

Orr, D (1993) 'Architecture as pedagogy', *Conservation Biology,* 7 (2): 226–8

Policy Studies Institute (2008) *HEFCE Strategic Review of Sustainable Development in Higher Education in England,* Policy Studies Institute, PA Consulting Group and Centre for Research in Education and the Environment, University of Bath, Report to the Higher Education Funding Council for England (HEFCE)

Richardson, J (2008) *Degree Attainment, Ethnicity and Gender: A Literature Review,* Equality Challenge Unit, York: Higher Education Academy

Ryan, A (2009) *2008 Review of Education for Sustainable Development (EfS) In Higher Education in Scotland*, HE Academy EfS Project in collaboration with the Scottish Funding Council and Universities Scotland

Ryan, A (2012) *ESD and Holistic Curriculum Change: A Review and Guide*, York: Higher Education Academy

Skelton, A (1997) 'Studying hidden curricula: developing a perspective in the light of post-modern insights', *Pedagogy, Climate and Society*, 5 (2): 177–93

Smith, B (2005) 'The role of national UK organisations in enhancing the quality of teaching and learning', in Fraser, K (ed.) *Educational Development and Leadership in Higher Education: Development an Effective Institutional Strategy*, London: Routledge, pp. 16–29

SQW Consulting (2009) *Education for Sustainable Development and Global Citizenship (EFSGC): Analysis of Good Practice in Welsh Higher Education Institutions*, Report to the Higher Education Funding Council for Wales (HEFCW)

Sterling, S (2004) 'Higher Education, sustainability, and the role of systemic learning', in Corcoran, P and Wals, A (eds) *Higher Education and the Challenge of Sustainability: Problematics, Promise, and Practice*, Dordrecht: Kluwer Academic Publishers, pp. 47–70

Sterling, S (2011) 'Transformative learning in sustainability: sketching the conceptual ground', *Learning and Teaching in Higher Education*, 5: 17–33

Sterling, S (2012) *The Future Fit Framework: An Introductory Guide to Teaching and Learning for Sustainability in HE*, York: Higher Education Academy and Centre for Sustainable Futures, Plymouth University

Tilbury, D (2007) 'Monitoring and evaluation during the UN Decade of Education for Sustainable Development', *The Journal of Education for Sustainable Development*, 1 (2): 239–54

Tilbury, D (2011a) 'Higher education for sustainability: a global overview of commitment and progress', in GUNI (ed.) *Higher Education's Commitment to Sustainability: From Understanding to Action*, Higher Education in the World 4, Barcelona: GUNI, pp. 18–28

Tilbury, D (2011b) *Education for Sustainable Development: An Expert Review of Processes and Learning*, Paris: UNESCO

UNECE (2011) *Learning for the Future: Competences in Education for Sustainable Development* (ECE/CEP/AC.13/2011/6). Available at: www.unece.org/fileadmin/DAM/env/esd/6thMeetSC/Learning%20for%20the%20Future_%20Competences%20for%20Educators%20in%20ESD/ECE_CEP_AC13_2011_6%20COMPETENCES%20EN.pdf (accessed 25 May 2012)

Wals, A and Jickling, B (2002) '"Sustainability" in higher education: from doublethink and newspeak to critical thinking and meaningful learning', *Higher Education Policy*, 15: 121–31

Willis, P (1978) *Learning to Labour*, New York: Colombia University Press

# 8 Sustainability research

A novel mode of knowledge
generation to explore alternative
ways for people and planet

*Rehema M White*

## Introduction

Research is not merely the gathering of facts, but is a process by which we collect, analyse and interpret knowledge in order to answer a research question, resolve a problem or contribute greater understanding of a phenomenon. Knowledge is not just the acquisition of facts but has deeper meanings; it is commonly defined as justified true belief. The role of research at sustainable universities is to enhance knowledge generation, mobilisation and implementation for a more equitable, healthier and happier society and to understand and develop environmental integrity.

Sustainability research shares attributes with the concept of education for sustainability. It focuses on sustainability issues, which can be globally and contextually defined, and is achieved using approaches commensurate with sustainability principles, comprising some or all of the following aspects: interdisciplinarity, participation, co-production, building capacity and awareness, contributing to theory while having local impact and global relevance, synthesising different forms of knowledge, encouraging reflexion, linking with teaching and learning. It alters our conceptualisation of what research itself is or might be. This new mode of knowledge generation involves interdisciplinary teams engaged in tackling our complex, 'wicked' problems, and also individuals working within disciplines to contribute new thinking on issues of sustainability. Sustainability science is an emerging disciplinary area that partially addresses the requirements of sustainability research, but which suffers constraints imposed by its positioning from a science base.

Sustainability research can be promoted through initiatives that stimulate partnerships, interdisciplinary debate and innovation, links between theory and practice and institutional self-analysis. However, it may be threatened by external structures such as reductionist methods of assessing research, territorial research funding and the audit culture. Internal university divisions, disciplinary boundaries and management units may limit research communication and incentives.

The debate on the purpose and nature of research and how it relates to our pursuit for sustainable development is wide ranging and diffuse. This chapter explores the concept of 'sustainability research' within the context of knowledge generation, mobilisation and implementation in sustainable development. Understanding how research can promote sustainability and integrate with other areas of scholarship and practice is essential in theorising and creating a sustainable university.

This chapter will first question what research is, or is expected to be, then define how research relates to university function and our pursuit of sustainable development, exploring in particular the relationship between sustainable development education and sustainability research. It will examine how sustainability research is determined through a focus on sustainability issues and on the nature of the research process. The role of research in the sustainable university will be analysed, including the impact of the emerging field of sustainability science. Finally, recommendations for how sustainability research can be nurtured in a sustainable university will be made. This perspective is drawn from my own disciplinary background, a shift from biology towards human geography, and informed especially by my research and experiences as dean of research in a South African university, and as developer and Director of the Sustainable Development Programmes at the University of St Andrews.

## What is research?

Research is a core activity in any university. It is the process of enquiry by which academics generate knowledge, and critically respond to both fundamental questions and real-world issues. It has instrumental, conceptual, strategic and process uses (Nutley *et al.*, 2007). In a sustainable university, research is crucial in allowing the institution to contribute to the imagining and implementing of a better world, in facilitating critical, reflexive and engaged learning processes and in improving practice.

Research can also be thought of in more functional terms: as an individual project, an activity undertaken by faculty, postgraduate or undergraduate students or a programme developed to meet particular goals. In this more functional meaning, research can be envisaged as a process of steps by which information is collected and then analysed (Creswell, 2008). Creswell (2008) defines the steps broadly as: to develop a question, collect data to address the question, then offer an answer to the question. Research is thus not merely the gathering of facts, but is a process by which we systematically collect, analyse and interpret data in order to answer a research question, resolve a problem or contribute greater understanding to a phenomenon (Leedy, 1997).

Knowledge is often defined as justified true belief, as opposed to merely opinion, but there is extensive debate over definitions of knowledge (see Lehrer, 1990). Research can be basic or applied (for example, Creswell, 2008); the former generally implying a theoretical generation of knowledge in which application is not (yet) apparent, and the latter the production of knowledge for immediate, practical use. Foucault challenged us to distinguish general modes of thinking (epistemes) to understand society. As we will see below, sustainability research offers us opportunities to develop a new knowledge mode.

## Defining sustainability research

### *Visions of sustainable development*

There are many definitions of sustainable development, most based on the understanding that 'Sustainable Development is development that meets the needs of the present without compromising the ability of future generations to meet their own needs' (World Commission on Environment and Development, 1987). In this chapter I propose that sustainable development constitutes a recognition that social justice and environmental integrity are interdependent; a process offering multiple pathways towards alternative futures; a plurality of perspectives that offers a new model of knowledge generation, mobilisation and implementation. This optimism for a new way of being is crucial; it acknowledges that '[g]enuine sustainability ... will come not from superficial changes but from a deeper process akin to humankind growing to a fuller stature' (Orr, 2002: 1457). This definition assumes that scientific advancements alone cannot fix our 'sustainability problems', but rather sees that '[v]alues are both the ends and means of the Great Transition' (Kates *et al.*, 2006: 1) and hence the humanities and arts will also be required to contribute. Sustainability research needs to permit civil society to ask not just how but 'why' (Brand and Karnoven, 2007).

### *Sustainable development education*

There has been substantial debate on the concept and practice of 'education for sustainable development', 'education for sustainability', 'sustainable development education' and 'sustainability literacy', but relatively little on the parallel and related concept of 'sustainability research', or other such terms. Let us thus start with a summary of sustainable development education to frame the context for research.

While the terms 'education for sustainability' and 'education for sustainable development' are widely used in theory and in practice (for example, Huckle and Sterling, 1996; HEA, 2012), there are critiques that the preposition 'for' implies dogma, and a predetermined direction (Jickling, 1992; Sterling, 2001). Some people thus prefer the term 'sustainable development education' (SDE). Whichever term is used, SDE, as I will call it, implies not just obtaining facts about sustainable development, but also a form of personal transformative learning (Sterling, 2003). As Stibbe and Witham suggest, this learning can help us reconfigure society and escape our current trajectory towards 'a planet which is a less hospitable place for human life, and the lives of countless other species' (2009: 1).

*Sustainability research*

Research contributes to the generation of knowledge, and its mobilisation, including learning. Academics are by nature sceptical of evangelical intent, and at the University of St Andrews we have found distrust of terms such as 'research for sustainable development'. The term 'sustainable development' itself is also problematic for some of our colleagues. Many scientists view sustainability as the main goal, but they perceive it as a largely environmental issue, without fully reconciling the potential contribution of the 'development' of humanity. On the other hand, 'development' is difficult for some in the social sciences and humanities, who object to the current dominant economic paradigm and the modes of progress that have been imposed on 'developing countries' by colonial and, more recently, corporate interests. 'Sustainable research' implies research that will be sustained, or continue into the future, rather than a specific connection to sustainable development. Given this contested context, I have chosen to use the term 'sustainability research', with the understanding that this is an area valid to all disciplines and epistemologies, including natural and physical sciences, social sciences and arts and humanities.

In partnership with SDE, sustainability research is about more than merely knowledge domains and a collation of facts to be analysed in relation to sustainability issues. Sustainability research also incorporates a particular approach to research, and an expectation that the researcher and research participants may themselves be affected by the research process.

## Sustainability issues

Let us first consider the *sustainability issues* that might be addressed in sustainability research. Academics and university managers may either believe sustainability to be very limited, usually in relation to their own area (for example, a geologist might envisage it as only the drive towards better understanding energy and mineral reserves) or existing only in relation to the areas of others (for example, an international relations academic might see it only as an environmental concern, of interest to him/her only in relation to environmental governance). Alternatively, people are confused because sustainability research seems to touch on everything, and we consistently state that sustainability scholarship is relevant to all disciplines. The list of sustainability issues to be tackled at any given sustainable university will be derived from local imperatives, national goals, global priorities and areas of research specialisation. Table 8.1 offers example frameworks of sustainability issues, two derived from debates in sustainability science and one employed at the University of St Andrews to inform teaching and research.

These elements will normally be contextualised, in relation to regional/global imperatives, academic strengths of the institution/organisation and other parameters. The first two frameworks are taken from Kajikawa (2008) and Parris and Kates (2003).

*Table 8.1* Different suggested sustainability research issues, themes or frameworks

| Sustainability science – journal focus (Kajikawa, 2008)* | Sustainability science – goals (Parris and Kates, 2003) | Sustainable Development Programmes (University of St Andrews) |
| --- | --- | --- |
| Climate | **Human needs goals** | **Priority areas**\*\* |
| Biodiversity | Improve health | Sustainable production and consumption |
| Agriculture | Provide education | Natural resource protection and environmental enhancement |
| Fishery | Reduce hunger | Building sustainable communities |
| Forestry | Reduce poverty | Climate change and energy |
| Energy and resources | Provide housing | Cross cutting themes |
| Water | **Life support goals** | **Research and education** |
| Economic development | Reduce emissions of atmospheric pollutants | Policy, decision-making and governance |
| Health and lifestyle | Stabilise ocean productivity | Changing behaviour |
| | Maintain fresh water availability | Values, philosophies and ethics |
| | Reduce land use cover change | Gender and equity |
| | Maintain biodiversity | |
| | Reduce emissions of toxic substances | |

\*    Theme identification drawn from only three journals: *Sustainability Science, Proceedings of the National Academy of Sciences of the United States of America* and *Sustainability: Science, Practice and Policy.*
\*\*   Priority areas defined partly in relation to Department for Environment, Food and Rural Affairs, UK.

Many of these issues represent 'wicked problems' (Rittel and Webber, 1973), which demand long-term attempts to develop adaptive responses and build even clumsy solutions. Wicked problems have no definitive formulation; solutions are not 'true-false' but 'better-than' or 'as-bad-as'; each is unique and there is often no trial or error possible, yet the outcome of decision-making is critical (Rittel and Webber, 1973). A key aspect of addressing wicked problems is being able to identify appropriate questions.

## Sustainability research process

The second aspect of sustainability research is the *research process*. As is well acknowledged for SDE, the process is as important as the issues covered and the outcome. Sustainability research will demonstrate the research attributes shown in Table 8.2 and discussed below, although not all attributes may be exhibited in every piece of research.

Sustainability issues, as illustrated by the examples above, transcend disciplinary boundaries. *Interdisciplinary research* is thus required; it avoids partial framing of the problem and research questions, contextualises environmental and technological constraints and opportunities, and enhances the potential for stakeholder interactions. However, interdisciplinary research is enhanced by maintaining specialism areas: academic rigour within different disciplines. Hence, while we need interdisciplinary teams, we also need

*Table 8.2* Proposed attributes of sustainability research

| Attribute | Brief description |
| --- | --- |
| Interdisciplinarity | Engagement between different disciplines, often from different epistemological positions; may include holistic synthesis such as systems thinking |
| Linking theory and practice | Blue sky and applied research, but not always both in the same project |
| Local impact with global relevance | Embedded within local region and/or engaging in global debates and exchanges |
| Participatory approaches | Stakeholder engagement, often in research design as well as process |
| Linked to learning | Informing and informed by teaching; enhancing capacity building of participants; recognising the blurred boundaries between participatory action learning and participatory action research |
| Employs different knowledge forms | Integrates academic, practitioner, indigenous and local knowledges as appropriate |
| Includes knowledge mobilisation | Includes flows of knowledge to and from stakeholders and through learning |
| Reflective process including self-assessment | Demands analysis of self and reflection on why as well as how research is undertaken |

individuals working within disciplines to contribute new thinking on sustainability issues.

In contrast with integrated interdisciplinary approaches, multidisciplinarity implies a non-integrated engagement between two or more disciplines. Transdisciplinarity has two meanings, in different literatures: first, a form of interdisciplinarity in which theory from one discipline informs another, and second a form of interdisciplinarity that involves coordination between theoretical and practical aspects within and across disciplines (Max-Neef, 2005). Not all forms of interdisciplinarity are equal; in some programmes one discipline leads and the contributions of others are marginal. Interactions across epistemological distance, for example physics and social anthropology, are likely to be more challenging. Interdisciplinary research offers a different form of research, a 'triangulation of depth, breadth and synthesis' (Klein, 1996). It tends to question the assumptions of disciplines themselves, as well as address the research questions.

*Systems thinking* offers a holistic framing, recognising that issues need to be studied in context, and in relation to other issues (Meadows, 2008; Seddon, 2008). Design thinking can be a form of synthesis. For many scientists, socio-ecological systems form a research focus (Berkes and Folke, 1998).

There is a normative basis to sustainable development that requires engagement with real-world issues; hence we need excellent ways to *link theory and practice*. This is not to say that we do not do theoretical research and only do applied research. On the contrary, we also need philosophers, physicians and biologists to explore, for example, the role of humanity, new energy options and the functional potential of plants.

There is support for locally derived, culturally specific implementation of sustainable development as well as a need to address planetary phenomena and an awareness of globalisation. Sustainability research thus faces the dual challenges of contributing to *local processes* while acknowledging *global debates and impacts*.

Applied research requires collaboration in research design, process and dissemination with those involved in the implementation of sustainable development, including policy-makers, practitioners and communities. This philosophy of *engagement* with non-academics illustrates a significant principle of sustainable development; participation. *Participation* is believed to strengthen democratic principles, enhancing social and political equality by giving power to those who do not have power (for example, for economic or racial reasons, or because of gender, age or religion); to facilitate empowerment; and to enhance project outcomes because of the plurality of perspectives involved and potentially improved acceptance of ideas (Blackstock *et al.*, 2007). *Participatory action research* (PAR) enables a critical engagement with feminist and post-modern concepts, although insensitive applications may actually exacerbate power imbalances (Kindon *et al.*, 2008). Modified versions of Arnstein's ladder (Arnstein, 1969) used in practice by agencies and institutions recognise the categories of Inform, Consult, Involve and Empower as increasing the amount and intensity of participation. While there is usually a cost in terms of time, money and/or effort in increasing participation, there is also increasing benefit to participants (Blackstock *et al.*, 2007). Empowerment may include capacity building, social learning and development of trust or relationships across actors. In this way, the role of research moves beyond that of merely providing solutions to sustainability issues to that of building capacity in people to themselves understand and respond to sustainability issues.

The conceptualisation of *capacity building* has recently shifted towards 'action capacity development', in which learning is achieved across multi-actor groups of people while they are pursuing a particular goal, often aided by knowledge brokers (Acquaye-Baddoo *et al.*, 2010). Often a knowledge advisor or broker is required to build trust across these actors. Social learning is 'learning that occurs when people engage one another, sharing diverse perspectives and experiences to develop a common framework of understanding and basis for joint action' (Schusler *et al.*, 2003). It includes cognitive (new knowledge) and normative (new ways of acting) aspects (Webler *et al.*, 1995).

Higher levels of participation require meaningful involvement in the planning as well as the process of research. This *co-production of knowledge* is becoming more widely accepted in sustainability research (Pohl *et al.*, 2010) and in research implementation (Boyle and Harris, 2009). This requires a mindset in which scientific or academic knowledge is not privileged over other *forms of knowledge*, and an openness to new ontologies and ideas.

*Academic knowledge* is produced, usually within academic institutions, according to accepted methods of research practice. It is published under peer review in journals to engage with academic debate. In sustainability research, the interdisciplinarity of much of the research has prompted the emergence in recent years of a suite of new journals aiming to facilitate debate focused on sustainability rather than discipline.

Academic knowledge is divided into disciplines and basic and applicable knowledge, and is distinguished by epistemological belief. Ontology, the set of assumptions underlying a theory or system of ideas, may vary from realism to nominalism. Epistemology, how knowledge is derived and our assumptions about how we know the world, ranges from positivism to anti-positivism. An academic may believe that people's actions are largely determined by their upbringing and social strategy, or may consider that individuals have greater powers to change their lives (see Burrell and Morgan, 1979). A researcher may choose to use quantitative or qualitative methods to address our research questions. These differences are crucial, since interdisciplinary sustainability research will often need teams of academics working across the epistemological spectrum, tackling the tensions that emerge and developing novel ways of combining the advantages of quantitative and qualitative methodologies (Table 8.3).

*Table 8.3* Typical differences between quantitative and qualitative research method-ologies (adapted from Bryman, 2006).

| Quantitative | Qualitative |
| --- | --- |
| Objective – reality exists | Subjectivity recognised |
| Numbers (numerical) | Words (non-numerical) |
| Point of view of the researcher (researcher sets the questions and agenda) | Points of view of the participants (researcher seeks the respondent's agenda) |
| Researcher distant | Researcher close |
| Theory testing (deductive) | Theory emergent (inductive) |
| Static (hypotheses set and tested) | Process (research focus and themes develop as research continues); observation |
| Structured | Unstructured (less structured) |
| Generalisation | Contextual understanding |
| Hard reliable data | Rich deep data |
| Macro | Micro |
| Behaviour (pattern) | Meaning (understanding) |
| Artificial setting | Natural setting |

*Practitioner knowledge* can add significantly to understanding (Sternberg and Iorvath, 1999). It is usually qualitative, contextualised, in-depth; and may include incorrect assumptions of association. It tends to be experiential and non-theoretical, observation- and manipulation-based. Attempts to capture practitioner knowledge and combine it with scientific knowledge have been made.[1]

*Indigenous knowledge,* or Traditional Ecological Knowledge, offers more than just specific facts regarding conservation in a particular area; indigenous perspectives offer new ontologies that can change the way in which we see the world around us. For example, the Cartesian separation of people from nature is foreign to many belief systems around the world, in which people consider themselves within nature. Indigenous knowledge is often qualitative, based on local knowledge, usually of a group of people; specific to a particular region; represents combined knowledge from across several generations; and is culturally embedded (Snively and Corsiglia, 2001). Not all knowledge from the global South is indigenous, and greater engagement with academics from countries in the global South is suggested (Raven, 2002), with opportunities to better link scientific diasporas for sustainability (Seguin *et al.*, 2006). While recognising that we can learn much from such countries, it is also suggested that significant scientific and technological capacity-building will be required there (Clark *et al.*, 2002).

*Local knowledge* is distinct from indigenous knowledge in that it is not associated with a particular culture and belief system and has not necessarily been passed down over generations. It tends to be locally specific, rooted in place and more detailed than scientific knowledge. It is thus valuable in the implementation of management theories, but can also be important in synthesising with scientific knowledge to enhance the potential for management strategies. For example, a computer model to predict the movement of red deer across the Scottish Highlands was greatly improved through the addition of assumptions offered by local stalkers using a participatory GIS process (Irvine *et al.*, 2009).

Constructivist thinking suggests that knowledge and belief are not strictly separable (Coburn, 1971). Nygren argues that indigenous and local knowledges are less distinct from other forms than is sometimes believed. Rather than seeing knowledge as the combination of discrete knowledge forms, she suggests that '[k]nowledge production is seen as a process of social negotiation involving multiple actors and complex power relations' (1999: 267). She proposes knowledge dichotomies based on tacit versus scientific, folk versus universal, indigenous versus Western, and traditional versus modern. A major challenge for sustainability research is recognising that these different forms of knowledge exist, identifying which to integrate in a particular piece of research and understanding how to do so, while attempting to address these power issues.

The notion of transformative learning is also prevalent in sustainability research, especially in approaches such as PAR, for participants and for the researchers themselves. *Research-led teaching* enables students to be excited by new frontiers in academic knowledge. However, recognition of the validity of different perspectives and knowledge forms, and an openness to self-analysis, offer new possibilities for *teaching-led research*. For example, in the Sustainable Development Programmes at the University of St Andrews, students combine partner subjects with the study of sustainable development. A student of international relations thus has the ability to offer new insights during class discussions to, for example, a biologist lecturer teaching on a core interdisciplinary module.

*Reflection* is rarely considered a part of the research process, yet reflective learning is increasingly being seen as important in education (Moon, 2004). The normative nature of sustainable development and the importance of the sustainability researcher being open to their own experiences and fallibility, as well as to the inputs of others, suggest that *reflexivity* will also be a characteristic of sustainability research.

There is by implication here an assumption that a successful sustainability researcher will be a decent human being, and that a dogmatic epistemological position will impede openness and respect for different disciplines, forms of knowledge and roles (for example, willingness to develop ideas through student contact) that is necessary for sustainability research. Such ideas offer a return to the Greek concept of Paideia for 'mastery of self' in scholarship (Orr, 2004). The goals of the sustainability researcher will have to shift from output focused (number of publications, amount of grant income) to process and outcome focused (learning, personal development, impact).

Finally, in sustainability research the boundaries between knowledge generation and *knowledge mobilisation* are blurred. Knowledge transfer gave way to knowledge exchange and is now being supplanted by knowledge mobilisation (see Canada's research impact site at www.researchimpact.ca/home/index.html). Nutley *et al.* (2007) describe a continuum of research use, differentiating among users and types of knowledge. In sustainability research this concept captures the nature of knowledge movement across networks of different stakeholders, through both formal mechanisms and relationships. Networks offer different roles for individuals involved in knowledge exchange, permitting integration of different forms of knowledge (Roome, 2001). This process again enhances capacity and relationships, and feeds back into the research process. However, knowledge mobilisation attempts need to recognise that action potential is embedded within power structures that reflect cultural biases and inequalities among and between researchers and other stakeholders (Van Kerkhoff and Lebel, 2006).

Many of these attributes are not exclusive to sustainability research. They describe a new model of research similar to the shift from Mode 1 to Mode 2 research described by Gibbons *et al.* (1994).

---

**Box 8.1: Social and natural sciences working across theory and practice: enhancing collaborative management of red deer in Scotland through participatory action learning**

Red deer range across Scottish estates, with some estates desiring high deer densities to support stalking enterprises, while others require lower densities to facilitate regeneration of the Caledonian Forest. Establishing deer number targets is further complicated by biodiversity regulatory requirements, economic contributions, tourism and road accident incidents. A research programme[2] first assessed stakeholder perspectives of these issues, and then developed a project[3] to help build the capacity of those people actually involved in establishing and running Deer Management Groups.

This project demonstrated the need for natural and social scientists to work together; the former understanding deer ecology and the latter the principles and practice of participation. It also showed how good theoretical research could lead to action-based projects that both had practical impact and led to peer-reviewed publications on the process of collaborative natural resource management. This research focused on a sustainability issue and also demonstrated attributes of sustainability research.

**Box 8.2: Forms and of knowledge and capacity building: forest utilisation in South Africa and carbon emission reduction in Scotland**

A research project[4] exploring the utilisation by local communities of a large state forest in South Africa was changed through collaborative working between academics and indigenous people. The community group demonstrated the decline in community forests over time, shifting the research focus from utilisation of state forest to the regeneration of community forests. The participatory action research approach enabled community members to understand and reflect on the underlying reasons for community forest degradation and to begin to implement management changes. The community steering group 'owned' the data, and used flipcharts that had been produced in research meetings, empowering them to persuade the rest of the community to engage in a management shift. In this sense, the academics acted more as facilitators of a capacity-building process than as knowledge extractors.

A project to co-design a strategy with local authorities to reduce community carbon emissions in Scotland[5] brought together representatives from academia, local government, communities and NGOs. A series of day-long seminars on energy, food, transport and community awarded equal input to different forms of knowledge, without privileging academic knowledge. We used Open Space Technology, Knowledge Cafés, ice-breakers, networking opportunities, a market place for projects and ideas, and other tools to facilitate knowledge exchange and relationship building. The research was overseen by a Steering Group, embedded in the local authority, engaged in ongoing initiatives and aiming to produce clear outputs. Outcomes included increased understanding of regional carbon emission reductions, an extensive scoping document, seminar reports and resource sheets for communities and development officers to act on, academic presentations and papers, a strategy for the local authority to implement, and policy input to national government in relation to the Public Engagement Strategy; but perhaps most important were the trust and relationship building and the realisation of the richness of the knowledge within different stakeholder groups. This catalysed a shift in approach, especially in the local authority, and has facilitated subsequent interactions in relation to climate change and other sustainability initiatives.

In the case of the first project, funding was awarded to tackle a specific problem (unsustainable utilisation of a state forest). The latter project was funded under a scheme to explicitly deliver not only project-specific benefits, but also collaborative learning across the network and development of partnerships between universities and local government that would catalyse further joint research. Sustainability research can be nurtured at project and programme level, through topic and approach.

## Disciplinary contributions

*Sustainability science* is an emerging disciplinary area that partially addresses the need for research in relation to sustainability (see Kates *et al.*, 2001; Parris and Kates, 2003; Clark, 2007), but which has constraints imposed by its positioning as a science that engages with other disciplines, as opposed to an open space for new debate. It does imply a shift from 'knowing' to 'learning and adaptation', but it has not resolved the epistemological tensions resulting from the engagement of traditional scientific methods with value-based stakeholder subjectivity (Burns *et al.*, 2006). The focus issues proposed tend to be functional rather than visionary (Table 8.1). Kajikawa (2008) views it as a potential emerging discipline, but Clark and Dickson (2003) instead propose a broader field of enquiry in which theory and practice, local and global perspectives and diverse academic disciplines can be accommodated. It is dangerous to propose that this is the only area of research focusing on sustainable development; the opportunity for other interdisciplinary research, including non-science study, and for single discipline research on sustainability issues should be fiercely defended. However, the emergence of sustainability science does offer new focus for debate on sustainability research and has the potential to evolve.

## The limits of science

Many of the issues facing us in sustainable development are not scientific in origin; they are moral and ethical and so can be informed but not solved by science (Lubchenco, 1998). If sustainable development is the recognition of the interdependence of social justice and environmental integrity, sustainability research has to focus on social as well as environmental issues, and on interactions between the environment and society. However, Rittel and Webber (1973: 155) note that '[t]he search for scientific bases for confronting problems of social policy is bound to fail because of the nature of these problems'. In fact, Redclift (2005) argues that although Agenda 21 promotes scientific understanding in the pursuit of sustainable development, belief in a global science is highly disputed. Modern science is largely a construct from the global North, with possibilities for indigenous types of science, viewed through different cultural perspectives, yet to be acknowledged (Snively and Cosiglia, 2001). The current dominant mode of science perpetuates visions of environment as distinct from society, and indigenous perspectives can challenge this view (Ingold, 2000).

While naturalists claim that research in the social sciences is similar to that of the natural/physical sciences, anti-naturalists claim a distinction because social science deals with desires and beliefs (Bhaskar, 1978). Karl Popper distinguished science claims as being empirically refutable, testing hypotheses against falsification rather than verification. While some social science may take non-science, qualitative approaches, the arts and humanities include the branches of knowledge concerned with human culture, including the liberal arts, and are thus non-science in nature. The ontology of science promotes an objective view of the world, with a positivist epistemological position. This view limits the adoption of other realities, and the integration of qualitative data with quantitative scientific outputs is problematic for sustainability research (Modvar and Gallopin, 2005; Bryman, 2006). In addition, science itself has become fragmented and specialised, with many disciplines not explicitly addressing the interaction between people and the environment (Redclift, 2005).

However, there is a new vision of science emerging. It has been suggested that the criteria of truth deriving from science need reconsideration (Modvar and Gallopin, 2005). The consilience of knowledge production formats, a synthesis focus with porous disciplinary boundaries, a multi-scale approach, and cultural and genetic evolutions may lead to a form of post-normal science, with increasing democratisation of the science method (Costanza, 2003; Carolan, 2006). Feminist perspectives also question the objectivity and ownership of science (Harding, 1991). Yet, even as science may adapt to new societal demands, there needs to be a place for a strong contribution from the arts and humanities to help us understand and express our cultures and intent in relation to sustainability.

## Sustainability research within all disciplines

*All disciplines* can contribute to sustainability research, either through contributions from within the discipline or through interdisciplinary collaboration. Consideration of ethical issues has become both norm and requirement in all research, and sustainability issues and attributes should likewise be a consideration. While it may seem logical how some disciplines are involved, others have less obvious contributions. For example, a study of early Russian philosophers illustrates how views of the biosphere emerged in the East and how these insights might influence future sustainability thinking (Smith, 2011). As discussed above, those not actively engaged in sustainable development debates may have a limited perception of its scope; and those not within a particular discipline are unlikely to be aware of the key debates. Both a sustainability specialist or a generalist able to work across multiple areas, and a disciplinary specialist may be required to interpret ideas in the light of sustainability issues and make contributory areas of enquiry explicit.

## The role of research in a sustainable university

All research, whether focused on sustainability issues or undertaken in a manner commensurate with sustainability principles, can play a role in making a university sustainable. Research activities offer excellent means by which a sustainable university can engage with the external community. Good research facilitates critical reflection and can aid institutional attempts to operationalise sustainability, raise sustainability awareness and build university community.

The cross-structural interactions (interdisciplinary, between staff and students) help build social capital. Research-led teaching stimulates critical and contemporary analysis of local and global issues. Finally, the participatory research approaches outlined above involve non-academic stakeholders, improving the potential for community engagement.

## Nurturing sustainability research

### *Mainstreaming, funding and publishing*

Over the last decade especially, the *field of enquiry of sustainable development* has become more mainstream. There has been an increase in journals devoted to sustainable development (for example, *Sustainable Development, Sustainability Science*), an acceptance of sustainability in leading journals (for example, the sustainability science forum in *Science*), vibrant debates within established journals and an increase in the number of interdisciplinary journals accepting sustainability and other articles. Many universities now have sustainability institutes, research centres or research themes. Some funding calls address sustainability issues or encourage research with sustainability attributes. Degrees in sustainable development have multiplied. New and established researchers can be made aware of these opportunities. However, mainstreaming also presents challenges, as the meaning of sustainable development is further diluted. For example, funding calls promoting enquiry into 'sustainable economic growth' without acknowledging the impact of the consumer culture on resources and our limits to growth (Jackson, 2009) are paradoxical. Further, while the rhetoric around participatory research and stakeholder engagement may now be widespread, there is rarely recognition of the time and effort taken to run genuinely participatory projects.

### *The impact agenda*

The impact agenda, a call for research to have effects, initially tended to focus on policy impacts, although broader debates later identified instrumental (influencing policy, practice, legislation and behaviour), conceptual (reframing debates and understanding) and capacity-building (personal and skills development) impacts (Nutley *et al.*, 2007). We can see impact as a process, or a change in mindset, and need to recognise its relationship to power. In the UK, research councils now demand a consideration of impact within proposals, and impact reports after projects end. Academic response to this has been mixed. On the one hand, many agree that some, although not all, research should influence society (Pain *et al.*, 2011), and certainly sustainability research attempts to address societal and planetary problems. On the other hand, impact is at times framed inadequately; for example 'impact on economic growth'. Measurement of impacts is problematic; academic impact is often long term and indirect, serendipitous and additional to other impacts (Meagher *et al.*, 2008). Direct causality is thus often difficult to define. It can be tracked forward from research outcome or backwards from research user communities (Davies *et al.*, 2005). Maximising impact from research demands stakeholder engagement, knowledge brokers and resource input.

The impact agenda seems to be driven in part by modern notions of accountability, exacerbated by the current era of financial austerity. Such accounting processes misunderstand the nature of research as a capacity-building, mutual-learning, evolving process of engagement between scholars and other stakeholders, tending to see it instead as a clear set of solutions to distinct problems. In fact, the pursuit of impact may lead to 'perverse incentives and dysfunctional consequences' (Walter *et al.*, 2004: 4), and may raise ethical questions as we attempt to change lives or make wide-ranging recommendations. However, sustainability research, with its articulated linkage of theory and practice, can benefit from this debate, providing it maintains a critical position with regard to the meaning of impact.

### Crossing disciplinary and structural boundaries

Internal university divisions, disciplinary boundaries, management units and career path norms can limit research communication and incentives to engage across disciplinary and structural boundaries. Current UK research assessment schemes also actively disincentivise interdisciplinary research and the production of creative conceptual hybrid scholarship (Brand and Karnoven, 2007). Instead, specialism and reductionism are promoted.

Sustainability research requires generalists to work across and specialists within disciplines. Role specialists may also be required, with knowledge mobilisation and stakeholder engagement demanding particular skills. Brand and Karnoven (2007) propose an 'ecosystem of expertise' for sustainable development, in which contemporary forms of 'experts' connect with outreach experts who connect with non-experts, interdisciplinary experts, meta-experts who broker claims across different kinds of experts, and the civic expert who engages in democratic process across experts and non-experts. These new spaces in knowledge generation and mobilisation encourage the development of *modern polymaths* who can promote excellent and high-level cross-discipline innovation (Young and Marzano, 2010). Specialism and reductionism seems to be particularly rife in the global North. Ntuli (1988) describes how a move to the UK required his poet, painter, sculptor, musician and actor work to be segmented into 'different compartments'.

Interdisciplinarity requires the building of trust and relationships among academics, translation across language and terminology differences, a respect for different epistemological positions, funding opportunities, and the time and space in which to overcome these additional research hurdles. The practical recommendations below suggest how to enhance communication and nurture interdisciplinarity and sustainability research within the constraints of broader academic structures.

## Stakeholder engagement and research in partnership

Stakeholder engagement, the development of partnerships and co-production take time and effort (Australian Government, 2008; Sarkissian *et al.*, 2009). This time investment may provide benefits in terms of outcomes, capacity building and potential for future research questions to be relevant, but these benefits may not be evident to colleagues and institutions. As the discourse regarding participatory research becomes more prominent, and there are political attempts to rework relationships between state, commerce, community and academia currently underway, sustainability researchers can take advantage of these opportunities. This involves not only gaining funding, but also being able to learn from, reflect on and publish experiences from the process of sustainability research, critically analysing research attributes as well as sustainability issues. Stakeholders could include: government (national or local), local communities, business, future employers, practitioners, NGOs, communities of interest and/or other possibilities. Bringing in partners develops mutual awareness and capacity building for those involved, as well as developing relationships. For example, working with potential employers allows them to understand the benefits of the 'green' economy and causes them to demand sustainability-literate graduates, reinforcing the need for students to engage in sustainability research and education.

## The audit culture

The audit culture promotes targets, quality assurance, specialised focus and individual accountability, whereas systems management proposes that a value-based approach to management can actually increase effectiveness and mitigate against the illogic of poorly interpreted targets (Seddon, 2008). In academia the former translates as a drive for more outputs (papers, graduates, funds), which are rewarded through resource flows such as staff and departmental budgets internally, and through research assessment ranking externally. Currently, despite requirements for a narrative on 'research environment' and 'impact' in the UK, for the majority of staff, research assessment rewards those who deliver targets and not those who retain a broader vision of research as a learning, capacity-building, relationship-strengthening process. A more systems-based approach would also help transcend some of the internal university divisions and management units.

## Recommendations for the promotion of sustainability research

We can promote sustainability research by facilitating attributes of sustainability research, and by encouraging research initiatives that address sustainability issues or themes, within and across disciplines. Some practical suggestions are as follows:

1  Develop a sustainability research strategy for your institution, embedded in real-world priorities, building on your strengths, capitalising on your

external links and providing a vision for where you wish to go. This may include the emergence of a sustainability institute and a locus for communication, resources and information, and the identification of sustainability issue themes.

2   Hold sustainability theme-based workshops and half- or one-day seminars to which people from different disciplines are invited to contribute. Inspire the university community through world-class external speakers.

3   Offer seed funds, accessed through a formal competitive process, to stimulate new and established academics to develop ideas for research. Even a few small grants can inspire new ideas, which if not funded internally may then be taken further and leverage external funds.

4   Support partnership working by formally acknowledging the time and effort put into developing trust and relationships locally and further afield, with academic and non-academic partners. Invest in personal relationships and spend time getting to know what the priorities of local or interest groups really are – for example, sit on committees, engage in local community events.

5   Offer social and professional networking opportunities for researchers – for example, cheese and wine after a seminar, invitations for lunchtime walks, invitations to exhibitions or new institute launches. Such events build trust as well as knowledge of what other parties are doing. Spend your coffee budgets!

6   Reward academically excellent initiatives that are interdisciplinary, work in partnership or open new areas of enquiry, through promotion, research leave or other incentives. You may even consider cross-departmental appointments (for example, in sustainable development and biology; sustainable development and history).

7   Recognise that students often nurture the seeds of radical new ideas, and encourage them to develop these within an academically excellent context. Link your research and teaching programmes to emphasise and enhance the sustainability research contributions.

8   Support cross-department PhD scholarships, or partial scholarships, focused on sustainability research. Encourage debate in sustainability research from all PhD students, and enable them to learn research approaches commensurate with sustainability principles during their research training.

9   Contribute in a practical manner to society and planet, but also support the development of new ideas, offering alternative theoretical insights for how we might live. In a practical sense, engage with your institutional operations, with awareness and action; with your local community and region; with national debates; but also with global discourse on what sustainability might be and how we might move towards it.

10  Recognise that academics are people. Keep them interested, allow them personal development, encourage them to pursue 'mastery of self', and provide them with a pleasant, stimulating environment. Some pressure may be positive for academics, but stress impedes creativity and the joy of academic enquiry. Be open to the personal transformative changes that sustainability research may invoke.

## Conclusions

Sustainability research addresses sustainability issues, and employs approaches commensurate with sustainability principles. These principles infuse new life into more general emerging models of knowledge generation. While others have promoted interdisciplinary and participatory approaches and knowledge mobilisation, the combination of all of these and other attributes, plus the acknowledgement that the researchers themselves can change, makes this a new mode of knowledge generation. No sustainable university can afford to ignore the power of sustainability research to transform its thinking, operation and practice. A sustainable university also has a responsibility to ensure that sustainability research contributes to new understanding and visions of how our societies can flourish and our environments remain healthy and resilient.

*Acknowledgements*

The funders of the specific research projects described are gratefully acknowledged: RELU (RES 227-025-0014), ESRC (RES-811-25-0002), ESRC/ SFC/LARCI (RES-809-19-0021) in the UK and DWAF, South Africa. Many thanks to the numerous participants and to my research colleagues on these and other projects for the insights gained.

### Transition Research Network

The Transition Research Network (TRN) is a self-organising group of researchers and practitioners active in the Transition Movement. The Transition Movement (TM) is a grassroots movement addressing the triple challenges of climate change, peak oil and economic instability, with almost 1,000 local initiatives worldwide (see Higgins *et al.*, this volume; www.transitionnetwork.org; Bailey *et al.*, 2010; Hopkins, 2011, 2010; Maxey and Gillmore, forthcoming). Drawing on collaborative approaches such as those promoted by the National Coordinating Centre for Public Engagement, TRN replaces hierarchical, 'extractive' research models with empowering, co-designed initiatives in which practitioners and researchers jointly set research questions and develop, undertake and report on research projects. Academic interest in Transition is growing because it is a significant social movement addressing sustainability and because funders increasingly require research to demonstrate policy and practical impacts. TRN is supporting the TM's capacity to engage with and make optimum use of this research interest. It also works with funders to develop best practice, moving away from collaboration as a 'tick box' exercise driven by metric assessments, towards approaches such as the UK

Research Council's 'Connected Communities' programme, which fosters deep collaboration.

Modelled after a Transition Initiative, TRN took shape at meetings in late 2011 and early 2012, in which 60 participants collectively designed its strategy using the principles and methods of permaculture[1] (Macnamara, 2012; TRN, 2012). TRN is now setting up working groups in a number of thematic areas, including food, energy and the evaluation of Transition Initiatives. It hosts an open email list and welcomes proposals for further working groups, research projects and themed meetings, as well as general expressions of interest.

## References

Bailey, I, Hopkins, R and Wilson, G (2010) 'Some things old, some things new: the spatial representations and politics of change of the peak oil relocalisation movement', *Geoforum*, 41 (4): 595–605

Hopkins, R (2010) *Localisation and Resilience at the Local Level: The Case of Transition Town Totnes (Devon, UK)*, PhD thesis, Plymouth University

Hopkins, R (2011) *The Transition Companion: Making Your Community More Resilient in Uncertain Times,* Totnes: Green Books

Macnamara, L (2012) *People and Permaculture*, East Meon, Hampshire, UK: Permanent Publications

Maxey, L and Gillmore, H (forthcoming) 'Transformation through Transition: Transition Tours as fieldwork partners delivering sustainable education in GEES', *Planet 27*

Transition Research Network (2012) *Transition Research Network: New Knowledge for Resilient Futures Meeting Report.* Available at: www.transitionresearchnetwork.org (accessed 30 June 2012)

### For more information

www.transitionresearchnetwork.org
email: transitionresearchnetwork@gmail.com
tweet@TransitionRN

## Notes

1   See, for example, the Conservation Evidence site at http://conservationevidence. regulus.titaninternet.co.uk/Default.aspx.
2   *Collaborative Frameworks in Land Management: A Case Study on Integrated Deer Management.* Funded by Rural Economy and Land Use Programme, 2006–09.
3   *Building Capacity Through Participatory Action Learning for Implementation of Sustainable Natural Resource Management.* Funded by Economic and Social Research Council, UK, 2010–12.

4   *A Monitoring System for Community Forestry: Combining Scientific and Local Knowledge in the Eastern Cape.* Funded by Department of Water Affairs and Forestry, South Africa, 1999–2001.
5   *Enhancing Local Authorities' Community Engagement: Co-designing and Prototyping Strategies for Carbon Emission Reductions.* Funded by Economic and Social Research Council/Scottish Funding Council/Local Authority Research Council Initiative, 2009–10.
6   Permaculture is a sustainable design approach, philosophy and set of practices, which informs the Transition Movement.

## References

Acquaye-Baddoo, N-A, Ekong, J, Mwesige, D, Nass, L, Neefjes, R, Ubels, J, Visser, P, Wangdi, K, Were, T and Brouwers, J (2010) 'Multi-actor systems as entry points to capacity development', *Capacity.org,* 41, 26 December 2010. Available at: www.capacity.org/capacity/opencms/en/topics/multi-actor-engagement/multi-actor-systems-as-entry-points-to-capacity-development.html (accessed 12 March 2012)

Arnstein, S (1969) 'A ladder of citizen participation', *Journal of the American Institute of Planners,* 35: 216–24

Australian Government (2008) *Stakeholder Engagement: Practitioner Handbook,* Department of Immigration and Citizenship, Belconnen, Australia: National Communications Branch of the Department of Immigration and Citizenship

Berkes, F and Folke, C (eds) (1998) *Linking Social and Ecological Systems: Management Practices and Social Mechanisms for Building Resilience,* New York: Cambridge University Press

Bhaskar, R (1978) 'On the possibility of social scientific knowledge and the limits of naturalism', *Journal for the Theory of Social Behaviour,* 8: 1–28

Blackstock, K, Kelly, G and Horsey, B (2007) 'Developing and applying a framework to evaluate participatory research for sustainability', *Ecological Economics,* 60: 726–42

Boyle, D and Harris, M (2009) *The Challenge of Co-production,* London: New Economics Foundation

Brand, R and Karnoven, A (2007) 'The ecosystem of expertise: complementary knowledges for sustainable development', *Sustainability: Science, Practice and Policy,* 3: 21–31

Bryman, A (2006) *Social Research Methods,* Oxford: Oxford University Press

Burns, M, Audouin, M and Weaver, A (2006) 'Advancing sustainability science in South Africa', *South African Journal of Science,* 102: 379–84

Burrell, G and Morgan, G (1979) *Sociological Paradigms and Organisational Analysis,* London: Heinemann Educational Books

Carolan, M (2006) 'Science, expertise, and the democratization of the decision-making process', *Society and Natural Resources,* 19: 661–8

Clark, W, Buizer, J, Cash, D, Corell, R, Dowdeswell, E, Doyle, H, Gallopin, G, Glaser, G, Goldfarb, L, Gupta, A, Hall, J and Al, E (2002) 'Science and technology for sustainable development: consensus report of the Mexico City Synthesis Workshop', *Initiative on Science and Technology for Sustainability,* Mexico City, 20–23 May 2002

Clark, W (2007) 'Sustainability science: a room of its own', *Proceedings of the National Academy of Sciences of the United States of America,* 104: 1737–8

Clark, W and Dickson, N (2003) 'Sustainability science: the emerging research program', *Proceedings of the National Academy of Sciences of the United States of America,* 100: 8059–61

Coburn, R (1971) 'Knowing and believing', *The Philosophical Review,* 80: 236–43

Costanza, R (2003) 'A vision of the future of science: reintegrating the study of humans and the rest of nature', *Futures,* 35: 651–71

Creswell, J (2008) *Educational Research: Planning, Conducting, and Evaluating Quantitative and Qualitative Research,* Upper Saddle River: Pearson

Davies, H, Nutley, S and Walter, I (2005) *Approaches to Assessing the Non-academic Impact of Social Science Research,* Report of the ESRC Symposium on Assessing the Non-academic Impact of Research, St Andrews: Research Unit for Research Utilisation, University of St Andrews.

Gibbons, M, Limoges, C, Nowotny, H, Schwartzman, S, Scott, P and Trow, M (1994) *The New Production of Knowledge: The Dynamics of Science and Research in Contemporary Societies,* London: Sage

Harding, S (1991) *Whose Science? Whose Knowledge?,* Ithaca, New York: Cornell University Press

HEA (Higher Education Academy) (2012) *Education for Sustainable Development.* Available at: www.heacademy.ac.uk/education-for-sustainable-development (accessed 25 September 2012)

Huckle, J and Sterling, S (1996) *Education for Sustainability,* London: Earthscan

Ingold, T (2000) *The Perception of the Environment: Essays on Livelihood, Dwelling and Skill,* London: Routledge

Irvine, R, Fiorini, S, Yearley, S, Mcleod, J, Turner, A, Armstrong, H, White, P and Van Der Wal, R (2009) 'Can managers inform models? Integrating local knowledge into models of red deer habitat use', *Journal of Applied Ecology,* 46: 344–52

Jackson, T (2009) *Prosperity Without Growth: Economics for a Finite Planet,* London: Earthscan

Jickling, B (1992) 'Viewpoint: why I don't want my children to be educated for sustainable development', *Journal of Environmental Education,* 23: 5–8

Kajikawa, Y (2008) 'Research core and framework of sustainability science', *Sustainability Science,* 3: 215–39

Kates, R, Clark, W, Corell, R, Hall, J, Jaeger, C, Lowe, I, Mccarthy, J, Schellnhuber, B, Dickson, N, Faucheux, S, Glallopin, G, Grubler, A, Huntley, B, Jager, J, Jodha, N, Kasperson, R, Mabogunje, A, Matson, P, Mooney, H, Moore, B, O'Riordan, T and Svedin, U (2001) 'Sustainability science', *Science,* 292: 641–2

Kates, R, Leiserowitz, A and Parris, T (2006) 'Great Transition values: present attitudes', in Kriegman, O and Raskin, P (eds) *Frontiers of a Great Transition,* GTI Paper Series No 9., Boston, USA: Tellus Institute

Kindon, S, Pain, R and Kesby, M (2008) 'Participatory action research', in *International Encyclopedia of Human Geography,* Amsterdam and London: Elsevier, pp. 90–5

Klein, J (1996) *Crossing Boundaries: Knowledge, Disciplinarities and Interdisciplinarities,* Virginia, The University Press of Virginia

Leedy, P (1997) *Practical Research: Planning and Design,* Upper Saddle River: Prentice Hall

Lehrer, K (1990) *The Theory of Knowledge,* Boulder, Colorado: Westview Press

Lubchenco, J (1998) 'Entering the century of the environment: A new social contract for science', *Science,* 279: 491–7

Max-Neef, M (2005) 'Foundations of transdisciplinarity', *Ecological Economics,* 53: 5–16

Meadows, D (2008) *Thinking in Systems: A Primer,* London: Chelsea Green Publishing Company

Meagher, L, Lyall, C and Nutley, S (2008) 'Flows of knowledge, expertise and influence: a method for assessing policy and practice impacts from social science research', *Research Evaluation,* 17: 163–73

Modvar, C and Gallopin, G (2005) *Sustainable Development: Epistemological Challenges to Science and Technology,* Report of the workshop 'Sustainable development: epistemological challenges to science and technology' Santiago, Chile, 13–15 October 2004, Santiago, Chile

Moon, J (2004) *A Handbook of Reflective and Experiential Learning: Theory and Practice,* Oxon, UK: RoutledgeFalmer

Ntuli, P (1988) 'Orature: a self portrait', in Owusu, K (ed.) *Storms of the Heart: An Anthology of Black Arts and Culture,* London: Camden Press

Nutley, S, Walter, I and Davies, H (2007) *Using Evidence: How Research Can Inform the Public Sector,* Bristol: The Policy Press

Nygren, A (1999) 'Local knowledge in the environment-development fiscourse: From dichotomies to situated knowledges', *Critique of Anthropology,* 19: 267–88

Orr, D (2002) 'Four challenges of sustainability', *Conservation Biology,* 16: 1457–60

Orr, D (2004) *Earth in Mind: On Education, Environment and the Human Prospect,* Washington, Island Press

Pain, R, Kesby, M and Askins, K (2011) 'Geographies of impact: power, participation and potential', *Area,* 43: 183–8

Parris, T and Kates, R (2003) 'Characterizing a sustainability transition: goals, targets, trends, and driving forces', *Proceedings of the National Academy of Sciences of the United States of America,* 100: 8068–73

Pohl, C, Rist, S, Zimmermann, A, Fry, P, Gurung, G, Schneider, F, Speranza, C, Kiteme, B, Boillat, S, Serrano, E, Hadorn, G and Wiesmann, U (2010) 'Researchers' roles in knowledge co-production: experience from sustainability research in Kenya, Switzerland, Bolivia and Nepal', *Science and Public Policy,* 37: 267–81

Raven, P (2002) 'Science, sustainability, and the human prospect', *Science,* 297: 954–58

Redclift, M (2005) 'Sustainable development (1987–2005): an oxymoron comes of age', *Sustainable Development,* 13: 212–27

Rittel, H and Webber, M (1973) 'Dilemmas in a general theory of planning', *Policy Sciences,* 4: 155–69

Roome, N (2001) 'Conceptualizing and studying the contribution of networks in environmental management and sustainable development', *Business Strategy and the Environment,* 10: 69–76

Sarkissian, W, Hofer, N, Shore, Y, Vajda, S and Wilkinson, C (2009) *Kitchen Table Sustainability: Practical Recipes for Community Engagement with Sustainability,* London, Earthscan

Schusler, T, Decker, D and Pfeffer, M (2003) 'Social learning for collaborative natural resource management', *Society and Natural Resources,* 15: 309–26

Seddon, J (2008) *Systems Thinking in the Public Sector: The Failure of the Reform Regime and a Manifesto for a Better Way,* Axminster: Triarchy Press

Seguin, B, Singer, P and Daar, A (2006) 'Scientific diasporas', *Science,* 312: 1602–03

Smith, O (2011) 'The ecology of history: Russian thought on the future of the world', in Bergmann, S and Eaton, H (eds) *Ecological Awareness,* Berlin: Lit-Verlag

Snively, G and Corsiglia, J (2001) 'Discovering indigenous science: implications for science education', *Science Education,* 85: 6–34

Sterling, S (2001) *Sustainable Education: Revisioning Learning and Change*, Totnes: Green Books

Sterling, S (2003) 'Sustainable education: putting relationship back into education', *Convergence* magazine, Dublin: Cultivate

Sternberg, R and Horvath, J (1999) *Tacit Knowledge in Professional Practice: Researcher and Practitioner Perspectives*, Mahwah, New Jersey, USA: Lawrence Erlbaum Associates, Inc.

Stibbe, A and Witham, H (2009) 'Introduction', in Stibbe, A (ed.) *The Handbook of Sustainability Literacy: Skills for a Changing World*, Totnes: Green Books

Van Kerkhoff, L and Lebel, L (2006) 'Linking knowledge and action for sustainable development', *Annual Review of Environment and Resources*, 31: 445–77

Walter, I, Nutley, S and Davies, H (2004) *Assessing Research Impact*, St Andrews: Research Unit for Research Utilisation, University of St Andrews

Webler, T, Kastenholz, H and Renn, O (1995) 'Public participation in impact assessment: a social learning perspective', *Environmental Impact Assessment Review*, 15: 443–63

World Commission on Environment and Development (1987) *Towards Sustainable Development*, Chapter 2 in *Our Common Future: Report of the World Commission on Environment and Development*, UN Document A/42/427. Available at: www.un-documents.net/wced-ocf.htm

Young, J and Marzano, M (2010) 'Embodied interdisciplinarity: what is the role of the polymath in environmental research?', *Environmental Conservation*, 37: 373–5

# 9   The student experience

Campus, curriculum, communities
and transition at the University of
Edinburgh

*Peter Higgins, Robbie Nicol, David Somervell and
Mary Bownes*

## Introduction

> Each generation has the duty not to engage in the wishful thinking that the
> problem can be left for descendants to solve.
>
> (Tremmel *et al.*, 2009: 87)

The management of any significant change in large institutions requires a multi-
level approach that appeals to a wide range of interests. Further, addressing
sustainability, with all its connotations, contested concepts and its interface with
personal values, needs institutional clarity of purpose. Planning and
implementing such change in universities presents both challenges and
opportunities.

Pro-active estate management to reduce environmental impact and contain
costs is common to many large organisations, but in universities the 'community'
of staff and students has a significant impact on any such developments. This
can be through analysis and critique, but also through individuals and groups
within this community developing their own initiatives and driving the agenda.
Importantly, the experience of change towards a more sustainable university
offers important programme-related and personal learning opportunities for
students and staff.

This chapter analyses the processes of change towards a holistic promotion of
social responsibility and sustainability that acknowledges the significance of
campus, community and curriculum within universities. In doing so it considers
the University of Edinburgh as a case study and reflects both the historical
commitment and current approaches to management of its estate and
procurement, and the particular role of the student body in developing initiatives
such as the Transition Edinburgh University project. The strategic planning
necessary to facilitate such developments in a 'traditional' university with a
prominent research profile is discussed, and future prospects are considered in
the context of a critique of recent student- and institutionally-led developments.

As with all universities, the community (staff and students) interacts through
curricula, research, administration, and within the campus and the city. This
chapter recognises that these circumstances offer particular opportunities to

develop within that community of practice, an understanding of and values orientation towards social responsibility and sustainability, and to demonstrate good practice in those facets of university life. Indeed, many would consider this a duty of all universities within their wider local, national and even international communities.

The student experience is central to that task and by this we mean the totality of the student interface with the campus, their community of practice and curriculum. The student-centred, transformative focus of this approach is, of course, nothing new and can be found in the writings of Rogers (1983) and his work in person-centred therapy, learner-centred teaching and client-centred therapy together with his role as an international ambassador in the resolution of inter-group conflict. It was evident, too, in the work of Dewey (1963) and the transformation he wanted to see in American society; and in the work of the Brazilian Paulo Freire (1996) whose educational approach involved a purposeful educational engagement to transform social, economic and political relationships. The transformational nature of these approaches (among others) may be seen as central to contemporary approaches such as problem-based learning, real-world learning and action learning.

The experience of students in UK universities, and increasingly in other countries, is that of a client paying for a service. The introduction of a free market from 2012 – requiring universities in England and Wales to charge fees – led to the transfer of the cost of higher education from the state to many individual learners, and this will undoubtedly change the relationship that students have with their institution.

At the same time students expect, and universities increasingly acknowledge the value of approaches to learning and teaching beyond the lecture theatre (for example, experiential learning, problem-based learning and work-based learning). Real-world issues such as sustainable development demand these approaches and acknowledge the central role of students in their own learning experience; and this too may alter the student–institute relationship. In short, we can expect to see more discriminating and financially aware students and some universities are already adjusting their central mission statement in response. Plymouth University, for example, began its *Students as Partners* programme in 2012, embedding a focus on students and prioritising student experience in every sphere.

Bone and Agombar's (2011) study for the Higher Education Academy strongly supports this view. For first-year students sustainability issues were significant in their university choices. They stated that curricula should be modified accordingly and permeate their courses, and over 80 per cent of respondents believed skills in sustainable development are significant for employability and for future employers. This perception and the financial issues noted above combine to send a clear message to universities concerning student choices regarding places and programmes.

In any university, particularly those long-established, the complex relationships between students and staff, the university and its curricula, and the institution and its local community have developed through many generations. The centrality of such relationships resonates with the view of the Scottish polymath, Sir Patrick Geddes (1854–1932) that communities are built around 'place-work-folk' (see Meller, 1990: 34). Geddes' insights are particularly apposite as he is considered by many to be the founding philosopher of the concept of sustainability and the notion of 'think global, act local'. He spent much of his career in Scotland as a professor at the universities of Edinburgh and Dundee, exploring the issues associated with sustainable development and education (Higgins and Nicol, 2011).

For students, the efforts that the university puts into sustainability in its estate and programmes may be particularly visible, as is the *absence* of such efforts. As Eisner (1985) reminds us, the implicit curriculum (the way in which things are done) and the null curriculum (what is deliberately or inadvertently left out) are often as visible and significant to learners as the curriculum that is explicitly included.

In this chapter we consider these tensions, responsibilities and opportunities, using the University of Edinburgh as a case study. This is particularly appropriate as the University is large, diverse and urban, with a complex estate developed over centuries since founding in 1583. All these factors provide challenges and opportunities in the practice and promotion of sustainability and education for sustainable development (ESD).

## The University of Edinburgh context

The University of Edinburgh is a significant feature of the city, with over 29,000 students (from over 120 countries) and 8,000 teaching, research and support staff (3,300 of whom are academic) engaged in over 101 undergraduate and 259 taught postgraduate programmes with an annual budget of over £650 million. While any consideration of sustainability in this context is challenging, the University has prioritised this agenda since 1990 – embedding social responsibility and sustainability (SRS) in its last two strategic plans as an underpinning theme.

In some ways making such commitments seems straightforward. For several decades, in common with many other universities, sustainability has been an issue of great interest to staff and students. The University of Edinburgh has been at the forefront of international research into the science and social science of many sustainable development (SD) issues, and also has a longstanding academic tradition in environmental education, social justice and related areas, through individual courses and degree programmes (undergraduate and postgraduate) and student research opportunities (for example, doctoral studentships).

Sustainability is problematic for an institution to deal with as it has a clear interface with personal values. In the 1990s the University pursued an initiative called 'Curriculum Greening', an earnest effort to introduce sustainability into many subject areas across the university. Several subsequent initiatives have made some progress in ensuring that sustainable-development-oriented courses and programmes are available to students. However, throughout this period any assumption that sustainability should be incorporated in a widespread fashion into the curriculum has, quite properly, been questioned and critiqued, and this is a key reason for the elective nature of programme developments. This is not unique to Edinburgh. For example, a research study of at the University of Plymouth found staff considered aspects of the language of ESD inaccessible, and were not universally supportive of widespread introduction (Cotton *et al.*, 2007).

Building a community consensus (students and staff) that supports such developments is central to demonstrating the *practice* of sustainability (Eisner's [1985] 'implicit curriculum') in an institution. Tensions arise when an individual perceives a discrepancy between their values and the explicit or implicit values and purpose of an institution. For the institution, clarity of purpose is important, but so is respect for individual values. The University of Edinburgh's approach to this has been to develop a guiding principle through its 2008–12 and 2012–16 Strategic Plans (University of Edinburgh, 2008, 2012a). Making 'a significant, sustainable and socially responsible contribution to Scotland, the UK and the world, promoting health and economic and cultural wellbeing' is one of the four elements of the corporate mission outlined and this is being implemented through a detailed *Social Responsibility and Sustainability Strategy 2010–20* (University of Edinburgh, 2010a).

The University has signed up to a number of national and international agreements on sustainable development including the Universitas21 (2009) *Statement on Sustainability* and in 2010 the equivalent Universities UK statement of intent, committing the University to a leadership role in progressing global sustainable development and ESD. This theme and approach also resonates with the University's Internationalisation Strategy, *Edinburgh Global* (University of Edinburgh, 2009).

In addressing sustainable development and also considering this as an educational issue, the University takes its role in providing opportunities and leadership seriously. It also recognises that in a global educational market there are very significant opportunities in providing programmes and conducting research that addresses such issues.

## The Scottish Government context

Sustainable development is one of the Scottish Government's key national performance outcomes and features in many aspects of government work. A 'greener and fairer' nation is one of its overarching strategic objectives (Scottish Government, 2012). With all-party support it has set extremely ambitious targets for reductions in greenhouse gas emissions through the Climate Change (Scotland) Act 2009, including a 42 per cent $CO_2$ reduction by 2020 against 1990 levels. There is also significant policy development aimed at addressing the social inequality within Scottish society. In these processes the government has signalled the importance of societal change towards a sustainable future and highlighted the role of education in informing that process. One of the reasons why such demanding emissions targets can be set is that the landscape of Scotland has considerable potential for renewable energy generation (wind, wave, tidal, hydro and solar) and a relatively modest population of 5.2 million within the 62 million UK population. If these targets are met it will make Scotland a world leader in supporting its own energy demands through sustainable sources, an agenda to which the University of Edinburgh is committed to contributing.

Such commitments will mean that the technological and social aspects of SD will need to feature in schools and further and higher education institutions (FE/HE). Unsurprisingly, there are clearly related policy imperatives in education, and SD is one of four strategic themes in the Scottish Funding Council's (2009) Corporate Plan[1] for universities and colleges. The pressure to meet renewables and social justice targets must act as a stimulus for ESD. For example, the Scottish Government's economic strategy includes six priorities, one of which is newly included – the 'Transition to a Low Carbon Economy'. This is even more ambitious than the legislation, and suggests that all of Scotland's demand for electricity should be met by renewables by 2020, and that a greener economy could support 130,000 jobs by 2020 (Scottish Government, 2011: 53).

The Scottish Government has taken seriously the UN Decade of Education for Sustainable Development (UNDESD, 2005–14). It has stimulated activity in Scotland through policy initiatives, and the core justification for these has been laid out in a series of documents, the most recent of which, *Learning for Change,* sets out expectations of each education sector, including FE/HE, highlighting that 'creating a sustainable future for Scotland will require widespread understanding and huge cultural change – and the key to achieving this is education for sustainable development' (Scottish Government, 2010: 1). The University has established a 'Learning for Change Task Group' to explore the implications of this second action plan.

For Edinburgh's Moray House School of Education, such policy initiatives have additional consequences, as it is here that many of Scotland's teachers are trained, and they need to be prepared to address the implications of *Learning for Change* in their future work in schools. One highly significant further development is that in 2011 the incoming Scottish Government made a manifesto commitment to explore 'One Planet Schools' – signalling an intent to help schools move towards a 'one-planet future' – where they will gradually reduce their use of resources and develop a values orientation that addresses sustainability. The intention is to take a whole-school approach to this through the integration of three equally important facets – ESD, Development Education and Outdoor Learning. This, too, will become a feature of pre- and in-service teacher education, and will be supported by the General Teaching Council for Scotland's (GTCS, 2012) review of 'Professional Standards for Scottish and Global Sustainability', which sets new expectations of the profession to address these issues in its teaching. Such developments are also in accord with national commitments through UNESCO, and its active programme to 're-orient teacher education to address sustainable development' (UNESCO, 2012). This action was one of the key recommendations of Agenda 21, the main formal outcome of the 1992 UN Conference on Environment and Development in Rio de Janeiro, and has been reaffirmed at every successive world summit.

## Campus

The University has a strong record in practically implementing SD, particularly through its estates and procurement functions, and the work of the Edinburgh University Students' Association (EUSA). In terms of the day-to-day experience of students and staff it is perhaps the work on issues such as institutional energy efficiency, a commitment to Fairtrade procurement, and the establishment of Transition Edinburgh University that are most obvious. These efforts have led to national recognition and a number of awards[2] (University of Edinburgh, 2011: 2). In this area the concept of the university as a living laboratory – which can provide for and benefit from action research projects on campus – is gaining momentum.

As noted earlier, the physical scale of the University is substantial. With over 300 buildings on five campuses, sustainable estate management is a challenge, and has necessitated developing a comprehensive sustainability programme. Promoting a 'more sustainable university and reducing the university's environmental and carbon footprint' involves

> making most effective use of natural resources, promoting whole life costing to contain utility costs, supporting continuous improvements in campus infrastructure, contributing to the university's missions of promoting excellence in research and teaching, and liaising with others such as the Scottish Environment Protection Agency to promote best practices on campus.
>
> (Somervell, 2006: 11)

Such developments have reaped financial benefits. Since 1989, 'more than 5% of the university's utilities spend has been invested each year in energy efficiency projects delivering cumulative savings of £10 million' (Somervell, 2006: 11; University of Edinburgh, 2010b, 2011).

Since 2003 £12 million has been invested in combined heat and power energy centres in the central campus and two other sites. Communication has become an increasing feature of such developments, informing the university and city communities of the initiatives and their benefits in both environmental and economic terms. Together, the three energy centres generate 80 per cent of the power on these campuses[3] reducing $CO_2$ emissions by over 8,500 tonnes, and generate over £1.5 million of savings annually (see University of Edinburgh, 2010b).

The University of Edinburgh was accredited as Scotland's first Fairtrade University in 2004, and through the efforts and commitment of staff and students this status continues to be maintained. This success was an early indicator of the move towards sustainable procurement, a policy that is now embedded in the university's culture. As a result of strong student and staff support and a commitment among senior staff, the introduction of this policy was largely unproblematic. Corporate guidance now supports the procurement of goods and services to ensure efficiency and effectiveness, while minimising social and environmental risks (University of Edinburgh, 2012b).

The university's approach to social responsibility and sustainability is guided by a Sustainability and Environmental Advisory Group (SEAG), established in 1999 as a development from a ten-year-old Environmental Committee. The group comprises senior academic and administrative staff, members of EUSA and other representatives from across the university. It holds a pivotal position in these aspects of university life by proposing developments, commenting on or endorsing related proposals and advising the university's central management group. SEAG has been active in many aspects of estate and infrastructure management, encouraging changes in the way people travel to work and in energy usage.

Since 2007 SEAG has refocused to include more academic involvement and to ensure expertise in social responsibility and sustainability, in its broadest sense, were included through SEAG events and promotional activity. The aim – to develop a coordinated university strategy addressing research, teaching, behaviour and the estate – led to the ten-year, cross-university *Social Responsibility and Sustainability Strategy* adopted in 2010. Key operational task groups (for example, on climate action, engagement, fair trade and curriculum) feed into SEAG to ensure the strategy is delivered. Central to this is the convener's role as Vice-Principal External Engagement in conveying ideas upwards and providing leadership for the agenda.

The university's procurement office has an important interface with SEAG and EUSA, notably through its drive to promote acceptance of Fairtrade products and services on campus. More widely, procurement staff encourage sustainable procurement and adoption of whole-life costing to deliver low-carbon, low-maintenance solutions. While this process is not an end in itself, the visibility and success of such initiatives and developments in the estate and infrastructure demonstrate, as *per* Einser's 'implicit curriculum' (1985), that the university is serious about sustainability.

## Curriculum

Progress towards a more sustainable university offers important programme-related and personal learning opportunities for students and staff. This is true within disciplines, but particularly in sustainable development issues that demand consideration of a range of disciplinary areas involving both the natural and social sciences. As well as a growing awareness of the necessity of such an approach, the global trend towards more student choice in course selection, particularly in the early years of degree programmes, has aided such developments.

Under the auspices of SEAG a number of academic staff across the university have carried out investigations to determine current provision. For example, a 'sustainability in the curriculum working group' reported to the University Senate on current provision and its recommendations for future development are now being implemented across the university. Representatives of EUSA have evaluated student interest in this area, too, and concluded that there is a growing demand from students for cross-disciplinary education in subjects relating to sustainability, and a growing market for graduate employment in corporate social responsibility, environmental education, environmental management, environmental auditing, environmental politics, risk management, sustainable citizenship, and sustainable development.

Students also recognise that such interdisciplinary programmes foster the personal and transferable skills necessary to succeed in their studies and subsequent employability. There are clear indicators of demand such as the growth in sustainability-related optional courses (see below) and rising numbers on interdisciplinary degrees, such as the MSc in Global Challenges offered online by the three Global Academies outlined below.

There is always a genuine tension between the need for academic freedom and student choice when trying to embed sustainability into the curriculum. This is being addressed by interactions and discussions with students at the appropriate level in the University where curriculum is decided. It is not likely that sustainability will be a compulsory part of all courses, though options for all should be possible and will be embedded where appropriate.

One example of this is that staff across the university (primarily from Education and Geosciences) have written and delivered *Sustainability, Society and Environment*, a course available to almost all undergraduates. Similarly, Biomedical Sciences staff launched *Our Changing World* in 2009 (University of Edinburgh, 2012c), introducing first-year students to a 'range of difficult, complex and inter-related issues that impact human wellbeing' (food, energy and water security, the spread of infectious diseases, developments in technology and medicine, and climate change) and relate these to their own subject discipline. Lectures are open to the public and available as podcasts online. Both courses are taken by a wide range of students, although the sciences predominate.

Such developments have mostly focused on *courses* at undergraduate level due to a general reluctance about introducing new *degrees* beyond the very wide range already offered at Edinburgh. However, as with other universities such as St Andrews, Edinburgh has recently introduced an undergraduate MA in Sustainable Development. Significantly, this emerged from an academic staff proposal developed initially by members of SEAG and then a wide range of academic and administrative staff from across the university. The interdisciplinary nature of this proposal required this broad-based collaboration, while enthusing and engaging the staff involved in development and delivery.

There are rising applications for most established Masters programmes (for example, Environmental Sustainability; Environment and Development; Culture, Ethics and Environment; Outdoor Environmental and Sustainability

Education) and the successful introduction of new programmes (for example, Carbon Management; Carbon Finance). Further developments are likely to follow, some broadening the range of specialist degrees, others offering options across several existing programmes.

One key process stimulating this is the development of a new structural framework in the university called the 'Global Academies'. Rather than replacing existing structures, this fosters interdisciplinary discussion and developments. Currently the 'Global Environment and Society Academy' (GESA) and others for 'Health' and 'Development' all have clear relevance to sustainability. New Masters programmes will be offered through each of these academies, and a Masters programme across all the academies called 'Global Challenges' will be available by distance learning from 2012. This increases flexible study choices while reducing travel (saving $CO_2$ emissions and costs). Relationships with other programmes are likely to develop as students develop an interest in taking courses as options that are delivered through these cross-disciplinary 'academy' programmes. Distance education will be an increasing feature of these programmes and this, too, will affect and be influenced by student expectations.

Government requirements shape both delivery and subject coverage of teacher training. Given the significance and difficulty of teaching SD in schools, it is paradoxical that Scottish providers of initial teacher education (including the University of Edinburgh) are not required to include ESD in their programmes. However, the review of 'Professional Standards' noted earlier is likely to recommend inclusion (GTCS, 2012). Notwithstanding this, Edinburgh, along with other providers, offers relevant options, and here the focus is on the values context implicit in teaching SD, as much as it is on the content.

As this case study illustrates, the range of subject areas where social responsibility and sustainability may feature is substantial and growing. Research opportunities abound, stimulating increasing interdisciplinary doctoral training and partnerships. Here again, the global academies are predicted to have an important role.

## Communities

University of Edinburgh students, all at different stages of their learning experience, come from around the globe and study a very broad range of disciplines. Some are just leaving home and beginning to take control of all aspects of how they live, while others are leading research in novel directions that can directly impact how sustainability and social responsibility are implemented and perceived.

In this section we look at how the diverse skill set and the enthusiasm of the student body can be harnessed to interact with equally diverse and talented staff and the wider community. The public needs to see direct benefit from having so many talented people in its midst. These interactions need to be equally diverse, with staff and students engaging with wider communities in many ways.

Some activities serve various communities directly, such as through student volunteering. Some impact locally and informally, through different lifestyles, attitudes and approaches to everyday things like recycling, sustainable travel choices, conservation of energy and generally being good neighbours. While students may be here for a short time individually, their community is here long-term and they make a significant positive impact on the economic, cultural and social development of the city, and this is increasingly being explicitly articulated.

Edinburgh's staff includes research leaders in many areas that significantly impact how we live now and how we may live in the future. Aspects of research, for example in innovations in medicine, the sciences (including sustainable energy) and ESD, raise significant ethical issues, which the wider public with their varied experiences, cultural and religious beliefs will have views on, and these need to be heard by researchers.

Similarly, in any area of research expertise, broader linkages facilitate the university community being part of and not separate from the wider community. The huge developments in sustainability-related engineering (for example, carbon capture and storage and innovative marine energy developments) will impact resource and energy use, with implications for most aspects of modern life from industrial processes to building design. Climate change mitigation, for example, requires widespread discussion of issues such as the likely need for behaviour change to conserve water and other resources, rather than relying on engineering solutions. In the Highlands and Islands of Scotland, the impact on fisheries and coastal agriculture has significant technical and social implications. In such realms, academic, business and social communities' interaction and the role of policy-makers is crucial.

The university's social science community has a wide-ranging role in understanding and engaging staff and students and wider communities on sustainability and social equity as a construct and as a values issue, and is engaged in topics that spread through philosophy, divinity, education, politics, economics and business studies.

Forthcoming priorities include gathering the community of researchers to do more together across disciplines and to share their ideas with the public. One of the most important aspects of getting the wider university community to engage with the sustainability and social responsibility agenda in an academic environment is finding ways to encourage and support people to develop their own plans. Individuals and small, enthusiastic groups are often more efficient at effecting change than top-down directions that people resist or evade. Increasingly, this is happening, for example, through the Global Academies' programme, research-related initiatives and the development of new centres such as the collaborative Edinburgh Centre for Carbon Innovation.

While individuals and groups within the University community may develop their own initiatives and drive the agenda, some of these may be in conflict or there may need to be choices made based on available resources. Here decisions are made at the school, college or university level, depending on the nature of the decision and the resource implications. Rather than appearing to take over activities or detract from the responsibility and enthusiasm people have for their own agendas, cooperative and facilitative approaches are emphasised (see Shiel, this volume). This is peer-referenced insofar as approaches that fail to conform to this approach risk being resisted or ignored.

Two areas of difficulty and complexity are the balance between the enthusiasts and the rest of the community, and the conflicting views of the best way to tackle specific global issues. Many are put off by any approach that says 'you must do this' and may react negatively rather than logically. It is a management responsibility to lay out the issues, where there are legal and government policy compliance requirements, or simply to encourage staff and students to be good citizens, allowing them to join in willingly. Outcomes are often dependent on how issues and programmes are presented to the University community.

It is also crucial to acknowledge and not underplay conflicting agendas, including the tension between the need to cut the institutional carbon footprint with the need to be an international university. The latter may allow recruitment of an international student population that can produce international leaders in many disciplines and lead projects on poverty and disease alleviation. How do we balance Fairtrade with supporting local communities and businesses? These issues, and the strategic direction proposed by the University need to be highlighted, with key input from all staff, students and wider communities. These crucial issues cannot be underestimated if we are to take seriously the notion attributed to Einstein that 'we can't solve problems by using the same kind of thinking we used when we created them' (Harris, 1995), and Orr's (1992) conclusion that the sustainability crisis is driven by graduates educated in modernist notions of progress, economic growth and the urge to dominate nature.

The place where students study also significantly impacts on how they will behave as citizens when they graduate. Many, such as medics, vets, teachers, social workers and nurses, have always been committed to altruistic work. Others have impacts on how we live, for example architects, engineers and chemists. Everyone, whatever their chosen career path, is a citizen and can have a real impact on how the world develops. University students spread out across the globe, and their education must equip them to be forward thinking and responsive to the needs of the wider communities they find themselves in – and able to think about and take action on the significant issues of the day.

## Transition Edinburgh University

While EUSA sabbatical officers have had an important role on SEAG, it is through the association's wide range of societies that it has significant impacts in driving and maintaining student commitment to SD. For example, the Edinburgh University People and Planet society was the driving force, along with estates and other staff, behind the establishment in 2008 of *Transition Edinburgh University* (TEU) (2011, 2012).

A study by five summer interns in 2009 identified that the total greenhouse gas emissions of the 37,000-strong community of staff and students was as much as 335,000 tonnes of carbon dioxide ($CO_2$) equivalent (Cooper and Lander, 2009). It also suggested that 85 per cent (285,000 tonnes) of this came from students' and staff homes, travel habits, consumables and leisure, and that these could be reduced through adopting a Transition Initiative modelled on the 946 Initiatives worldwide (Transition Network, 2012). Following that study, the Scottish Government supported the initiative through its Climate Challenge Fund with grants of nearly £360,000 between 2009 and 2011.

Working with EUSA and many others across the University, six TEU staff and several dozen undergraduate interns spoke face to face with 14,000 of the 37,000-strong University community about opportunities to cut personal carbon emissions. In the type of 'virtuous circle' development that universities are well suited to, research students of the Moray House School of Education and other schools participated in action research programmes evaluating this and related initiatives, often as part of their Masters' dissertations. Altogether 1,400 students and staff signed up for regular bulletins with ideas for practical changes and news of related events and became involved in a series of complementary initiatives.

Five themed projects encouraged participants to reduce their personal carbon footprint. These included a Big Green Makeover – where 90 student volunteers were provided with training as home energy auditors by staff from Changeworks, the Edinburgh-based Energy Saving Scotland Advice Centre. These students then visited the homes of their friends and others to identify simple measures that cut carbon footprints by an estimated 300 tonnes of $CO_2$ and made the occupants warmer (Transition Edinburgh University, 2011).

Sixty-two students and staff and others were trained as Carbon Conversations facilitators, able to lead a series of six informal, but carefully structured sessions which address fears about climate change and introduce practical measures to cut personal carbon footprints. Nearly 200 people participated in these, with an average projected annual reduction of over four tonnes of $CO_2$ each (Transition Edinburgh University, 2011).

A Food Pledge, combined with initiatives from the Hearty Squirrel Food Co-op, provided weekly organic vegetable bags to people on the two main campuses. A related series of events, aimed at raising awareness of peak oil and climate change and laying the foundations for personal change, was continued as a popular co-curricular offering of weekly events, entitled Global Challenges, in 2011.

It should be noted that the Climate Challenge Fund's emphasis on measuring reductions in carbon emissions rather than recognising the community capacity-building achieved, proved problematic and the TEU went into hibernation after continuation funding was refused in March 2011 (see Luna and Maxey, this volume, for further critique of carbon metrics). Fortunately, the University retained many of the key staff who then applied themselves successfully to electricity saving and emissions reductions on campus – typically achieving 5 per cent consumption reduction through harnessing the ideas of staff and students concerning the buildings they occupy. TEU re-emerged in spring 2012 as a network of groups linking active students and staff.

## Edinburgh Sustainability Awards

In 2010 the University's Sustainability Office joined with EUSA and the National Union of Students (NUS) to raise the profile of social responsibility and sustainability on campus. Following EUSA's successful Teaching Awards, Edinburgh Sustainability Awards (ESA) were launched to recognise student, academic and support staff contribution to innovation and change supporting the wider sustainability agenda.

In their second year the Awards were additionally promoted by 30 EUSA Student Ambassadors who were trained as volunteers to support teams in completing an online workbook devised with the NUS. Students then carried out on-site audits of the submissions and their involvement was recognised at the annual awards ceremony (attended by over 100 staff and students) where three Gold, eight Silver and six Bronze categories were awarded along with three Laboratory Awards and six Special Awards.

## Other student-led initiatives: *OurEd* and *The Edinburgh Manifesto*

The student community has been engaged in other ways, too. Social media is an increasing facet of student life, with a very high proportion using Facebook and increasingly Twitter. Responding to calls for a student-facing, online resource, a focus group helped identify how such a website could best engage with students. With the support of the university's web development programme, an online resource titled *OurEd* (University of Edinburgh, 2012d) was developed to inspire current and prospective students to become involved in sustainability and social responsibility issues.

The website has been popular among students, and student societies are particularly supportive of the platform as they can communicate their activities to fellow students across the university. As with many online resources, the impact of the initiative goes well beyond the university and 'should inspire students to embrace sustainability and social responsibility during their time at the University of Edinburgh and in their future careers' (Howell *et al.*, in press). It is now being rolled out as a service to other universities and colleges in Edinburgh.

Arising from ideas generated by these initiatives, a student forum entitled Autonomous University Edinburgh met during 2011 to share ideas on subjects ranging from ecological economics to permaculture. The ensuing discussions led to *The Edinburgh Manifesto,* outlining the student vision of learning for change. This calls for 'Education for Sustainable Development to be at the very core of learning and teaching at the University, of the University's relations with and responsibilities towards the community, both locally and globally, and of the running of the University itself' (Autonomous Edinburgh University, 2012: 2).

The students' request is based on

> a belief that sustainable development is a core element of University teaching and research; that Education for Sustainable Development involves including students and staff at all levels in tackling both global and local challenges; and that if we talk the talk, we have to walk the walk – and work towards sustainability within the University community.
>
> (Autonomous Edinburgh University, 2012: 2)

Encouragingly, these ideas have been seriously considered by senior academics framing the next strategic plan and so the direction of travel of this large institution slowly evolves.

## Future prospects

While there have been some successes, in common with other universities there is much more to do to continue to move progressively towards the goal of becoming a sustainable university. Many challenges have arisen, and while some feel that the pace of change is too slow, it is important that the difficulty of this process is acknowledged, respecting that policy-makers and senior colleagues have different, but valid, organisational drivers, which are in turn delimited by both what already exists and the imperative for consensus. These restrictions and lack of dialogue have occasionally prevented or delayed change. Additionally, administrative loads continue to limit efforts to improve the quality of student experiences generally and sustainability initiatives specifically.

While such issues are endemic and institutional, ways must be found to move towards a more sustainable future for universities, not just because of the scale of their direct environmental impact, but also because of the influence of both *what* they teach and the *way* they teach it. Before institutional change can be stimulated, there must be a sense of community engagement in this process, stimulating both acceptance of change and commitment to facilitating it by all members of the community.

The transitory nature of student involvement within the life-cycle of their studies places restrictions on their long-term individual involvement. Most students remain at university for between one and four years. They do not design whole courses, nor are they around long enough to sustain the ideas they may set in motion. Consequently, it is not always possible to deploy iterative processes where students are able to reflect on what they started with a view to improvement. This is not to say that the student voice is unimportant, it is more a case of understanding situational factors so that student involvement is seen as essential but at the same time to be realistic.

However, if universities are to successfully engage students they must acknowledge their potential as agents of change who have different spheres of influence. Students' opportunities for engagement are spatially diverse (involving university life but also off-campus life) and determined by individual and collective agency (the capability to take action that will have a social and/or environmental impact). This means universities do not simply lecture about the importance of sustainability while assuming students are 'citizens in waiting'. Rather, everyone in universities can engage in transformative approaches to citizenship that begin now and involve capacity building and action. One example of a change of culture in this regard is that *The Edinburgh Manifesto* (EUSA, 2012) has been endorsed by SEAG and aspects will inform the University's strategic direction.

The University has committed to sustainability through written policy and strategy documents (as noted above) and these have been approved by the University court. These actions are consistent with Tilbury and Wortman's (2004: 2) view of the importance of 'envisioning' and being able to imagine a better future. They have also indicated the importance of critical thinking and reflection, and the University has been proactive in ensuring that it subjects itself to internal scrutiny through increasingly rigorous reports and analyses of these at SEAG and engagement with UK and international initiatives that require demonstrable institutional commitment, for example Universitas 21 (2009).

Furthermore, the setting up of a specific committee over twenty years ago to advise the university senior management team with regard to sustainability, was a visionary concept. Because the membership was drawn from diverse operations within the university (teaching, research, estates, procurement) it meant that a holistic and 'warts and all' approach to systemic management could be adopted. It would be inaccurate to state that this rational approach has been easy. Because sustainability is inherently values-based, and its meaning rooted in contested definitions, it is unrealistic to assume that change through consensus-building is straightforward.

The empirical findings of Fullan suggest that 'educational change is technically simple and socially complex' (1991: 65). It is this complexity that leads to contestation and stress for change agents and change resistors alike. Fullan would argue therefore that stress is inherent to this process and a necessary precondition for change. In other words, if stress is not present it is likely that nothing is changing. It is also in keeping with Tilbury and Wortman's acknowledgement of the importance of systemic thinking while 'acknowledging complexities, looking for links and synergies in trying to find solutions to problems' (2004: 2).

In common with others, the University of Edinburgh has begun a process that has involved building partnerships towards a more sustainable future both within the university and beyond. It has meant responding to positive political circumstances (for example, the stimulus of UNDESD, interest in 'green' jobs and global challenges) and promoting dialogue and negotiation internally and externally while recognising the need to work collaboratively with staff, students and other communities towards participative decision-making (Tilbury and Wortman, 2004). These challenges are not peculiar to universities, but apply to any organisation that is serious about sustainability. Arguably, of all institutions a university, and especially one as diverse as Edinburgh, should find this process both natural and positive.

## Going greener: the birth of a Transition University movement

The concept of Transition Universities first emerged circa 2008 when student campaigners from People and Planet – the UK's largest student environmental network – started to question the pace of progress their universities were making towards sustainability. After five years of campaigning for the basic building blocks of sustainability (environmental policies, environmental management staff, etc.), it was clear that many institutions were still far from being resilient, low-carbon or community-led.

Inspired by the growing Transition Town movement, students from across the UK came together to propose a radical shift in campaign strategy – a move away from centralised campaign objectives to bottom-up practical transition projects on campuses. The aims – pioneered by the student-staff initiative, Transition Edinburgh University – were simple: raise awareness, build community, cut carbon emissions and create the institutional change required to 'future-proof' our universities from the ground up. It required a shift from a traditionally more confrontational campaigning approach to a much more collaborative, open-source effort that brought staff and students together more consistently.

Launched in 2009, the Transition University model developed by People and Planet has spread rapidly, and today the charity supports over 30 groups of students and staff nationwide in working towards a transition vision for their universities. The group offers training and resources to help university communities explore what they want to achieve at the local level and the skills to make transition projects a reality.

National events such as Go Green Week have helped to engage a wider audience in climate change and peak oil issues, whilst a Transition Universities website allows groups across the country to share project ideas, resources and communicate with each other. Progress is measured by the People and Planet Green League – a comprehensive and independent league table that compares universities' campus operations, community engagement, sustainability policies and carbon emissions as well as their approach to greening the curriculum.

Although independent, People and Planet has worked closely with the Transition Network from the outset. At the local level, Transition University initiatives are starting to work more closely with their local town initiatives and in some cases have been initiated by members of the Transition Network in collaboration with People and Planet.

For more information, visit: http://peopleandplanet.org/goinggreener
Transition Universities Wiki: http://peopleandplanet.org/transition-ideas
People and Planet Green League Table: http://peopleandplanet.org/
greenleague

*Louise Hazan is Climate Change Campaigns Manager for People and
Planet, the UK's largest student campaigning network. She founded the
People and Planet Green League – an annual green ranking of universities
that is credited with putting climate change on the desk of every vice-
chancellor in the country.*

## Notes

1   The Scottish Funding Council distributes funding to all Scottish further and higher
    education colleges on behalf of the Scottish Government.
2   These include Green Energy Award 2004, National Energy Efficiency Award 2008,
    Carbon Trust Standard 2009, GO Procurement Team of the Year 2012, Green
    Tourism Gold Awards for halls of residence 2011 and Healthy Working Lives Gold
    Award 2010.
3   Of the remaining energy bought in by the university, 40 per cent is from 'green
    sources'.

## References

Autonomous Edinburgh University (2012) *The Edinburgh Manifesto: Learning for
    Change: Students' Visions.* Available at: www.oured.ed.ac.uk/2012/03/students-
    vision-edinburgh-manifesto/ (accessed 20 April 2012)
Bone, E and Agombar, J (2011) *First-year Attitudes Towards, and Skills in Sustainable
    Development,* York: The Higher Education Academy
Cooper, O and Lander, R (2009) *Footprints and Handprints: The Edinburgh University
    Community's Climate Impact and How We Begin Reducing It,* Transition Edinburgh
    University. Available at: www.transitionedinburghuni.org.uk/files/Keynote%20
    summary.pdf (accessed 19 May 2012)
Cotton, D, Warren, M, Maiboroda, O and Bailey, I (2007) 'Sustainability development,
    higher education and pedagogy: a study of lecturers' beliefs and attitudes',
    *Environmental Education Research,* 13 (5): 579–97
Dewey, J (1963) *Experience and Education,* London: Collier Books
EAUC (Environmental Association of Universities and Colleges) (2009) Available
    at: www.eauc.org.uk/universities_uk_statement_of_intent (accessed 20 April
    2012)
Eisner, E (1985) 'The three curricula that all schools teach', in Eisner, E (ed.) *The
    Educational Imagination,* New York: Macmillan, pp. 87–108
EUSA (2012) *Learning for Change: Students' Visions: Edinburgh Manifesto.* Available at:
    www.oured.ed.ac.uk/wp-content/uploads/2012/03/Learning-for-Change-Edinburgh
    Manifesto.pdf (accessed 20 April 2012)

Freire, P (1996) *Pedagogy of the Oppressed*, London: Penguin

Fullan, M (1991) *The New Meaning of Educational Change*, London: Cassell

General Teaching Council for Scotland (2012) *Professional Standards for Scottish and Global Sustainability*, Edinburgh: GTCS

Harris, K (1995) *Collected Quotes from Albert Einstein*. Available at: http://rescomp. stanford.edu/~cheshire/EinsteinQuotes.html (accessed 7 June 2012)

Higgins, P and Nicol, R (2011) 'Professor Sir Patrick Geddes: *"Vivendo Discimus"*: by living we learn', in Knapp, C and Smith, T (eds) *A Sourcebook for Experiential Education: Key Thinkers and their Contributions*, New York: Routledge, pp. 32–40

Howell, R, Wisdahl, M, Farthing, J, Somervell, D and Bownes, M (in press) 'OurEd: creating online social responsibility and a sustainability community', *International Journal of Environmental, Cultural, Economic and Social Sustainability*

Meller, H (1990) *Patrick Geddes: Social Evolutionist and City Planner*, London: Routledge

Orr, D (1992) *Ecological Literacy: Education and the Transition to a Postmodern World*, New York: Albany, State University of New York Press

Rogers, C (1983) *Freedom to Learn*, London: Merrill Publishing Company

Scottish Funding Council (2009) *Corporate Plan 2009–12*. Available at: http://ww7. global3digital.com/exreport/pdfs/viewer.jsp?ref=4&page=2&zoom=std&view=list& layout=double#1 (accessed 20 April 2012)

Scottish Government (2010) *Learning for Change: Scotland's Action Plan for the Second Half of the UN Decade of Education for Sustainable Development*, Edinburgh: Scottish Government

Scottish Government (2011) *The Government Economic Strategy*, Edinburgh: Scottish Government

Scottish Government (2012) *Strategic Objectives*. Available at: www.scotland.gov.uk/ About/Performance/scotPerforms/objectives (accessed 20 April 2012)

Somervell, D (2006) 'University of Edinburgh's sustainable future: combined heat and power the key to low carbon strategy', *District Energy*, 4th Quarter, pp. 10–14, Westborough, MA: International District Energy Association

Tilbury, D and Wortman, D (2004) *Engaging People in Sustainability*, The World Conservation Union Commission on Education and Communication, Geneva: IUCN

Transition Edinburgh University (2011) *Transition Edinburgh University: CCF Funded Project Phase 2 Final Report*. Available at: www.teu.org.uk/wp-content/ uploads/2010/02/TEU-Project-527-Final-Report-April-2011.pdf (accessed 20 April 2012)

Transition Edinburgh University (2012) *Transition Edinburgh University*. Available at: www.teu.org.uk (under Resources) (accessed 20 April 2012)

Transition Network (2012) *Transition Initiatives Worldwide*. Available at: www. transitionnetwork.org/initiatives (accessed 20 April 2012)

Tremmel, J, Page, E and Ott, K (2009) 'Editorial: special issue on climate change and intergenerational justice', *Intergenerational Justice Review*, 9 (3). Available at www. intergenerationaljustice.org/images/stories/IGJR/igjr_09.pdf (accessed 20 April 2012)

UNESCO (2012) *Re-orienting Teacher Education to Address Sustainable Development*. Available at: http://www.unesco.org/en/esd/networks/teacher-education (accessed 20 April 2012)

Universitas21 (2009) *Universitas21 Statement on sustainability*. Available at: www. universitas21.com/relatedfile/download/165 (accessed 20 April 2012)

University of Edinburgh (2008) *Strategic Plan 2008–12*. Available at www.ed.ac.uk/schools-departments/governance-strategic-planning/strategic-planning/strategic-plan-2008-12 (accessed 20 April 2012)

University of Edinburgh (2009) *International Strategy: Edinburgh Global*. Available at: www.ed.ac.uk/about/edinburgh-global/about-us/overview (accessed 20 April 2012)

University of Edinburgh (2010a) *Social Responsibility and Sustainability Strategy 2010–20*. Available at: www.ed.ac.uk/about/sustainability/our-approach (accessed 20 April 2012)

University of Edinburgh (2010b) *SRS Highlights 2009/10*. Available at: www.ed.ac.uk/about/sustainability/includes/news-includes/srshighlights2010-2011 (accessed 20 April 2012)

University of Edinburgh (2011) *SRS Highlights 2010/11*. Available at: www.ed.ac.uk/about/sustainability/includes/news-includes/srshighlights2010-2011 (accessed 20 April 2012)

University of Edinburgh (2012a) *Strategic Plan 2008–12*. Available at www.ed.ac.uk/schools-departments/governance-strategic-planning/strategic-planning/strategicplan-2012-16 (accessed 10 November 2012)

University of Edinburgh (2012b) *Sustainable Procurement*. Available at: www.ed.ac.uk/schools-departments/procurement/sustainableprocurement/sustainable-procurement (accessed 20 April 2012)

University of Edinburgh (2012c) *Our Changing World*. Available at: www.ocw.ed.ac.uk (accessed 20 April 2012)

University of Edinburgh (2012d) *OurEd*. Available at: www.OurEd.ed.ac.uk (accessed 20 April 2012)

# 10 Well-being

## What does it mean for the sustainable university?

*Sarah Sayce, Judi Farren Bradley, James Ritson and Fiona Quinn*

### Introduction

The Brundtland definition of sustainability (WECD, 1987) has been interpreted predominantly in relation to the physical resource base, in climate and material terms. However, consideration of the sustainable university would be incomplete without due regard being paid to social issues, notably health and well-being. Indeed, the well-being of some 2.3 million students in higher education institutions (HEIs) is critical to the mainstream agendas of student recruitment, experience and retention and must be embedded within an institution's approach to sustainability (Dooris and Doherty, 2010: 6). HEIs are well positioned to have a significant impact on social sustainability and the development of habits that lead to good work–life balance – and resilience to stress among the staff and student body could have significant ramifications for the future well-being of society.

Although no official government scheme has been developed to support and promote well-being within universities specifically, the development of the 'healthy university' has been evident over the past decade with the informal English National Healthy Universities Network offering the opportunity for HEIs to share ideas, practice and resources (Dooris and Doherty, 2010: 12).

However, as students themselves and the economic climate in which they sit undergo renewed changes, so the mechanisms employed to support their well-being need to adapt if HEIs are to become more socially responsible.

Orme and Dooris argue that 'it is widely recognized that health is determined by a range of environmental, social and economic influences and that the health of people, places and the planet are interdependent' (2010:425). Without a workforce that is healthy, both physically and mentally, the education of students will be impaired. The converse is also true: an education system that either neglects or negatively impacts upon the health and well-being of staff and students will prove both less effective in output terms and make more problematic the interactions between staff and students.

A state of mutual interdependency, in which behaviours and attitudes of one group will likely impact on the other, exists; over time, lack of social well-being will compromise academic standards and economic sustainability. For this reason, it is perhaps surprising that the schemes promoted to universities in the UK to enhance their sustainability, such as the Higher Education Funding Council for England (HEFCE)'s sustainable development strategy (2005, 2008), the influential People and Planet Green League Table (People and Planet, 2012) and other schemes such as EcoCampus (2012) have no explicit criterion based on the health and well-being of staff and students.

This chapter considers some of the key considerations that relate to the well-being of staff and students in the light of the continuing pressures on the UK higher education system as it attempts to respond to the imperatives of a global education market while subject to the vagaries of radical shifts in national policy. Exhortations to deliver 'more' and 'better' are set within an apparently diminishing resource base, which is currently characterised by both cost constraints and uncertainty of income. From one perspective, this is likely to reduce the resources that can be devoted to well-being, both in terms of time and finance; from another, it could be argued that, unless such matters are taken seriously, the economic bottom line of the organisation will be adversely affected, reputation will be harmed, and the life chances of graduating students may be compromised.

More importantly, any university that purports to pursue sustainability as a priority must be governed by policies and practices derived from a culture of 'beyond legal compliance'. Such policies and practices relate to staff well-being, to factors surrounding the student body, and to the physical environment impacts on health and well-being.

## The well-being of staff

The need to protect workers' health has been recognised by the World Health Organization (WHO), who have produced and endorsed a Global Plan of Action for Workers' Health (WHO, 2007). This plan recognises a need to provide healthy workplaces that promote physical and mental good health. However, it makes no explicit link between pressure, generally regarded as 'good', and stress, which is viewed both as a negative influence on well-being and as a cause of illness.

Within the UK, the legal framework governing the well-being of employees and the responsibilities of employers is contained within a raft of legislation. Non-compliance is often only revealed in matters brought before employment tribunals or following actual or potential accidents. Although no specific legal requirements apply to employees within a university setting, funding authorities such as HEFCE, and agencies concerned with maintaining and enhancing quality, such as the Quality Assurance Agency (QAA), have codes, standards and guidance that require institutions to define and demonstrate their commitment to their employees.

Universities normally have well-developed policies in relation to the well-being of both staff and students. As one university's website claims, 'The health and wellbeing of all our staff and students is of the upmost importance to the university' (Kingston University, 2012). The well-being of staff may be presented as a matter of ethics and corporate social responsibility.

When considering university staff well-being, differentiation is required between academic staff and those who support the academic functions, and between functional categories within each group (Winefield *et al.*, 2003; Tytherleigh *et al.,* 2005). Support or general staff typically face the same well-being issues as any other office-based workers; academics, however, face specific pressures due to the nature of their work (Safaria *et al.*, 2011) but may also exhibit higher levels of tolerance to otherwise stress-inducing factors (Watts *et al.*, 1991) for reasons which are 'speculative' (Tytherleigh *et al.*, 2005: 55). An essential aspect of the academic role is to make professional judgements relating to academic performance, research ethics and curricula design, and is beset with the tensions recognised within other professional groups, as the status of professions is challenged and managerial systems are introduced to limit professional autonomy, centralise authority and increase systematisation of activities (Sayce and Farren Bradley, 2011).

Doyle and Hind (1998) argue that some of the stresses and pressures of academic work are actually or potentially stimulating and an integral part of the inevitably varying workload patterns; further, claims Buchholz (2011), they are good for you. Most academics are motivated by their disciplines and passionate about communicating their work through teaching, research and publication. Some level of pressure is a given within an organisation such as a university, which has built up a longstanding pattern of distinctive annual cycles in which periods of intense activity, such as around the commencement of the academic year and periods of major assessments, have been balanced by periods of comparatively less pressure, in which academic staff can refresh themselves and undertake activities that demand controlled long-term concentration, such as the preparation of research papers. However, the 'long summer vacation' as the in-built period of refreshment has gone, needing to be replaced with structured, managed periods of leave to ensure that staff take time to balance their lives and refresh themselves. In many institutions, the once-standard provision of paid sabbatical leave has been removed from academic contracts. This has been substituted by limited provision for 'teaching relief', related to specific approved research outcomes. Even greater emphasis is, therefore, placed on acquiring externally funded research projects, which are often reliant upon a substantial previous track record of research outputs.

Staff engaged primarily or exclusively in research may experience a different pattern of potential stressful activities related to funding and project deadlines; also, many are on temporary contracts linked to their ability to win external funds (Court and Kinman, 2008). For them, job security is likely to be a higher stress predictor than for teaching staff, although no longer are teaching staff protected from redundancy, other than in some selector universities, as UK student applications decline and the competition for overseas students intensifies.

However, while pressure can be regarded as positive, it may turn quite easily into stress, defined as 'the adverse reaction a person has to excessive pressure or other type of demand placed upon them' (HSE, 2011) and, therefore, needs to be combatted in an ongoing way (Smith *et al.*, 2000). Stress may not impact solely on an individual, but may be systematic within an organisation, caused by an inability to undertake adaptive action at the rate of change demanded by outside circumstances. Here, the consequence may be that the entire workforce is impacted negatively, regardless of the role within which they are employed. The issue for the sustainable university is to actively protect its staff from undue stress. For this, it must first accurately monitor the environment in which it operates to ensure that it is not predisposing its staff to collective stress. Second, it must take steps to promote behaviours that are sustainable, as these have been found to produce co-benefits in terms of health and well-being (UCL Institute of Health Equity, 2010).

Theories abound on stress causation, from Demand-Control Models, which postulate a combination of high demand and a feeling of the inability to be in control (see, for example, Karasek, 1979) to Demand-Resources Models (Bakker *et al.*, 2003), which relate high workloads to inadequate support resources (social, organisational, physical or psychological). Given that many lecturing staff still retain significant autonomy over their working patterns, it is more likely that academic stress will arise under the latter model's conditions. Sparks *et al.* (1997) argue that a long-hours culture is significant, a point not lost on Gill (2009), who points to a conspiracy of silence by academics of unacceptable work pressures rooted in a passionate commitment to their work.

Work stress is the subject of an extensive literature (see, for example, Cooper, 1998; Cartwright and Cooper, 2011; Sparks *et al.*, 1997). Stress leads to physical and mental illnesses, the incidence of which can be both tracked and managed. The need, both economic and social, to manage the mental well-being of employees in the early stages is critical as it is estimated that mental health alone could be costing the economy as much as £28.3 billion a year (NICE, 2009) or even more (Naylor and Appleby, 2012), but that good management could lead to a 30 per cent saving (Sainsbury Centre for Mental Health, 2007). Developing key mechanisms for the promotion of well-being should, therefore, be a mainstream agenda for the sustainable university.

While long recognised in relation to human service occupations, especially teachers (Brouwers and Tomic, 2000; Farber, 2000), long-term stress, or burnout, is less acknowledged within the higher education academic community, although, more than fifteen years ago Abouserie (1996) identified it as a concern among academics, notably the pressures placed by research deadlines. More recently, Watts and Robertson (2011) concluded that the incidence of burnout within the academic community was comparable with professions known to be at high risk, namely healthcare workers and schoolteachers. Burnout has three dimensions:

1   emotional exhaustion, leading to feelings of inadequacy particularly in undertaking pastoral roles;

2  cynicism, leading to a depersonalisation approach as a defence mechanism against exhaustion, which leads to difficulty in interpersonal relationships;
3  work-related dissatisfaction and a possible loss of efficiency.

It builds slowly and may go undetected for long periods until it has taken a strong hold (Maslach and Jackson, 1981; Maslach, 2006), by which time damage may have occurred to the staff member's colleagues and students.

The causes have been identified as reductions in funding and resources, increased workloads, pressure to obtain research funding, poor university management, lack of job security and insufficient pay and reward (Winefield *et al.*, 2003). Gill (2009) confirms these characteristics, also citing the 'hostile and dismissive judgements' produced in the peer review process before concluding that the 'seduction' of academic autonomy, with imposed business models, has led to working lives in which there is no boundary between work and anything else. Resultantly, Hastings (2002) identified that stressed staff may exhibit tendencies which exacerbate any poor behaviour among their student base – which, in turn, adds to staff stress, thus establishing a vicious circle.

Among the limited empirical studies of academics identified by Watts and Robertson (2011), only two were UK-based: Doyle and Hind (1998) and McClenahan *et al.* (2007). Doyle and Hind found evidence of greater susceptibility among female academics than male, while McClenahan *et al.* found the key predictors of burnout were associated with work demands, control and supervisory support. In their conclusions, they recommended that these three aspects be key considerations in any re-evaluation of job design, with control being ceded to the lowest level practicable. Tytherleigh *et al.*'s literature review (2005) of all categories of university staff concluded that while stress levels are higher than average across the board, academic and research staff experience the most stressed conditions, although the translation to physical and psychological disorders is lower than could be expected, the reasons for which, they suggest, require further research.

Finally, the Turkish study by Bilge perhaps best sums up the approach that a sustainable university might wish to adopt, namely that, in managing staff and helping to reduce stress and burnout, policies should recognise that 'academics who find their jobs meaningful, who find encouragement for personal development and who can assume responsibility for their jobs are more motivated to work and experience burnout less' (2006: 1157). However, particularly at times of change, such as now, this needs to be tempered by a careful management of workloads, as Boyd *et al.* (2011) found that autonomy becomes counter-productive if there are perceptions of injustice in processes and procedures.

In summary, the literature confirms that academic staff are at greater risk of suffering stress-related conditions than many other employees. Their role makes them particularly susceptible to burnout, with the attendant negative consequences for the student experience. Widening participation, fee increases, and increasingly pressured students may have further intensified this as their role is predicated on regular, intense interactions with large numbers of, mainly young, increasingly demanding and often vulnerable people, for whom support may make the difference between success and failure. While the literature is quite clear in identifying stress and burnout as the conditions to which academics are at high risk, little specialist literature exists on how they might be prevented. The sustainable university must address this.

## A changing higher education environment

The environment within which academics operate is in a process of continual change. The UK, for example, has seen an increasing audit culture and changes in the quality framework and the governance of universities, leading to the strengthening of managerial controls and bureaucratic processes aimed at better assuring an evidence base and control of consistency of standards between and within universities. These processes have inevitably led to greater systemisation and loss of academic freedom and professional autonomy, combined with increased workloads and individual accountability, as evaluation is embedded within every layer of delivery.

We now consider three aspects that are of importance for staff well-being: staff–student ratios, the changing nature of the student cohort, and changing technologies.

### Staff–student ratios and their impact on workload

The UCU (2010) estimate that, over the last decade, average staff–student ratios (SSRs) rose to just under 17:1. However, this masked significant differences between institutions, with some post-1992 universities reporting figures nearly double this. It also masked significant differences between subject areas within institutions, a well-recognised factor related to funding models. Further, SSRs within the UK are consistently significantly higher than for other member states of the Organisation for Economic Cooperation and Development, where the average was 15.3:1 in 2007 (OECD, 2009). This is meaningful as it is against such better resourced countries that UK HEIs compete for research money and students.

In view of this, the sustainable university should keep its SSRs under careful review. It is standard practice for universities to run regular staff satisfaction surveys, which may pick up signs of institutional stress, and it is encouraging that they are incorporating a range of questions targeted at identifying mental health issues. Specific Stress Surveys, modelled on the HSE Management Standards approach, are used to help identify the prevalence of the problem; however, such surveys tend to happen infrequently and may not identify or explain specific and underlying factors. As an adjunct, advice and support services, ranging from counselling to reflexology, are being introduced, arguably as a less costly measure than addressing the fundamental causes, but these will not work if take-up is poor. SSRs are a very crude measure of workload as they have an inbuilt assumption that teaching is the prime or only activity of academic staff, which for most it is not. Even if it was, a number of factors have made the role of the lecturer more complex than only a decade ago, due to changes in the nature of the student cohort, prevailing pedagogic practices and technology-enhanced delivery mechanisms.

### The changing student cohort

As is the case internationally, in the UK, a sustained feature of government policy has been the expansion of student participation rates predicated on widening access. The social, cultural and educational background of students has, therefore, become more diverse. No longer can it be assumed that students entering higher education come from homogenous groups in society or are able to commit to programmes and modes of study that have traditionally formed the bedrock of higher education. Further, previous assumptions as to culturally-based value sets and knowledge bases are challenged. This potentially, and actually, changes the relationship between academic staff and their students.

While a widely acknowledged issue, pedagogic responses have met with only limited success and most experimentation in class delivery has resource implications. The final evaluation report to the Higher Education Funding Council for England (HEFCE) of the grant to C-SCAIPE[1] (one of the UK government's Centres for Excellence in Teaching and Learning) concluded that while student engagement and success can be enhanced through innovative practice, educational enhancements come at a price, supporting the earlier finding of Prosser and Trigwell (1999). Nonetheless, if handled well, changes in modes of delivery, engaging students in new ways, can be stimulating for both staff and students, empowering both groups in a more dynamic relationship.

Students may be less well prepared for the rigours of higher education and assumptions about study skills, such as data investigation, critical reasoning and essay writing, may be wrong. Further, increasing cohorts of overseas students bring the challenges of language skills and cultural understandings. The responsibility to make good these deficits falls to the academic community, if the academic standards of graduates are to be maintained. In response, many universities have set up, with some success, Academic Support Centres to complement the work of those delivering core materials by providing generic or subject-related support materials.[2]

Another consideration relating to the student base is that of funding. The introduction and escalation of fees has created a strongly pronounced consumer culture among students. This is monitored closely via student satisfaction measures, not just by the National Union of Students[3] but by evaluation at programme, module and, at its most extreme, individual class level.

Given that fees are due to rise sharply in most English universities from 2012, the level of experience that students will, with justification, seek to achieve, is likely to increase. However, the experience of students is the result not just of the calibre of teaching, but the efficiency of timetabling, form of learning environment and associated facilities. Academics feel that they have a lack of control over the range of matters on which they will inevitably be judged. In moving forward, the sustainable university will need to ensure that learning environments and timetabling systems work in a way that supports effective learning and teaching so as to ensure that reasonable student expectations can be set, monitored and achieved.

### Changing technologies

Initiatives for blended learning, via virtual learning environments (VLEs), present opportunities and challenges. HEIs have made significant investment in this area, both in terms of actual development and in dynamic interactive facilities to support them. These developments can be invigorating and help to keep the academic environment and interactions lively and engaging.

However, they impact both the skill base and physical working conditions of academic staff, and can be potentially threatening to the sense of autonomy and the development of the interpersonal dynamic between staff and students. The preparation of learning materials is demanding in terms of time and financial resources, and has the potential to increase staff workload (Gill, 2009). While much has been made of the need for clarity in terms of copyright and liability for digital media, academic staff are now required to provide material for intranets on a daily basis and also accept the recording of lectures, seminars and tutorials, which, while carried out within the codes and standards set out by institutions, are no safer than any other media from being uploaded to the public.

### Responses

It is important to place what could be seen as a catalogue of negatives into a more constructive perspective. The need for workplace well-being is widely recognised, not least on a business case level (Sainsbury Centre for Mental Health, 2007; NICE, 2009), and an examination of HEI facilities and training programmes would indicate strong awareness of the need for positive approaches to be taken. HEIs offer a range of initiatives, including advice to staff on work–life balance, through to a proliferation of volunteering schemes, community engagement activities and sports, music and other activities. For example, Kingston University, through its Sustainability Hub, undertakes a range of activities with Transition Town Kingston.[4] Sensitivity to the needs of students has increased opportunities for staff, as discussed below, while increased provision of social space and the setting aside of space as faith rooms bear testament to a genuine intent to assist staff well-being.

However, it is one thing to make facilities available; it is another for them to have effect. For this, there needs to be take-up (and here there are workload implications) and effective monitoring. Until indicators are developed to track the success of well-being promotion initiatives, it will be difficult to argue the case that more should be done. Additional thought could be given to the more careful use of appraisal and personnel management to evaluate whether the causes of stress are being recognised, acknowledged and addressed, rather than seeking to deal with the symptoms.

As with any other aspect of sustainability, a holistic approach is required for the organisation to move forward. And, for this, stronger measurement and monitoring would be a good next step.

## Health and well-being co-benefits of living sustainably: an example (case study) from Transition Town initiatives

The first Transition Town was launched in 2005 in Totnes, Devon (UK), acting as a catalyst for the local population to explore and prepare for life beyond cheap oil and gas, and to respond proactively to climate change. Transition Towns bring together local communities through specific projects, and in a social context, to take positive action, such as minimising waste, promoting cycling and growing local food. Central to the 'Transition' process is the concept of resilience. Transition Streets, one initiative launched in Totnes, is designed to support neighbourhoods to adopt energy and water efficiency, reduce waste and promote sustainable travel and food consumption.

In order to assess the potential health and well-being co-benefits of this initiative, we carried out a Health Impact Assessment (HIA), which is an evidence-based process that aims to predict the positive and negative impacts of a strategy, proposal or development. The HIA process provides an opportunity to promote sustainable communities, by ensuring that new strategies and developments are considered in the context of their contribution to the health and well-being of local populations.

The findings highlight the possible associated well-being benefits of engagement in such an initiative, and the potential for building social capital. The initiative emphasises positive changes in *lifestyles* through, for example, healthy eating and increased physical activity. Participants in the scheme reported associated changes in lifestyle, such as growing their own produce and sharing vegetables. An increase in *social cohesion* through neighbours meeting and socialising together is a positive impact for those engaged, with the potential for ongoing gains. Participants report *skills development* and *increased knowledge* of climate change issues, which in turn leads to behaviour change. This could be transferred into the wider community through skills sharing. One positive impact is the potential for *greater disposable income* due to cost savings from increased energy and water efficiency and referral to other housing repair funding schemes that provide insulation improvements. Increased warmth and reduction of damp could provide *health benefits*, particularly for those in fuel poverty.

This case study demonstrates the health and well-being co-benefits of living more sustainability. Transition initiatives are now being integrated into higher education, where these benefits could be experienced by students. For example, Transition University St Andrews includes a community garden and promotes active living through cycling and eating nutritious local food.

*Janet Richardson is Professor of Health Service Research at Plymouth University. Dr Amy Nichols is Lecturer in Nursing Studies at Plymouth University. Tina Henry is Head of Improvement/Urgent Care for NHS Devon.*

## The well-being of students

### *Changes in demographic and the student cohort*

As the social and demographic mix of students at HEIs has diversified, so, too, has the range of mental and physical well-being issues to which students are susceptible. Student stress and anxiety derive from 'a range of underpinning economic, lifestyle, academic, environmental and service-related factors'. Students often struggle with 'lack of money, time, information and advice, eating and living generally unhealthy lifestyles, and ... a wide range of symptoms such as headaches, irregular sleeping patterns, allergies and relationship problems' (Dooris, 2001: 56). It is now a decade since that analysis was conducted; the pressures since that time are likely to have worsened, not improved, although university support systems have burgeoned in recognition of the issue.

Kitzrow (2003), in tracking changes in student mental health, found that, from the initial problems of adjustment and individuation that were seen from the 1950s to early 1980s, the next two decades witnessed more severe concerns, including 'suicidality, substance abuse, history of psychiatric treatment or hospitalisation, depression and anxiety' (Pledge *et al.*, 1998: 387).

As a result of the macroeconomic situation, the demographics of undergraduate students in the HE sector will likely continue to change over the ensuing years. The NUS (2010: 32) predicted that the percentage of eighteen- to twenty-year-olds entering full-time programmes will drop by some 13 per cent from 2009 to 2019, while the number of older age groups, mainly on part-time undergraduate courses, is expected to rise over the same period. This potential change in study mode and age range is likely to contribute further to the breakdown of traditional cohort-based programmes, creating a lack of the informal social mechanisms that help to provide peer support during HE programmes. Combine this with the increasing social diversity of people achieving university places, and the breadth of student support needs will continue to increase significantly.

## *Financial pressures*

Over recent years, the financial pressures that impact the decision to study at university level and the student experience have changed significantly. Increased tuition fees and living costs, and the decline in graduate employment openings, add to potential student mental health issues. Further, the frequent financial need for part-time working can mix with timetabling pressures to result in more students missing classes or working late at night, increasing stress and risking illness and, sometimes, personal safety.

Research by NUS Scotland (2011:6) confirms that, after exams and assessments, career prospects are the most stressful aspect of studentship – with 75 per cent finding it 'reasonably' or 'very' stressful. Over half also experience stress due to financial pressures, while a third feel dealing with debts to be stressful. This study also highlighted that the main support for students comes from family, friends and their GP, rather than from within their institution, indicating the need for enhanced institutional responses.

HEIs, therefore, need to go beyond the narrow focus of students as future workers and develop a transformative approach to higher education that 'serves the dual purpose of enhancing both personal and collective well-being' (Dooris and Doherty, 2010: 14). This involves thinking beyond the traditional issues faced by students and looking to address their well-being holistically through new ways of engagement and support. Accordingly, the sustainable university will need to provide enhanced support to better assure students' physical safety and mental well-being.

## New systems of support

Community involvement through volunteering, such as horticultural and biodiversity engagement projects, have the potential to act as a destressing mechanism (Forest Research, 2010; Bell *et al.*, 2008; Guite *et al.*, 2006; Weldon *et al.*, 2007).

Volunteering on such projects not only helps students with their mental well-being but also gives them the opportunity to take part in physical activity (O'Brian *et al.*, 2010) and potentially, depending on the project, to source locally produced food, all helping to improve students' eating and living lifestyles. This multi-angled approach fits well with the sustainable university, as such community participation helps to improve social and environmental aspects, while simultaneously empowering those who use sites and services, helping to reduce maintenance costs and enhance user benefit.

With the sharp rise in student numbers over recent years has come a new process of 'studentification' whereby the vast number of students make up a significant proportion of the population of a town. This can lead to negative feelings between students and local residents, affecting the well-being of both groups and, in the long term, the sustainability of the university. Universities, along with other stakeholders, must manage and integrate students into neighbourhoods in order to address sustainability and well-being issues (Dooris and Doherty, 2010).

Physical wellbeing is also important, including safety, alcohol, drugs, smoking, healthy eating, and sexual health; one or more of these out of balance can impact mental health. The NUS (2011) highlight the prevalence of violence and harassment against female students with over two-thirds of respondents to their well-being survey reporting verbal or non-verbal harassment in and around their institution. The study also found that the majority of perpetrators were students studying at the same institution, highlighting the need for HEIs to address violence and harassment within their student body in order to improve the safety of all students (2011: 4).

Orme and Dooris suggest that universities have the power to 'change hearts and minds to make a real and recognisable difference to this and future generations' (2010: 434), and, as such, recommend they develop student values, knowledge and understanding in ways that that will have significant opportunities for long-term impact on public health and sustainability in families, communities, workplaces and society as a whole.

## The learning and working environment: a key contributor to health and well-being

A university is an environment in which students are not only educated, but also develop personally and socially (Dooris, 2001). Universities occupy a multiplicity of spaces that change as ideologies, processes and pedagogy change. The people who occupy them and the diverse nature of their activities each have their own resource, environmental and person requirements. That multiplicity ranges from costly and inflexible scientific laboratories to lecture halls and classrooms, social spaces and staff offices. The variety of working and learning environments can present a challenge to those responsible for their efficient management. The pace of change within which education delivery takes place – including current trends towards blended learning, off-site delivery and the refocusing that is occurring across a range of institutions towards reintroducing small group and individual tutor sessions – all place demands on the strategic approach to the development and configuration of the university estate.

One of the fundamental requirements for both the design and management of workspaces is to allow people to work and perform optimally and under comfortable conditions (Roelofsen, 2002). Health and occupant well-being is an important part of sustainability; however, it is one of the least considered aspects of sustainable design. In OECD countries, people spend 90 per cent of their time indoors, with the economically active population spending around 60 per cent of their waking life in the workplace environment (OECD, 2003). As a result, the importance of healthy university workspaces for both staff and students cannot be overestimated, yet it remains subordinate to aspects such as cost and specialist function.

Relationships between the built environment and health-related issues have been difficult to establish, but evidence is growing that the pathways and mechanisms by which the built environment affects health have increased (Rao *et al.,* 2007). The health implications of poor indoor environmental quality and problems are increasingly well documented. Yet, while these may be addressed at the point of initial construction, in practice, addressing such issues in the *management* of the workplace environment is more challenging, as occupancy requirements change and spaces may no longer be used for their original purpose or may be used inappropriately by their occupants.

The physical and environmental factors that affect the quality of the internal environment for the occupants include, *inter alia,* ergonomics, sound levels, air quality, temperature and humidity levels. Personal control over these factors can be critical in achieving satisfaction. Further, they relate to up to a quarter of the total average absenteeism (Roelofsen, 2002). Such improvements are co-beneficial: not only is health improved, but many are linked to climate change mitigation actions (Haines *et al.*, 2009).

The provision of office space for academics changed little over the past century: typically, individual offices feeding off corridors. Since 2000, however, attempts have been made to create new types of academic office environments, based on the now-accepted principles of open-plan office layouts. The aim of these new workspaces is to offer a more creative and collaborative working environment, thus encouraging the promotion of knowledge flow between occupants, to support greater creativity and innovation in research and teaching (Parkin *et al.,* 2011).

However, creative and cognitively demanding tasks require privacy and opportunity for quiet reflection. Given the nature of academic work, a single space may not be able to provide adequately for such a variety of activities. One solution is to provide a range of activity space within the academic office environment (Parkin *et al.*, 2011). This idea has been developed both from other types of creative workplaces and from other student academic spaces. Many learning resource centres have now introduced several different types of workspace to suit various work requirements and personal preferences. For individual study, silent or quiet work areas are allocated, each with shared work desks and single study booths. These are accompanied by several multi-activity meeting rooms and studio spaces for group work and meetings, including cafés. This may well become the pattern for the future sustainable university.

The relationship between buildings and health environments is not just a design consideration. The management and maintenance of workspaces is regularly identified as one of the leading problems (Tong and Wilson, 1990). Poorly resourced and ill-maintained workspaces can lead to a variety of psychological and mental health issues for staff, including staff feeling unappreciated and undervalued by their institution (Amabile *et al.*, 1996).

More widely in the general office sector, the provision of so-called 'green offices' has been found to be positively correlated with both productivity and well-being. Gensler (2005) argued that functionally flexible offices increase employee satisfaction and productivity, reduce stress and ill health, and increase staff retention. These findings are reinforced by Morgan and Anthony (2008), who found strong links between working environments and individual and organisational performance. More recently, Armitage *et al.* (2011) found, in an Australian study, that Green Star rated (i.e. energy efficient) buildings increase both user satisfaction and the health and well-being of employees. However, noting that the estates of many universities are old, even simple measures can be introduced to help employee well-being. Smith and Pitt (2009) found that the introduction of plants to remove or reduce indoor air pollutants provides psychological and physical health benefits for employees.

For students, the relationship with the quality and configuration of space is also critical; good spaces promote interaction and facilitate learning, whereas dull oppressive spaces do not. Kingston University's experimental C-SCAIPE project has proven, in student evaluations, to lead to feelings of well-being and the creation of a real sense of community. In light of this, the university plans to create other social learning spaces where students can develop a sense of belonging and engage in a variety of activities.

In summary, the link between buildings, their design, management, configuration and use has a perceived and actual influence, both on the physical and mental well-being of those who use them.

## Recommendations for the sustainable university

For the university seeking to pursue sustainability goals, the well-being of all its stakeholders should be a key responsibility – not just for managers, but shared. Every individual, being an organic part of the whole, should consider their own and others' well-being in recognition of the interdependency that exists between staff and students and of the institutional impacts that ensue. We now summarise and conclude under the three main headings considered above, but in reverse order.

### The estate

Almost as a prerequisite for any consideration of social sustainability, the physical environment must be considered. It needs to be designed, adapted and managed to generate feelings of well-being. This can be achieved through careful design, refurbishment and maintenance. The use of natural light, access to green spaces, the incorporation of environmental technologies, and individual controls are all important. Beyond this, appropriate and sensitive management and maintenance can enhance the occupational experience. Possibly, the most important way in which well-being can be promoted through buildings is by adapting and responding to user needs, and by recognising and respecting that universities are diverse organisations with complex and ever-evolving needs.

Academic staff members need social spaces in which to develop ideas and debate; they also need quiet spaces in which to think, write, research and tutor. If this is to be provided off-campus, through flexible working practices, then timetabling and other workload factors need to be carefully devised so that responsibilities for workspace provision are not merely exported. In addition to good quality formal teaching spaces and Learning Resource Centres, students respond well to the provision of small-scale learning spaces that foster community and collaborative learning. All users benefit from good interior design, including the creative and appropriate use of colour and smaller details, such as the incorporation of plants. Elements of 'green' technologies can promote perceptions of well-being, possibly connected to the continually heightening awareness of environmental concerns among the university's citizens. The final word on the provision of space should go to Oommen *et al.* who argue that

> if managers fail to address the psychological dimensions when planning facilities, complex issues like low job satisfaction and decreased work productivity will arise. Moreover, a workplace has to be a place where employees are satisfied when carrying out their work.
>
> (Oommen *et al.*, 2008: 42)

We would argue that this applies equally to stakeholders within the sustainable university.

## The new economics foundation's work on well-being and sustainability

The London-based think-tank, the new economics foundation (nef), has run a programme of work on well-being since 2001. The work builds on our belief that societies need to *measure what matters*, because headline indicators both frame what we regard as important and provide incentives for political action.

nef has looked at the evidence about what drives well-being for individuals. We have reviewed this evidence for policy-makers, and produced a set of evidenced-based messages about everyday actions that promote well-being. The Five Ways to Well-being are:

### *Connect...*

With the people around you. With family, friends, colleagues and neighbours. Think of these as the cornerstones of your life and invest time in developing them. Building these connections will support and enrich you every day.

### *Be active...*

Go for a walk or run. Step outside. Cycle. Garden. Dance. Exercising makes you feel good. Most importantly, discover a physical activity you enjoy.

### Take notice...

Be curious. Savour the moment, whether you are walking to work, eating lunch or talking to friends. Be aware of the world around you and what you are feeling. Reflecting on your experiences will help you appreciate what matters to you.

### Keep learning...

Try something new. Rediscover an old interest. Take on a different responsibility at work. Set a challenge you will enjoy achieving. Learning new things will make you more confident as well as being fun.

### Give...

Do something nice for a friend, or a stranger. Thank someone. Smile. Volunteer your time. Join a community group. Look out, as well as in. Seeing yourself, and your happiness, linked to the wider community can be incredibly rewarding and creates connections with the people around you.

More information about nef's work on well-being can be found at http://www.neweconomics.org

*Juliet Michaelson is Senior Researcher at the Centre for Well-being at nef.*

### Students

Students are the creators of tomorrow's and, in many cases, today's society. They need to undertake their studies in an environment and with the support of academics who foster their creativity and learning, and equip them to be active citizens for a more sustainable society. For this, they need to be resilient and able to withstand pressure without that leading to undesirable stress and loss of well-being. As they have become more diverse in their backgrounds and their social and domestic pressures, so too many students face difficult financial situations, which can impact negatively on their health and well-being. As this happens, their chances of fulfilling their intellectual potential and entering into good jobs in which they can make a positive contribution to society are reduced.

Evidence shows that universities take very seriously their obligations in this regard, with many seeking to heighten awareness of the student voice and to empower them in their learning. Further, they are seeking to facilitate learning, both on and off campus through the enhancement of VLEs, the introduction of expanded study facilities and the greater availability of services. Specifically, in relation to supporting social and health-related facilities, many universities, such as Kingston, provide a wide range of support for students with special needs, as well as childcare facilities, stress and health counselling, and sports and gym facilities. However, what is more difficult to provide is a real sense of cohort, particularly as class sizes have increased and investment in blended learning has led to the need to increase time spent preparing 24-hour-a-day accessible materials at the expense of small group teaching and, in some cases, the demise of one-to-one tutorials.

Consequently, the student body has faced feelings of isolation and stress and, in the face of financial hardship, many have turned to long hours in paid employment to support its studies. Not only is this destructive of success but no amount of university, central-led support, in terms of handling money and coping with stress, will help if the student is either at work or feeling isolated and depressed. If this happens, they become in danger of becoming very disconnected, despite the prevalence of social media. Promotion of volunteering and other forms of human-based, extra-curricular activities can be crucial as a counter-balance but are not likely to be taken up by the time-poor.

The basic generic infrastructures are increasingly in place for the support of students. But we face a choice: do we rely increasingly on electronically-based support, or do we return to greater human contact between academics and students? Both require investment; possibly both are needed.

## *Staff*

The greatest emphasis in this chapter has been on the issues concerning academic staff well-being. Despite an extensive literature on the causes of stress and burnout in the workplace, only limited recent literature relates to academics in higher education. What research there is indicates that they are a high-risk group, comparable to teachers and health sector workers. Consideration of burnout is important in relation to academic staff, primarily because those suffering from burnout are likely to be unable to fully motivate students, even before their own efficiency becomes impaired.

The emphasis on positive well-being promotion, advocated by the new economics foundation (Aked *et al.,* 2008), and now seen at government level with the inauguration in 2011 of the National Happiness Survey, has been taken up by many organisations. Universities, too, have been very proactive in seeking ways to support all their staff. Training schemes providing support for professional, social and life skills abound. Occupational health departments work hard to support colleagues who are struggling, and reward schemes have been introduced to encourage and celebrate success.

Yet, despite all these mechanisms, two major concerns must be addressed: first, the issue of well-being remain difficult to measure and, in a world governed by metrics, it is hard to argue for investment in something for which the indicators of success are not well developed. Those seeking to persuade senior management to invest in well-being, then, may face challenges in securing funds. One recommendation here is that established HEI sustainability evaluations include well-being. Second, the causes of stress and burnout need addressing. Excessive workloads, feeling the loss of autonomy, and a perception of unsupportive management, particularly at times of change, feature highly in all analyses. Given the changes that UK HEIs are facing, most institutions are unable to reduce staff workload, and a strengthening of bureaucracy associated with quality processes and performance measurement exacerbates the situation. Against that, academic staff experience high levels of job satisfaction and relatively high levels of autonomy, both of which provide positive support to resilience against stress. However, if the sustainable university does truly hold to social sustainability principles, investment in countering 'workload creep', consequent on the changes outlined in this chapter, is a prerequisite.

So, finally, we finish by repeating the quotation that we set out at the beginning of this chapter: '[I]t is widely recognized that health is determined by a range of environmental, social and economic influences and that the health of people, places and the planet are interdependent' (Orme and Dooris, 2010: 425). If any university does not address this holistically, it is not a sustainable university.

## Notes

1   See: http://www.kingston.ac.uk/virtual-tour/penrhyn-road/c-scaipe
2   For example, at Kingston there has been a wide roll-out of discipline-led Academic Support Centres, operating on both a drop-in and formal session basis, which have become, year on year, further integrated to the student offer. However, as the Centres are academically staffed, with additional trained student support, academic staff workloads are increased, although they do provide a staff development opportunity to colleagues. The key to their success at Kingston has been closely related to context specificity.
3   The 2012 Annual National Student Satisfaction Survey (NSS) (www.thestudent survey.com/) of final-year undergraduates measures a wide range of indicators. The results have wide-ranging implications, from national league tables to the identification at a very local level of satisfaction to which individual academics may be called to account.
4   For more information on the Sustainability Hub at Kingston, see Taylor, this volume. For a discussion on universities and the Transition Town Movement, see Harder and White, this volume, and Higgins *et al.*, this volume.

## References

Abouserie, R (1996) 'Stress, coping strategies and job satisfaction in university academic staff', *Educational Psychology: An International Journal of Experimental Educational Psychology*, 16 (1): 49–56

Aked, J, Marks, N, Cordon, C and Thompson, S (2008) *Five Ways to Well-being: A Report to the Foresight Project on Communicating the Evidence Base for Improving People's Well-being*, London: new economics foundation

Amabile, T, Conti, R, Coon, H, Lazenby, J and Her, M (1996) 'Assessing the work environment for creativity', *The Academy of Management Journal*, 39 (5): 1154–84

Armitage, L, Murugun, A and Kato, H (2011) 'Green offices in Australia: a user perception survey', *Journal of Corporate Real Estate,* 13 (3): 169–80

Bakker, A, Demerouti, E, de Boer, E and Schaufeli, W (2003) 'Job demands and job resources as predictors of absence duration and frequency', *Journal of Vocational Behaviour,* 62: 341–56

Bell, S, Hamilton, V, Montarzino, A, Rothnie, H, Travlou, P and Alves, S (2008) *Green Space and Quality of Life: A Critical Literature Review,* Stirling, Scotland: Greenspace

Bilge, F (2006) 'Examining the burnout of academics in relation to job satisfaction and other factors', *Social Behaviour and Personality,* 34: 1151–60

Boyd, C, Bakker, A, Pignata, A, Winefield, A, Gillespie, N and Stough, C (2011) 'A longitudinal test of the job-demands resource model among Australian university academics', *Applied Psychology: An International Review,* 60 (1): 112–40

Brouwers, A and Tomic, W (2000) 'A longitudinal study of teacher burnout and perceived self-efficacy in classroom management', *Teaching and Teacher Education,* 16 (2): 239–53

Buchholz, T (2011) *Rush: Why You Need and Love the Rat Race,* London: Penguin

Cartwright, S and Cooper, C (2011) *Innovations in Stress and Health,* Basingstoke: Palgrave Macmillan

Cooper, C (ed.) (1998) *Theories of Occupational Stress,* Oxford: Oxford University Press

Court, S and Kinman, G (2008) *Tackling Stress in Higher Education,* London: University and College Union

Dooris, M (2001) 'The "health promoting university": a critical exploration of theory and practice', *Health Education,* 101 (2): 51–60

Dooris, M and Doherty, S (2010) 'Healthy universities: current activity and future directions: findings and reflections from a national-level qualitative research study', *Global Health Promotion,* 17 (3): 6–16

Doyle, C and Hind, P (1998) 'Occupational stress, burnout and job status in female academics', *Gender, Work and Organization,* 5: 67–82

EcoCampus (2012) *EcoCampus.* Available at: www.ecocampus.co.uk/ (accessed 15 April 2012)

Farber, B (2000) 'Introduction: understanding and treating burnout in a changing culture', *Journal of Clinical Psychology,* 56 (5): 589–94

Forest Research (2010) *Benefits of Green Infrastructure,* Farnham: Forest Research

Gensler (2005) *These Four Walls: The Real British Office,* London: Gensler

Gill, R (2009) 'Breaking the silence: The hidden injuries of neo-liberal academia', in Flood, R and Gill, R (eds) (2009) *Secrecy and Silence in the Research Process: Feminist Reflections,* London: Routledge

Guite, H, Clark, C and Ackrill, G (2006) 'The impact of the physical and urban environment on mental well-being', *Public Health,* 120 (12): 1117–26

Haines, A, McMichael, A, Smith, K, Roberts, I, Woodcock, J, Markandya, A, Armstrong, B, Campbell-Lendrum, D, Dangour, A, Bruce, N, Tonne, C, Barrett, M and Wilkinson, P (2009) 'Public health benefits of strategies to reduce greenhouse-gas emissions: overview and implications for policy makers', *The Lancet,* 374: 2104–14

Hastings, R (2002) 'Do challenging behaviours affect staff psychological well-being? Issues of causality and mechanism', *American Journal on Mental Retardation,* 107 (6): 455–67

Higher Education Funding Council for England (HEFCE) (2005) *Sustainable Development in Higher Education,* Bristol: HEFCE

Higher Education Funding Council for England (HEFCE) (2008) *A Strategic Review of Sustainable Development in Higher Education in England*, Bristol: HEFCE

HSE (2011) *Working Together to Reduce Stress*, London: Health and Safety Executive (HSE)

Karasek, R (1979) 'Job demands, job decision latitude, and mental strain: implications for job redesign', *Administrative Science Quarterly*, 24: 258–308

Kingston University (2012) *Wellbeing Services*. Available at: www.kingston.ac.uk/health-and-counselling-service/wellbeing-services/ (accessed 15 April 2012)

Kitzrow, M (2003) 'The mental health needs of today's college students: challenges and recommendations', *NASPA Journal*, 41 (1): 167–81

Maslach, C (2006) 'Understanding job burnout', in Rossi, A, Perrewé, P and Sauter, S (2006) *Stress and Quality of Working Life: Current Perspectives in Occupational Health*, Charlotte: Information Age Publishing, pp. 99–114

Maslach, C and Jackson, S (1981) 'The measurement of experienced burnout', *Journal of Occupational Behaviour*, 2: 99–113

McClenahan, C, Giles, M and Mallett, J (2007) 'The importance of context specification work stress research: a test of the demand-control-support model in academics', *Work and Stress*, 201: 85–95

Morgan, A and Anthony, S (2008) 'Creating a high-performance workplace: a review of issues and opportunities', *Journal of Corporate Real Estate*, 10 (1): 27–39

Naylor, C and Appleby, J (2012) *Sustainable Health and Social Care: Connecting Environmental and Financial Performance*, London: The King's Fund

NICE (2009) *Promoting Mental Wellbeing at Work: Business Case*, London: National Institute for Health and Clinical Excellence (NICE)

NUS (2010) *Living Together, Working Together*, London: National Union of Students

NUS (2011) *Hidden Marks: A Study of Women Students' Experiences of Harassment, Stalking, Violence and Sexual Assault*, London: National Union of Students

NUS Scotland (2011) *Think Positive about Student Mental Health: Silently Stressed*, Edinburgh: National Union of Students Scotland

O'Brian, L, Williams, K and Stewart, A (2010) *Urban Health and Health Inequalities and the Role of Urban Forestry in Britain: A Review*, Farnham: Forest Research

OECD (2003) *Environmentally Sustainable Buildings: Challenges and Policies*, Paris: OECD Publishing

OECD (2009) *Education at a Glance*. Available at: www.oecd.org/dataoecd/41/25/43636332.pdf (accessed 15 April 2012)

Oommen, V, Knowles, M and Zhao, I (2008) 'Should health service managers embrace open plan work environments? A review', *Asia Pacific Journal of Health Management*, 3 (2): 37–43

Orme, J and Dooris, M (2010) 'Integrating health and sustainability: the higher education sector as a timely catalyst', *Health Education Research*, 25 (3): 425–37

Parkin, J, Austin, S, Pinder, J, Baguley, T, and Allenby, S (2011) 'Balancing collaboration and privacy in academic workspaces', *Facilities*, 29 (1): 31–49

People and Planet (2012) *Green League*. Available at: http://peopleandplanet.org/greenleague (accessed 17 April 2012)

Pledge, D, Lapan, R, Heppner, P, and Roehlke, H (1998) 'Stability and severity of presenting problems at a university counselling center: a 6-year analysis', *Professional Psychology Research and Practice*, 29 (4): 386–89

Prosser, M and Trigwell, K (1999) *Understanding Learning and Teaching: The Experience in Higher Education*, Birmingham: SRHE/OUP

Rao, M, Prasad, S, Adshead, F and Tissera, H (2007) 'The built environment and health', *The Lancet,* 370: 1111–13

Roelofsen, P (2002) 'The impact of office environments on employee performance: the design of the work place as a strategy for productivity enhancement', *Journal of Facilities Management*, 1 (3): 247–64

Safaria, T, bin Othman, A and Wahab, M (2011) 'The role of leadership practice on job stress among Malay academic staff: a structural modelling analysis', *International Education Studies*, 4 (1): 90–100

Sainsbury Centre for Mental Health (2007) *Mental Health at Work: Developing the Business Case,* London: Sainsbury Centre for Mental Health

Sayce, S and Farren Bradley, J (2011) 'Educating built environment professionals for stakeholder engagement', in Rogerson, R, Sadler, S, Green, A and Wong, C (eds) *Sustainable Communities: Skills and Learning for Place Making,* Hatfeld: University of Hertfordshire Press

Smith, A, Johal, S, Wadsworth, E, Davey Smith, G, Peters, T (2000) *The Scale of Occupational Stress: The Bristol Stress and Health at Work Study,* Norwich: HSE Books

Smith, A and Pitt, M (2009) 'Sustainable workplaces: improving staff health and well-being using plants', *Journal of Corporate Real Estate*, 11 (1): 52–63

Sparks, K, Cooper, C, Fried, Y and Shimon, A (1997) 'The effects of hours of work on health: a meta-analytic review', *Journal of Occupational and Organizational Psychology*, 70: 391–408

Tong, D and Wilson, S (1990) 'Building related sickness', in Curwell, S, March, C and Venables, R (eds) *Buildings and Health,* London: RIBA Publications

Tytherleigh, M, Webb, C, Cooper, C and Ricketts, C (2005) 'Occupational stress in UK higher education institutions: a comparative study of all staff categories', *Higher Education Research and Development,* 24 (1): 41–61

UCL Institute of Health Equity (2010) *Fair Society, Healthy Lives: The Marmot Review.* Available at: www.instituteofhealthequity.org/projects/fair-society-healthy-lives-the-marmot-review (accessed 3 June 2012)

UCU (2010) *UCU Policy Briefing: Student–teacher Ratios in Higher and Further Education.* Available at: www.ucu.org.uk/media/pdf/q/6/ucupolicybrief_ssratios_may10.pdf (accessed 15 April 2012)

Watts, S and Robertson, N (2011) 'Burnout in university teaching staff: a systematic literature review', *Educational Research,* 53 (1): 33–50

Watts, W, Cox, L, Wright, L, Garrison, J, Herkimer, A, and Howze, H (1991) 'Correlates of drinking and drug use by higher education faculty and staff: implications for prevention', *Journal of Drug Education*, 21: 43–64

WECD (1987) *Our Common Future: Report of the World Commission on Environment and Development*, New York: United Nations

Weldon, S, Bailey, C and O'Brian, L (2007) *New Pathways to Health and Well-being: Summary of Research to Understand and Overcome Barriers to Accessing Woodland,* Edinburgh: Forestry Commission Scotland.

Winefield, A, Gillespie, N, Stough, C, Dua, J, Hapuarachchi, J and Boyd, C (2003) 'Occupational stress in Australian university staff: results from a national survey', *International Journal of Stress Management*, 10: 51–63

World Health Organization (WHO) (2007) *The World Health Report 2007: A Safer Future: Global Public Health Security in the 21st Century.* Available at: www.who.int/whr/2007/en/index.html (accessed 15 April 2012)

# Part III
# Institutional change

# 11 Whole institutional change towards sustainable universities

## Bradford's Ecoversity initiative

*Peter Hopkinson and Peter James*

This chapter adds to existing discussion of whole institutional change for sustainability (for example, by Bartlett and Chase, 2004; Corcoran and Wals, 2004; Hopkinson, 2010; Jones *et al.*, 2010) by examining the University of Bradford's Ecoversity initiative. This began in 2005 with the objective of achieving significant progress towards sustainability in all aspects of the university's life simultaneously, including the estate, the curriculum, and the staff and student experience of campus life. The initial four sections below describe four stages: *Genesis* (2000–05), *Making Connections* (the first two years of the initiative, in 2005–07 when it was estates-focused), *Transformation* (the core period of 2007–10, when the university received funding for a national flagship project to make the entire student experience more sustainable so that Ecoversity's centre of gravity shifted towards academic and student matters), and lastly, *Consolidation* (the post-2010 response to diminished funding and new institutional challenges). A final section relates the initiative's experience, achievements and failures in relation to key questions within the education for sustainable development (ESD) literature, such as whether ESD is a coherent, intellectually robust body of knowledge or a moralising crusade which does not belong in higher education (Knight, 2005) and/or a high-level concept that masks significant differences between curriculum areas and is therefore difficult to implement consistently (Hopkinson and James, 2010); whether ESD must be transformational to be successful (Orr, 1994, Sterling and Gray-Donald, 2007); whether top-down or bottom-up approaches are more likely to be successful in ESD implementation (Wade, 2008); and what the relationship is between ESD-related curriculum change and other aspects of university life (M'Gonigle *et al.*, 2006). Firstly though, the following section discusses the theoretical background to Ecoversity's activities.

## Theory and practice

As noted below, a key feature of Ecoversity was the attempt to implement the precepts of the education for sustainable development (ESD) literature within the university. However, as space is limited, and other chapters discuss this literature more fully, we focus here on socio-cultural theory, which proved helpful in understanding many of the early problems experienced by Ecoversity, and in shaping subsequent actions. As we discuss in the final section, the implication of this approach is that the implementation of ESD is less unique than its theoretical literature often suggests, and that more might be achieved in practice if it is conceptualised as a curriculum change initiative with many similarities with others. We also believe that, although most socio-cultural-theory-influenced work on higher education has focused on curriculum change, it can also be applied without modification to consider change for sustainability in general within universities.

Socio-cultural theory 'takes as its unit of analysis *social practice*, instead of (for example) individual agency, individual cognition, or social structures' (Bamber *et al.*, 2009: 8). Analyses of social practices relevant to curriculum change within higher education have highlighted factors such as discipline, profession, role and other factors (Alvesson, 2002; Bamber *et al.*, 2009; Trowler, 2008). They have also noted that universities are generally 'change averse' and that barriers which are often highlighted in the ESD literature, such as disconnected structures and decision-making processes, 'turf wars', disciplinary resistance, lack of prioritization and a preference for talking about actions rather than actually achieving them are generic to most curriculum change (Fullan and Scott, 2009; Weick, 1976).

Socio-cultural theory highlights the importance of national, regional and local context in shaping university life (Bamber *et al.*, 2009). In the case of Bradford, one important contextual factor was serious industrial decline from the 1960s on, so that the city now has a high proportion of its population on low incomes. The university also has a large proportion of its intake from ethnic minorities, reflecting the city's social composition. These factors result in a relatively high proportion of students living at home. (This distinctive undergraduate pattern is also due to the university's pioneering efforts to improve participation beyond traditional university recruits in the UK.) The University of Bradford itself is small to medium sized (with 11,000 students) and therefore tries to differentiate itself from peers through an emphasis on graduate employability, and on focused research and teaching areas. The university is also relatively unusual in having 70 per cent of its students on professionally accredited courses, and around half on STEM (science, technology, engineering and maths) courses.

Bamber *et al.* (2009) argue that successful change requires: a) a 'ladder of implementation' in which actions are occurring simultaneously at all levels of the institution, and thereby reinforcing each other, and b) 'frameworks for action' among key players, which take contextual and social practice factors into account when planning and implementing actions (see Figure 11.1 for a schematic of these and related topics highlighted by Bamber *et al.*). They identify four key aspects of these frameworks – understanding change; theories of change; enhancement identities; and reflexive questioning.

*Understanding change:* (by those who are driving it) – this includes awareness of the culture, social practices and activity systems (such as course management

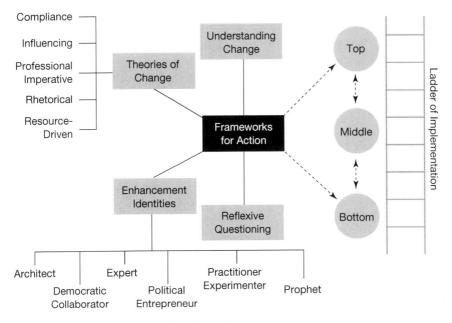

*Figure 11.1* Key elements of socio-cultural theory

approval mechanisms) within the institution and how these shape attitudes and behaviour; identifying and understanding the communities of practice related to activity systems; recognising that significant change involves a 'ladder of implementation' with varying actors and actions at different levels within the organisation; and understanding the importance of non-formal learning and knowledge production.

*Theories of change*: These are often implicit but considering them explicitly can help to better understand and overcome barriers, and to identify opportunities that might otherwise be overlooked. Bamber *et al.* (2009: 186–7) highlight four practice-based exemplars:

1   influencing (where core individuals or groups influence others through 'contagion');
2   resource-driven/dependent (where rational actors respond to changes in incentives or disincentives);
3   rhetorical support (where people respond to authority figures and structures);
4   professional imperative (where people respond to external expectations that form an important part of their social identity, e.g. as a chemist, or as an educator).

There is also a fifth implicit theory of change, which is that of compliance with requirements, regulations, rules etc. that are introduced by authority figures or groups that are perceived to have the legitimacy to promulgate and enforce them. As will be seen, all of these were utilised in the Ecoversity initiative, with a different mix over time, and between different areas.

*Enhancement identities*: These are mindsets, visions of the world and ways of interacting with others. Bamber *et al.* (2009) identify six – prophet, expert, democratic collaborator, political entrepreneur, practitioner experimenter and architect – but stress that there can be more, and that identities are not mutually exclusive, as individuals can move between them in different circumstances or over time. This was very much the experience of Ecoversity, which required a large element of prophecy and entrepreneurship to get off the ground, but a great deal of collaboration and experimentation as it began to be implemented.

*Reflexive questioning*: Dialogue and openness is essential in understanding and influencing the social practices of others, requiring empathy, and an ability and willingness to change in response, by the change originator (Revans, 1980). Bamber *et al.* (2009) suggest a number of questions to encourage this, which, with some adaption, provided a very helpful checklist to Ecoversity practitioners.

The following sections describe Ecoversity's 'frameworks for action'.

## The Ecoversity initiative: genesis, 2000–05

Sustainability was not a high priority within the University of Bradford as a whole, and among most of its staff, during the 1990s and early 'noughties'. Generally, it was not seen as relevant to either the pressing problems of unemployment, poverty and community relations within the locality, or to the institution's focus on employability, or to its STEM-based culture.

Nonetheless, several (largely unconnected) initiatives did occur as a result of the enthusiasm and interest of individuals acting as 'experts' or 'practitioner experimenters' (for example, in adapting curricula to new requirements), and through more or less forced responses to the growing external 'steering' with regard to sustainability from government pronouncements, regulations and other sources (Broadbent *et al.*, 2010). These subsequently provided important foundations for Ecoversity. They included:

- council-led Local Agenda 21 activities (arising from the UK government's commitments at the 1992 Rio conference) in which the university was involved as a major economic player in the city area;
- modification of curricula to reflect sustainability concerns in disciplines that had obvious connections with them (for example, Civil Engineering, Environmental Science, Peace Studies);
- analysis of the university's energy consumption and environmental footprint, initially through student projects and subsequently through estates-led initiatives such as participation in the Carbon Trust's Higher Education Carbon Management Club and the work of a newly appointed Energy Manager;

- HEEPI (Higher Education Environmental Performance Improvement) – a national project funded by HEFCE (Higher Education Funding Council for England) to support environmental change in UK higher education, based on the experience of the co-authors of doing similar work in the private sector. This helped raised internal awareness of good practice elsewhere, and led to the university's first Environmental Manager appointment;
- growing interest amongst key estates personnel in the potential for campus regeneration and green buildings;
- proposals for an 'eco-mill' using derelict warehouses on the campus periphery to provide a base for environment-related teaching and innovation and to demonstrate the potential for sustainability-led regeneration.

In 2004–05, a number of factors enabled these initiatives to be linked into a broader vision of sustainability-led change, which connected with strategic organisational concerns and thereby mobilised additional interest and support. One was the publication of a HEFCE (2005) consultative report, *Sustainable Development in Higher Education*, which stressed the need for senior managers to pay more attention to the topics, and highlighted actions that could be taken. One consequence was the embedding of sustainability as one of six core values in the university's 2005–09 corporate strategy.

Discussion of what sustainability meant also highlighted the links between sustainability and another area of pressure from government, HEFCE and other bodies – which was social inclusion in higher education. This was particularly relevant to Bradford, as a city with considerable deprivation, and a high proportion of ethnic minorities and international students with strong connections to the global South (which also provided almost 20 per cent of the university's student intake). The university also had a problem at this time of students rejecting offered places, or not applying in the first place, because of perceptions (shared by many existing students) that the main campus was physically unattractive, with limited social and study space and no student 'life'. There was much truth in this, as most buildings dated from the 1960s and were in need of refurbishment, and much space was devoted to car parks.

As a result, an increasing number of people came to feel that Bradford's social context, academic profile and growing competence in environmental management and green building made it uniquely positioned to be an exemplar of how to connect all these areas through an integrated approach 'cemented' by the concept of sustainable development. A key figure in both creating this vision, and persuading key decision-makers that it was feasible, was the newly promoted Director of Estates. Through a combination of personal and – through the involvement of other estates staff – collective action, he was able to simultaneously play the role of 'prophet' (graphically summarised by two slides, one showing the current, then unimpressive, campus, and another showing what it could be within a decade), 'expert' (on the opportunities for new build and refurbishment on campus), and 'architect' (especially in creating connections between campus transformation and the importance of student experience for both recruitment and achieving sustainability).

## The Ecoversity initiative: making connections, 2005–07

Ecoversity was launched in late 2005 at the national Parliament building in London by the Vice-Chancellor. It had a senior management 'sponsor', a programme manager, and four strands – environment, community, education for sustainable development and economy – each with its own champion. Despite this broad remit, its early years were most strongly identified with estates-related activity. Planned (and eventually realised) capital projects such as a new sustainable student village and a new atrium with enhanced social and informal learning spaces on the main campus building – were both tangible and time-consuming and therefore attracted much of the attention and time of key stakeholders. The programme manager was also an estates employee (aka the Environmental Manager) and there was initially no comparable role for the curriculum and other non-estates dimensions of the scheme.

By June 2007, estates had rapidly progressed a number of large-scale projects, including new and refurbished buildings (with adoption of stretching targets under the UK BREEAM green building assessment scheme to signal environmental commitment); movement of (reduced) car parking from the campus centre to its periphery; and landscaping. However, a review also revealed that while there was widespread interest and enthusiasm for the vision (alongside some scepticism as to whether it could be achieved), many staff and students saw Ecoversity as failing to engage with them and falling short of its goals. Some pointed to very visible and highly symbolic areas, such as limited recycling facilities or obvious areas of energy and other wastage, as evidence of the gap between marketing and reality. Ecoversity insiders also had a sense that multi-stakeholder engagement and enthusiasm, and concrete actions outside the estates area, were held back by bureaucratic processes. The result was a gradual sense of slippage and failure to communicate and share the vision with the wider university.

These problems were related to an implicit change model based around resources (an assumption that investment in the physical environmental would automatically create changes in attitudes and behaviour, and rhetorical support from senior figures). However, the latter was undermined by the disconnect with actual actions on the ground, which were held back on the academic side by limited linkages with professional imperatives, and by few examples of practice-based exemplars that could be adopted by others. There were also no real resource incentives/disincentives for academics to become engaged in the process. This situation was related to the continued centrality of the more top-down 'prophet', 'expert' and 'architect' roles and limited development of the 'democratic collaborator', 'political entrepreneur' and 'practitioner experimenter' ones, which are important in engaging with potentially suspicious academics and in creating connections across the university, and between the different stages of the implementation 'ladder'. The problem was exacerbated by the adoption of a very formal project management approach to the Ecoversity task groups. Such approaches can, of course, have great value for managing well-defined projects in settings such as estates or administration. However, they can easily be seen as bureaucratic and managerialist by academics and therefore unsuited for the complexity of curriculum change.

## The Ecoversity initiative: transformation, 2007–10

A number of factors combined during 2007 to transform the reach and impact of the programme. The university was successful in obtaining large-scale project funding from HEFCE for an Ecoversity StuDent project, which aimed to create a 'learning laboratory' to examine whether and how a relatively mainstream university could transform the staff and student experience with regard to sustainability. A new vice-chancellor was also appointed whose personal as well as rhetorical support for Ecoversity was given credibility by his background at the University of Plymouth, a UK leader in Education for Sustainable Development (ESD). Coincidentally, the previous senior management champion, the Registrar, retired and was replaced as programme sponsor by the Pro-Vice Chancellor (PVC) for Learning and Teaching, creating a more direct channel for influencing academics.

The Ecoversity StuDent project enabled creation of a Sustainable Education Directorate (SED) in May 2007. This had five full-time staff – the Director of ESD (as a 100 per cent role, compared to part-time previously), a Student Engagement Coordinator, two researchers undertaking action and longitudinal research respectively, and a secretary. Ecoversity became increasingly identified with the SED and its PVC champion, both internally and externally, with a resulting shift in its centre of gravity towards academics and students. This new direction was enabled and cemented by a three-day, off-site 'Change Academy' exercise involving seven senior figures in shaping a strategic vision for the initiative and developing a stronger and more integrated implementation plan. This was subsequently reflected in the embedding of ESD within the university's 2009–14 Corporate Strategy.

## Formal curriculum change

There are many examples of how curricula in a variety of disciplines can be adapted to engage with sustainability agendas (Blewitt and Cullingford, 2004; Dawe *et al.,* 2005; Herrmann, 2000; Roberts and Roberts, 2007) and foster sustainable literacy (Forum for the Future, 2004; Forster, 2006; Cade, 2007; SDC, 2007). However, Dawe *et al.* (2005) highlighted many barriers, including an overcrowded curriculum; perceived irrelevance by academic staff; limited staff awareness and expertise; and limited institutional drive and commitment. Some critics have also seen ESD as 'a moralising agenda' which attacks academic freedom and the cherished value of higher education (Butcher, 2007).

To minimise controversy and resistance, Ecoversity sought a holistic and widely accepted 'road map' of what sustainability is as the basis for its actions. It found this in a UNESCO-informed framework (Hopkinson *et al.*, 2008a). This proved to be extremely helpful in highlighting areas that were already being addressed, and encouraging open and interesting conversations about future opportunities for development. It was especially useful in ensuring that Ecoversity was not seen as a purely environmental initiative by raising issues such as poverty, equality, health and faith which were often directly relevant and meaningful to the daily lives of Bradford students, both on campus and in the immediate communities in which they lived.

Great emphasis was also placed on helping academics to consider the implications of sustainability, and to experiment with responses to it, through non-coercive means. Ecoversity and ESD became an element at all new staff inductions, and within the mandatory professional development course for new Bradford lecturers. Examples of practice-based exemplars from elsewhere were publicised internally, and in later stages were brought into the university through the hosting of several national and international events on ESD. Small-scale pilot projects were also set up to introduce more ESD into targeted subject areas. These gave insight into important issues such as the very different disciplinary cultures, how individual schools addressed teaching and learning improvement, and their course approval and review structures.

The pilots reinforced a sense that Ecoversity needed to create 'native' capacity within each of the seven schools. The result was an 'academic pioneer' scheme involving two staff members in each school being paid for 0.2 full-time equivalent (FTE) of time in order to review existing ESD-related teaching and learning within their school curricula and to take actions to increase it. In addition to direct contact with the ESD Director, the pioneers also met as a group – initially on an approximate two-monthly cycle to update and share progress, thereafter on a more ad hoc basis to monitor progress.

The outcomes and impacts of this exercise were varied, and depended on how pioneers interpreted their role, the size and complexity of the schools, their specific culture and disciplinary focus, and the extent to which the pioneer was supported or able to gain support for the initiative. All the pioneers had periods of success but also periods of frustration and feelings that nothing was happening, which became problematic when reporting on 'progress' at project boards. Group meetings with the pioneers, and bi-lateral meetings with pioneers in specific schools sometimes became unproductive and negative. For this and other reasons, a number of staff left the role, requiring more time to induct their replacements.

Although the emphasis was on voluntary action as much as possible, it was felt that this needed to be underpinned by an element of compulsion. Hence, the University's Course Approval and Periodic Review process (CARP) was amended in September 2008 to require all applicants to demonstrate that their course was addressing relevant sustainability topics. In most cases, the decision as to whether this was being met was delegated to school level, in the form of an associate dean working with a pioneer.

## Academic pioneers

An important feature of Ecoversity was the school-level champions initiative, known as 'academic pioneers'. They were very autonomous, resulting in diverse approaches and outcomes, as two examples indicate.

While an initial pioneer appointment in the School of Life Sciences proved ineffective, a successor had a very good understanding of the university teaching and learning culture, and had chaired academic review (CART) meetings. She consulted widely and created a school-level ESD statement for inclusion in all course documentation. She then initiated several successful curriculum projects, including one with a colleague in pharmacy, which used a new NHS Sustainable Development Strategy document as a vehicle for curriculum interventions around ESD and professional practice and responsibility (Lucas *et al.*, 2010).

In another school, the pioneer was energetic and a passionate Ecoversity supporter but was much less involved in CART and curriculum change. The resulting inability to influence teaching and learning structures, together with strong divisional silos, soon made her demoralised and frustrated. However, staff changes enabled a second colleague with good curriculum experience to work alongside her and the resulting combination of energy and knowledge resulted in rapid progress, including a school-wide ESD statement and course modifications. This original pioneer remains a committed Ecoversity supporter, and now acts as a very effective green champion who has helped to achieve reductions in her school's energy use.

## The informal and campus curricula

A key principle of Ecoversity was that ESD within the formal curricula needed to be reinforced by an 'informal curriculum' in which sustainability was made visible in existing activities such as clubs and societies, and new ones such as opportunities to become involved in campus change through attendance at special events, volunteering and other means. The importance of this – under the heading of 'service-learning' – has been highlighted in pedagogic discussion and practice in US universities (Jacoby, 1996) and at Leicester (Warwick, 2007). Hopkinson *et al.* (2008b) and Lipscombe (2009) have also highlighted its relevance to ESD.

This strand of Ecoversity involved developing links and good working relations with a wide range of student-focused service areas, and developing a team of Interns, Ambassadors or 'Actioneers' (who received some modest financial support). In addition to familiar models of student engagement, for example halls-based campaigns, competitions and student-union-led events, this inspired more original 'action research' activities such as a permaculture gardening project and leading a Peace Camp for Bradford schools. The activities were supported through social networking – not only Facebook but also a university-specific site called DevelopMe! (www.bradford.ac.uk/developme/) – and a newspaper (*The Seed*) with regular discussion groups using its content to raise awareness and stimulate new initiatives. These proved especially useful to international students and those outside halls of residence.

Both formal and informal curricula can also be strengthened if their features are embodied in the physical environment of learners, thereby creating a 'campus curriculum'. Orr (2004) describes how the Adam Joseph Lewis Center at Oberlin College in Ohio has been designed to link building occupants to natural systems, and to provide concepts and data that can illustrate topics such as full cost accounting and systems analysis. Temple (2007) also notes how good environmental conditions – such as lighting, heating and noise control – and a sense of being 'cared for', support feelings of belonging and psychological security, which are in turn linked to effective learning. The converse is when features of the built environment and its operation – such as broken and never-mended equipment or fittings, over-heated rooms and unnecessary waste – appear to send messages of lack of care and hypocrisy because of the disconnect between stated values and aspirations, and everyday realities.

Fortunately, such disconnects were minimised by growing evidence that Ecoversity was transforming the university campus, through actions such as:

- the refurbishment of the main Richmond Building, and the addition of a new atrium (opened in 2006);
- the reduction and repositioning of car parks, with ecologically sensitive and visually attractive landscaping of the 'liberated' space;
- a BREEAM Outstanding Sustainable Student Village, whose construction began in 2010.

The changes began to alter perceptions so that many people were no longer embarrassed by the campus, and found it a more pleasant place to work/study. Ecoversity also sought to optimise the learning potential of the development's sustainability features, for example displays in student-frequented areas showing electricity generation from new PV installations and rainwater harvesting. Other 'campus curriculum' activities included signage, information systems communicating activity and a Green Campus trail and brochure.

## Evaluation

Ecoversity StuDent's research strand found a significant increase in the number of students who were aware of and engaged with the project over its life. (Specific research activities and findings were reported as a series of fourteen Highlighting Ecoversity Research Briefings (HERBs): see www.bradford.ac.uk/ecoversity). In 2008–09, 68 per cent of student respondents to a university-wide survey mostly or definitely agreed that their unit/module had helped them to understand sustainability issues relevant to the subject, with 65 per cent reporting this at course level. Most students who had engaged with Ecoversity developed a deeper understanding of, and stronger personal commitment to, sustainability agendas, and made changes in their personal lives as a consequence. Interviews with recent alumni also found that engagement with Ecoversity helped them to develop professional and other skills that enhanced employability and that they were able to apply in a work context.

Perhaps the most important measure of success for the university as a whole was the perceptions of current and prospective students of the campus, and the Ecoversity-associated changes that had been made to it. Survey and informal feedback found a widespread appreciation of the new landscaped green spaces, the increased number of trees and habitats, and special areas like the Peace Garden and the Veg Out Patch. Many students reported awareness of recycling initiatives, green travel alternatives and the availability of ethically sourced food and drink, often leading to changes in attitudes and behaviours. These changed perceptions about the university campus were almost certainly a major factor in the improvement in the university's recruitment during the years of Ecoversity – with a 15 per cent reduction in the number of students citing the campus as a reason for rejecting the offer of a place.

In socio-cultural terms, the key feature of Ecoversity during this 2007–10 phase was the establishment of a 'democratic collaboration' structure with a central team liaising with academic experts in as non-coercive way as possible to develop their roles as practitioner experimenters and architects within their discipline and school. The ethos was to ensure that no one felt excluded by Ecoversity and that any offer or suggestion for change or action was taken seriously and supported if possible and feasible. A key mechanism for this was supportive and empathetic 'reflexive' questioning – for example, to highlight the progress, however small, that had already been achieved, and to find out what more could be done, and what was needed to achieve it. This often led to constructive dialogue, which established that the difficulties were in many ways the 'norm' for such initiatives and provided reassurance that the voices of individuals or groups were being heard and responded to. The corollary was adjusting and modifying Ecoversity project expectations, timescales and expected outcomes.

From the outside, this strategy may have seemed diffuse and anarchic but from inside it felt like a process of emergent change, bringing with it subtle changes in attitudes and support. As those involved in change management often testify, the most difficult trick is to create a new culture within the confines of an existing culture, without creating instability and anxiety – which often leads to the old culture snapping back in order to maintain control and order (Pennington, 2003; M'Gonigle *et al.*, 2006). All this was enabled by the strong rhetorical support (backed by 'hands-on' interest) from the VC and PVC, and the way that the latter played a high level political entrepreneur role, building good relations around the university that enabled 'space' for experimentation and initiatives to develop, while keeping them grounded in the everyday realities of institutional life.

## The Ecoversity initiative: consolidation, 2010

The end of Ecoversity StuDent funding in summer 2010 coincided with growing financial challenges for UK universities. Hence, whilst the estates strand retained considerable momentum as it took forward high-profile developments, the centrally funded Ecoversity team was reduced to two part-time roles – a maintained Director of ESD position, and a new one within the Strategic Planning Office. The former has a remit of assisting schools to continue developing the ESD component of their courses and other activities (which the maintenance of the CART requirement 'steered' them to do), and taking forward new or embryonic initiatives. The latter had the remit of maintaining strategic momentum by creating an annual reporting cycle on Ecoversity progress, overseen by a review group chaired by the VC. As part of this, the lead on student engagement was transferred to the Student Union, the lead on environmental performance to estates and related areas, and the lead on curriculum to a new Academic Development Unit (ADU).

Ironically, the increasingly visible redevelopment of the estate, the 'time lag' of people picking up on dissemination activities, and the winning of a number of prestigious awards, meant that external interest in Ecoversity peaked after actual ESD activity declined with the project's end (and the temporary exhaustion that often comes at the end of high-intensity periods with clear end points). This recognition intensified an internal discussion as to what the university should do next, especially after the PVC champion of Ecoversity left in spring 2011. Some argued that while sustainability should always be an important topic for the university, the recession and other factors meant that relatively greater attention should be devoted to other areas, notably research and league table positions. A related view was that while the Ecoversity 'brand' had been invaluable in galvanising action and enhancing the university's reputation, there was a risk that maintaining high-profile activity might mean being judged by impossibly high standards in the future.

Although this strategic discussion had no formal conclusion, a *de facto* change in the centre of gravity occurred with the emergence of new sustainability-focused research, learning and outreach activities. These are linked to a new £5.2 million Sustainable Enterprise Centre (opening in 2013), with substantial funding by the university (a significant sum for a medium-sized institution to commit to a visionary, albeit well researched, development). Its roles include developing new areas of research and teaching (especially by playing a 'democratic collaboration' role in linking and leveraging existing ones within different departments), business support and engagement (including business start-up and student enterprise), providing demonstration and training space for both internal and external audiences (for example, Bradford schools) and acting as a focal point for the university's continuing corporate commitment to

Ecoversity. The Centre's building, located in the heart of the campus, also makes an impressive statement, having achieved in total at the time of writing the highest number of points of any BREEAM scheme at its interim certification. This meant that the university is the only organisation to have two buildings (of seventeen in total) with the highest classification of Outstanding (at least equivalent to, and possibly more demanding than, the highest marked Platinum buildings in the US Green Building Council's LEED scheme).

The achievements and reputation of Ecoversity also helped to achieve a partnership with the high-profile Ellen Macarthur Foundation (launched in September 2010 with a focus on developing ideas about, and supporting actions to achieve, a 'circular economy'). This has involved an international conference of leading thinkers on the topic at Bradford in November 2011 and the development of a Bradford M-level course in the Circular Economy, working with B&Q, BT, National Grid, Cisco and Renault.

In early 2011, the UK National HE STEM programme also funded a Greening STEM project at the university. This aimed to deepen and extend Ecoversity's work on ESD and STEM curricula, especially in software engineering, computing, automotive engineering, civil and mechanical engineering and chemistry, and also aimed to undertake STEM infrastructure interventions to reduce their environmental footprint. This has included partnership in a new awards scheme and conference on the theme of sustainable laboratories which has been organised by the Bradford-based (and still continuing) HEEPI initiative (see James, 2012; and www.goodcampus.org).

## Discussion and conclusions

Ecoversity has been a considerable success in meeting its core objectives. The campus has been transformed, with more green space, new high performance buildings and effective refurbishment of existing ones. Environmental impacts have been reduced significantly. The university's carbon footprint (scope 1 and 2) is on track to achieve a 50 per cent reduction by 2015 compared to 2005, well ahead of sector and national targets and creating considerable savings in energy and other costs. An ESD component has been introduced into almost all courses, and new ones that directly address sustainability have been developed. The attitudes and behaviour of many staff and students has also changed. People began to use Ecoversity as a noun (Is Ecoversity represented at the meeting?), an adjective (How very Ecoversity!) or as a questioning or personification device (What does Ecoversity think about this?). All of this has had a positive impact on the overall perception of the university and the campus by current and prospective students, which has in turn been a significant contributor to improved student recruitment rates. The positive publicity has also enhanced the university's reputation in the higher education world, both nationally and internationally, allowing a relatively small institution to stand out from its peers in many respects. In addition, permanent capacity has been created which is now supporting new pathways for further progress.

Of course, some of the Bradford achievements might have happened without an Ecoversity-type initiative. In particular, the campus regeneration was overdue and some action was inevitable. However, linking it to a broader vision of sustainability through the Ecoversity vision did make it easier to justify extra difficulties or costs related to high environmental standards, and sometimes led to innovative actions that might not have occurred in other circumstances.

Naturally, the Ecoversity story has not been one of complete success. Effective cross-institutional links were slow to develop, and constant attention has been needed to maintain their effectiveness. The devolution of significant ESD responsibility to schools produced less progress than hoped in a number of areas, in part because the academic pioneer mechanism was only partially effective. Some initiatives aimed at staff and students did not succeed, or had early success and then fell back. Some senior managers were willing to prioritise Ecoversity while external pressures were strong but became more cautious as they weakened and/or were perceived to be outweighed by even greater pressures, for example to maintain financial viability. And some areas, such as IT and laboratory-based sciences, had a disconnect between more sustainability in their curricula and relatively little improvement in the environmental impact of their day-to-day operations.

The strong estates dimension to Ecoversity was also both advantageous and problematic. Constructing, refurbishing or demolishing buildings, and landscaping car parks, are very tangible demonstrations of change, especially when they are winning national accolades for their sustainability and changing staff and student attitudes. This undoubtedly gave Ecoversity an authority and rhetorical legitimation that could be used to create momentum in ESD in other areas. However, the different culture ('men of action' rather than male and female denizens of the ivory tower!) and practices (such as formal project management techniques) of estates do not connect naturally with academics, and so an equally strong pole is essential if there is to be productive relationships and mutual advance. The occasional failure of strategic estates visions to be reflected in day-to-day operations could also be of greater symbolic importance than junior estates staff might imagine, and therefore corrosive of staff and student credibility. Hence, it was only when Ecoversity was fully integrated into the university's strategic academic policies and discussions after 2007 that real transformation was possible.

Insofar as Ecoversity was successful, a major explanation of this was its identification of these difficulties (initially intuitively, but more analytically once the existing literature on university change was recognised) and the design of 'frameworks for action' to take account of them, albeit as the result of a trial and error process. Hence, the main pillars of its implementation – estates and ESD – proceeded with considerable autonomy, as did ESD implementation within schools. In particular, much effort was put into working collaboratively with academics in different disciplines to understand their existing social practices and repertoires, to identify points of connection with Ecoversity agendas to overcome responses such as 'students in my discipline aren't interested in sustainability', and to identify or develop practical exemplars that could be emulated. The converse of this was a willingness to de-emphasize some aspects of ESD if there were no points of connection with these practices, and to be as accepting as possible about local difficulties rather than sending the punitive authority messages that often accompany complex and difficult changes.

The concept of a 'ladder of implementation' also proved helpful. For a few crucial years, the Ecoversity initiative did achieve a self-reinforcing cycle in which senior management responded positively to both external pressures and bottom/middle-up demand, for example through rhetorical support and support for compliance/regulatory mechanisms which enforced action on middle and lower rungs. Staff on those levels both reacted to demands from above and, in many cases, felt empowered to develop their own ideas and solutions, and to explore best practice examples and links with professional imperatives in more open ways than they might otherwise have done. These actions then influenced students, some of whom were also being mobilised through the Ambassadors scheme and other measures. The sense of action below – and especially evidence that students were interested in the topic, and that this might therefore have the potential to enhance the university's external reputation – then bubbled back up to middle and senior levels, often with demands for specific supporting actions, which were accepted. The corollary of this was Ecoversity staff recognising and adapting to the constraints (sometimes by diplomatically holding back enthusiasm) at lower rungs of the ladder so that people were not set up for disappointment or failure. This was a factor in Ecoversity's areas of mixed performance. For example, the difficulties with academic pioneers were partially due to an over-reliance on the commitments of heads of school and insufficient recognition that the rungs below them, i.e. senior lecturers and lecturers, also had to be won over. A related explanation of the difficulties with IT and laboratories was an underestimation of the importance of technical support staff in the implementation ladder, and of the difficulties of reaching and persuading them.

A socio-cultural theory-based approach was particularly important in working with the STEM disciplines, which are so central to Bradford (Hopkinson and James, 2010). Many STEM academics and students, for reasons of both disciplinary tradition and personal psychology, are less receptive to ways of thinking that fundamentally challenge existing practices, especially when these do not appear to have immediate practical applications. Certainly, the authors' experience is that many STEM academics and students do find sustainable development to be a nebulous and partly ideological concept (as described in Knight, 2005), and if forced to respond to it can be antagonistic and negative. While some may see this as unfortunate, it is perhaps a reality which has to be taken on board within many institutions, and worked with or around in present circumstances. The experience of Ecoversity suggests that changes to STEM curricula are most likely to succeed when they clearly relate to core scientific and technical competencies such as analytical rigour, critical thinking, and empirical observation and testing. They are also more likely to be effective when they build on – rather than work against – existing disciplinary, departmental and teaching and learning cultures, and encourage students to take critical and reflective stances on the ideas/data and issues they are generating or being presented with. Examples include linking concerns about resource utilisation and wastage to good experimental practice procedures in disciplines such as chemistry, or to established approaches such as lean manufacturing in engineering.

The Ecoversity experience suggests that a monochromatic notion of sustainable universities, and ESD within them, which identifies only a small number of possibilities (for example, the radical/reformist dichotomy identified by Dryzek, 2005) may no longer be that helpful in guiding practical action, if it ever was. Similarly, teleological notions of the sustainable university and ESD as a stable end state of holistic and systemic thinking, and righteous action, can also be unhelpful in practice when confronted with the messy real-world challenges of university social practices and competing agendas. This means that progress towards sustainability is always likely to vary over time, with some periods of rapid progress, and others of stasis or even some backwards movement. In periods of heavy external pressure towards sustainability, significant progress should be expected. But when pressures wane, progress is likely to depend upon the ability to connect with local social practices and broader organisational agendas. Ecoversity was built on the ability to make connections between sustainability and university 'hot buttons' such as campus regeneration and student access. It developed momentum through making further connections with the institution's desire to enhance its reputation for employable graduates, and is now refocusing again on sustainability-related teaching and learning, and on making the IT and laboratory infrastructure into more of an informal and campus curriculum.

In conclusion, Bradford may have been particularly fortunate in 'being in the right place at the right time' through its unusual mix of campus regeneration needs, social inclusion strengths, and strength in disciplines that were facing external pressures for change. It was also fortunate in having a number of entrepreneurial and visionary individuals who could connect these both internally and to national agendas, which enabled first funding and then other linkages to develop. Nonetheless, we do believe that Ecoversity does highlight the value for most universities of academic-led 'middle-out' initiatives that can mediate between top-down policies and bottom-up enthusiasm and initiatives, and can respond flexibility to the diversity of social practices that are the reality of university life, and are likely to always be so. The following box pulls out ten key lessons from its experience for other organisations already following, or hoping to travel, a similar route.

**Ten pointers: how can we get started on, or maintain the momentum of whole institutional change if we don't have the level of funding that Ecoversity had?**

1   Check that you're not being too ambitious – it may be better to prioritise the most promising areas for progress and build out from these. If a project has stalled, someone leaves, resistance is too great, or funding is switched off, be prepared to accept the situation and move on to more promising areas.

2   Create a simple and compelling vision (as with the Bradford Estates Directors' slides) to inspire senior management and the troops, which links with mainstream organisational issues and goals and the external drivers that are shaping them. This may mean a focus on actions that can help finances or research/teaching performance in current conditions.

3   Look at the Harvard Green Campus initiative (Sharp, 2009), which Ecoversity learnt much from. This achieved great success from a limited initial resource through a 'bottom-up'/'middle-out' process, an emphasis on capacity building and a focus on changing the physical environment and engaging students rather than curriculum.

4   Maintain an alignment between different areas of campus sustainability as progress in one helps others, while poor performance in one can undermine the credibility of others.

5   Ensure that capacity (for example a champion or champions) is maintained in some shape or form with a brief of identifying and strengthening linkages between sustainability and emerging organisational developments. Don't let this capacity and remit be too closely identified with the estates department.

6   Maintain conversations and relationships that worked in the past and keep in as many internal loops as possible so that potential new opportunities can be identified at an early stage.

7   Create resources through benefit sharing – for example ringfencing a proportion of energy savings for general sustainability measures.

8   Use pilot projects to test ideas and build a case. Don't feel that actions always need to be rolled out across the university – focus on those where benefits are likely to be greatest.

9   Be persistent – many initiatives stall or erode as exhaustion or demoralisation sets in. It can take months to create momentum but years for most significant and tangible progress to be evident.

10  Keep writing, publishing and disseminating, both internally and externally, to maintain visibility.

## Strategic approaches to ESD and sustainability at the University of the West of England: another perspective

The University of the West of England, Bristol (UWE) is a large modern university, located in Bristol, one of Europe's greenest cities. It has approximately 30,000 students and 3,500 staff and provides a range of undergraduate, postgraduate and continuing professional development (CPD) courses plus extensive research and knowledge exchange activity across four campuses. In 2007, a 'cross-university sustainability board' was constituted, chaired by a pro-vice chancellor, reporting to the Academic Board and Board of Governors. Acknowledging unnecessary environmental impacts from the institution's operation, the board was charged with developing a five-year sustainability strategy and action plan. Crucially, the board recognised that the university's major impact on sustainable development would be in educating its students, whatever their discipline, and so education for sustainable development (ESD) would be at the core of the UWE approach.

The ambition for the initiative was given voice and commitment through a five-year plan, the 'Sustainability Strategy 2008/2012'. This articulated the vision that 'by 2012 UWE will be recognised by its students, staff and wider stakeholders as one of the leading universities in the UK for its sustainability performance'. This was a deliberately inspiring statement, which helped to galvanise the sustainability board and led to success and recognition in several key areas. This strategic goal was underpinned by objectives and performance targets in leadership and management in nine thematic areas. Regular meetings of the board, with reports against targets, kept up momentum; all the original targets were comprehensively revised in 2011.

In parallel to this high-level activity, two internal stakeholder networks were established to extend commitment and activity across the institution. On the operational side, sustainability leads were identified in each faculty and service to marshal action through convening regular representative sustainability groups. On the ESD side, a baseline study of sustainability content in all modules was undertaken to identify the extent, quality and nature of provision. The study also involved in-depth interviews with academics across the whole university. This report, with its ten recommendations, initiated an ongoing, university-wide ESD programme. This has seen the establishment, and embedding within the formal reporting structure, of an internal knowledge exchange group for ESD. It has also led to an annual sustainability conference for all staff linking ESD and operational sustainability; an ESD employability report; production of sustainability content for university-wide programmes; an induction in ESD for all new academic staff; incorporation of sustainable development into the new lecturer programme, and the development of a unique cross-faculty Masters programme, 'Sustainable Development in Practice'.

Recognition has been patchy, but not without considerable high points, such as being consistently rated as 'Good Honours' in the People and Planet Green League; a shortlisting by the *Times Higher Education* for Outstanding Contribution to Sustainable Development; gaining a Carbon Trust Standard mark of excellence; and winning the Green Gown award in 2010 for 'Continuous Improvement – Institutional Change'. On the other hand, there is work to be done in extending good practice, joining up better across the institution, and involving students in the initiative more fully. For example, a recent survey of all heads of department shed some light on weaknesses in monitoring and information flow of ESD activity across the organisation.

Coming to the end of the first five-year sustainability strategy is allowing UWE to take stock. There is strong commitment in the Vice Chancellor's Executive and Board of Governors, and UWE will rechallenge itself with a new five-year strategy. In reflecting on its achievements and remaining challenges the university will need to determine how it wishes to further embed sustainable development into its entire strategic and operational decision-making. The university has made great strides in developing its whole institutional approach but recognises that there is still much to be done strategically, operationally and in winning hearts and minds for the whole institutional change offer of the university.

*Marcus Grant is Associate Professor at the Institute for Sustainability, Health and Environment and member of the Sustainability Board, University of the West of England, Bristol.*

## References

Alvesson, M (2002) *Understanding Organizational Culture*, London: Sage

Bamber, V, Trowler, P, Saunders, M and Knight, P (eds) (2009) *Enhancing Learning and Teaching in Higher Education: Theory, Cases, Practices,* London: Open University Press/SRHE

Bartlett, P F and Chase, G W (2004) *Sustainability on Campus: Stories and Strategies for Change*, Cambridge, MA: MIT

Blewitt, J and Cullingford C (2004) (eds) *The Sustainability Curriculum: The Challenge for Higher Education*, London: Earthscan

Broadbent, J, Laughlin, R and Ghazma, A S (2010) 'Steering for sustainability: higher education in England', *Public Management Review*, 12 (4): 461–73

Butcher, M (2007) *Are You Sustainability Literate?*, Online report at www.spiked-online. com/index.php?/site/article/3821/ (accessed 31 October 2007)

254    *Peter Hopkinson and Peter James*

Cade, A (2007) *Employable Graduates for Responsible Employers*, Report to the Higher Education Academy, Loughborough: Student Force for Sustainability

Corcoran, P and Wals, A (eds) (2004) *Higher Education and the Challenge of Sustainability: Problematics, Promise, and Practice*, Dordrecht: Kluwer Academic Publishers

Dawe, G, Jucker, R and Martin, S (2005) *Sustainable Development in Higher Education: Current Practice and Future Developments*, York: The Higher Education Academy

Dryzek, J (2005) *The Politics of the Earth: Environmental Discourses*, Oxford: Oxford University Press

Forster, J (2006) 'Sustainable Literacy: Embedding Sustainability into the Curriculum of Scotland's Universities and Colleges', Report for the Scottish Funding Council, John Forster Associates. Available at: www.scwg.aaps.ed.ac.uk/docs/open/JFASustainable Lite10EB5B.pdf

Forum for the Future (2004) *Learning and Skills for Sustainable Development: Developing a Sustainability Literate Society: Guidance for Higher Education Institutions*, London: Forum for the Future for Higher Education Partnership for Sustainability

Fullan, M and Scott, G (2009) *Turnaround Leadership for Higher Education*, San Francisco: Jossey-Bass

HEFCE (2005) *Sustainable Development in Higher Education: Strategic Statement and Action Plan*, Bristol: HEFCE

Herrmann, M (2000) *Greening Teaching Greening the Campus*, Surrey: The Surrey Institute for Art and Design

Hopkinson, P (2010) 'The potential for sustainable development to reshape university culture and action', *International Journal of Environment and Sustainable Development,* 9 (4): 378–91

Hopkinson, P and James, P (2010) 'Practical pedagogy for embedding ESD in science, technology, engineering and mathematics curricula', Special Issue: Competences for sustainable development and sustainability: significance and challenges for ESD (guest editors: Zinaida Fadeeva and Yoko Mochizu), *International Journal of Sustainability in Higher Education,* 11 (4): 365–79

Hopkinson, P, Hughes, P and Layer, G (2008a) 'Education for Sustainable Development: using the UNESCO framework to embed ESD in a student learning and living experience', *Policy and Practice: A Development Education Review*, Spring 2008, 17–29

Hopkinson, P, Hughes, P and Layer, G (2008b) 'Sustainable graduates: a whole institutional approach', *Environmental Education Research,* 14 (4): 435–54

Jacoby, B (1996) *Service-learning and Higher Education: Concepts and Practices*, San Fransisco: Jossey Bass

James, P (2012) *The Effective Laboratory: Safe, Successful and Sustainable*, University of Bradford, HEEPI project. Available at www.goodcampus.org.

Jones, P, Selby, D and Sterling, S (2010) *Sustainability Education: Perspectives and Practice Across Higher Education*, London: Taylor and Francis

Knight, P (2005) 'Opinion: the Funding Council's circular on sustainability is pernicious, shameful and dangerous', *Education Guardian*, 8 February 2005.

Lipscombe, B (2009) *Education for Sustainable Development: Extra Curricular Education for Sustainable Development in Higher Education*, PhD thesis, University of Chester

Lucas, B, Hawksworth, G and Horncastle, E (2010) 'Why helping the public and planet is important for Pharmacy Education and practitioners of the future', *The Pharmaceutical Journal*, 284: 55–6

M'Gonigle, M, Starke, J and Penn, B (2006) *Planet U: Sustaining the World, Reinventing the University*, Gabriola Island, BC: New Society Publishers

Orr, D (2004) 'Can educational institutions learn? The creation of the Adam Joseph Lewis centre at Oberlin College', in Bartlett, P F and Chase, G W (eds) (2004) *Sustainability on Campus: Stories and Strategies for Change*, Cambridge, MA: MIT

Pennington, G (2003) *Promoting and Facilitating Change*, York: Higher Education Academy. Available at: www.heacademy.ac.uk/assets/documents/institutions/change_academy/id296_Promoting_and_facilitating_change.pdf

Revans, R (1980) *Action Learning: New Techniques for Management*, London: Blond and Briggs

Roberts, C and Roberts, J (eds) (2007) *Greener by Degrees: Exploring Sustainable Development Through the Higher Education Curricula*, Cheltenham: University of Gloucestershire

Sharp, L (2009) 'Higher education: the quest for the sustainable campus, Editorial', in *Sustainability: Science, Practice, Policy*, 5 (1), Proquest. Available at: http://sspp.proquest.com/archives/vol5iss1/editorial.sharp.html

Sterling, S and Gray-Donald, J (2007) 'Editorial', *International Journal of Innovation and Sustainable Development*, 3 (42): 241–8

Sustainable Development Commission (2007) *Skills and Sustainable Development: An Exploratory Dialogue Report from a SDC Stakeholder Event: Friday 20th July 2007.* Unpublished report.

Temple, P (2007) *Learning Spaces for the 21st Century*, Report for the HEA, York: Higher Education Academy. Available at: www.heacademy.ac.uk/assets/documents/research/Learning_spaces_v3.pdf

Trowler, P (2008) *Cultures and Change in Higher Education: Theories and Practices*, London: Palgrave Macmillan

Wade, R (2008) 'Education for sustainable development: challenges and opportunities', *Global Education Policy and Practice*: A development education review, 6, Spring 2008, Centre for Global Education

Warwick P (2007) 'Learning for life: incorporating into HE the transformative potential of ESD through service learning programmes', Paper presented at 2nd International Conference of Graduates as Global Citizens, 2007, Bournemouth

Weick, K E (1976) 'Educational organizations as loosely coupled systems', *Administrative Science Quarterly*, 21 (1) (March, 1976): 1–19, Johnson Graduate School of Management, Cornell University

# 12   Bottoms up for sustainability

## The Kingston experience

*Ros Taylor*

Our biggest challenge in this new century is to take an idea that seems abstract –
sustainable development – and turn it … into a reality for all the world's people.
(Kofi Annan, UN Secretary General, 2001)

[T]he people who will succeed fifteen years from now, the countries which will
succeed, are those which are most based on a sustainable vision of the world.
That is what we should be training people to do.
(Rt Hon Charles Clarke MP, UK Secretary of State for
Education and Skills, 2003)

Universities educate most of the people who develop and manage society's
institutions. For this reason universities bear profound responsibilities to
increase the awareness, knowledge, technologies, and tools to create an
environmentally sustainable future.
(Association of University Leaders for a Sustainable Future,
*The Talloires Declaration*, 1990)

## Introduction

There is no single way to achieve a sustainable university; context matters
(UNESCO, 2005) and will influence 'best practice' locally. Flagship projects,
such as Ecoversity at the University of Bradford, beneficially inspire the higher
education sector, but many institutions lack resources and support. Yet everyone
needs to act if the sector is to take a societal lead in achieving 'sustainability
literate' graduates (Forum for the Future, 2004; Higher Education Partnership
for Sustainability, 2004) and future leaders who understand the sustainability
challenge. This aspiration is arguably an important objective for universities. It
could be expected that higher education institutions (HEIs) would show
leadership by becoming exemplars of sustainability through their teaching and
research, their operations and their community interactions. Carbon reduction,
energy saving, waste minimisation, water management, biodiversity
enhancement, and fostering more sustainable communities (from promoting
staff and student well-being to encouraging thriving 'town–gown' relationships,
and contributions to wider societal debates) – these are all key objectives for a

sustainable university, in addition to appropriate curricula. The importance of these tasks cannot be overstated. As noted above, this has been widely recognised by political leaders and, more immediately for the UK HEIs, by the Higher Education Academy (HEA), the Higher Education Funding Council for England (HEFCE) and its partners throughout the UK, and by the support organisations for campus managers and professional staff, such as the Association of University Directors of Estates (AUDE) and the Environmental Association for Universities and Colleges (EAUC).

Recognition of this agenda is not new. In the early 1990s, several attempts were made to engage universities as leaders in the creation of sustainable societies. The Talloires Declaration (1990), the Association of Commonwealth Universities' Swansea Declaration (1993), and the Association of European Universities' Copernicus Charter (1994) all argued that sustainability policies and practices should be integrated and embrace these five core concerns:

1   ecological stability
2   climate change and biodiversity trends
3   societal considerations and stakeholder relationships and responsibilities
4   economic considerations
5   principles of intergenerational equity and justice

(Dawe *et al.*, 2003; Wright, 2002).

However, it was more than ten years later before initiatives such as the UNESCO Decade of Education for Sustainable Development, 2005–14 (2005) and, in the UK, sustainability consultations and benchmarking activities by HEFCE and the Department for Education and Skills (DfES), brought these ideas into prominence. In the UK, at least, few advances were made during the intervening period. Indeed, the anticipated prominence of environmental education, as discussed in the Toyne Report (1993), made little impact in practice and its recommendations were largely ignored. While it may be argued that more recent initiatives are more inclusive, more truly about sustainability, less dominantly environmentally-led, and reflect current government and wider society agendas, there is an important warning here. Sustainability in universities needs, itself, to be sustainable. In this context, the work of the HEA in promoting research into sustainability pedagogies and developing the Green Academy (Luna and Maxey, this volume) as a supporting forum and framework for institutional change; the HEFCE drivers and research investment in leadership for sustainable development (SD) (for example, sustainable procurement practice); and support developments such as LiFE (EAUC, 2011), are potentially important safeguards.

It is against this background that initiatives to achieve sustainable universities, whether enabled by top-level leadership or 'back-seat' driven by determined grassroots enthusiasts, must be examined. This chapter outlines the journey of one institution, Kingston University, towards becoming a sustainable university. It is the story of a collegiate grassroots initiative, embracing all aspects of the university's work, which, a decade later, has been fully acknowledged by the university's governors and executive. Sustainability is now embedded in Kingston's governance structures and policies as well as in its curricula, research, campus operations and community relations. Kingston's story is of a participatory, shared journey in that, from the start, academics, support staff and students from across the university worked together to change their HEI.

Similar reports of individual or small group actions triggering substantive change can be found throughout the sector – though typically these are often linked to more focused initiatives related to specific issues such as waste management, energy or travel, where the immediate payback benefits are most obvious. Their stories reveal many of the same key lessons learned at Kingston: the need for good communication, fun engagement, listening and sharing, clear goals and pathways, and effective monitoring.

## Kingston in context

Kingston University is a medium-sized UK academic institution located in southwest London (Box 12.1). It is a mainly urban, multi-site, post-1992 university. Building on its former polytechnic roots, it has strengths in subjects of direct professional relevance, such as engineering and applied sciences, art and design, architecture and planning, and business, education and healthcare sciences. It also has strong traditions in earth and environmental sciences, and an expanding curriculum in the humanities including politics, drama, media and film studies.

Indeed, in 2009, journalism and literature staff succeeded in launching Kingston University Press with *Sustainability in Practice* (Corrigan *et al.*, 2009) – collected papers drawn from the proceedings of Kingston's 2007 international conference of the same name – featuring among the four launch titles. In many ways, this typifies the interplay of departments and activities that characterises

---

**Box 12.1: Kingston University: facts and figures**

- 4 main campuses (plus foundation students at feeder colleges)
- 10 additional substantive residential and administrative sites
- 5 faculties
- 24,720 students: 20,251 full time, 3,975 part time
- 20,251 undergraduates; 4,469 postgraduates
- 4,124 non-UK students from 151 countries (2,431 are non-EU)
- 24 per cent of students are aged over 25 when commencing study
- 2,916 permanent staff
- All major world religions are represented and supported in a shared multi-faith centre.

(Data from December 2011)

the Kingston sustainability story. The production process raised awareness of sustainability among journalism and literature staff, with the printing company, and more widely, in-house and externally, at the gala launch. Several follow-up activities have been initiated by academic and technical staff in the humanities, including sustainability reporting and documentary making, as part of the assessed curriculum, and waste reduction and renewable energy initiatives in associated technical laboratories. The conference itself highlighted the commonality in challenges experienced in academia, business and local government sustainability endeavours.

The university is multicultural and ethnically diverse. Many students come from the local ethnic minority communities of west and central London and there is a strong presence of wider European and overseas nationals from wide-ranging backgrounds. This eclectic mix gives a vibrant and challenging environment.

## Starting out

Kingston's coordinated sustainability journey started in 2002 following a successful bid by two academics, a social geographer and an environmental scientist, to the university's HEFCE funds for cross-university initiatives. Success was unexpected; the mainstream bids supported staff development linked to e-learning, e-assessment, blended learning, etc., all very topical themes at the time. Support from the Head of Human Resources (HR) and the 'wow' factor – the distinctiveness of the bid – won the day.

The primary aims were to bring together people from across the university who were researching or teaching sustainability, and to establish baselines for curriculum and campus operations and developments. Secondary aims included staff development and training, development of business links, and improved networking around sustainability and environmental themes within the university and with external partners. It was envisaged that these activities could promote new opportunities for staff in terms of professional development, research and consultancy activity. Furthermore, these developments could lead to establishing the university as a 'hub' for information exchange and professional advice within the southeast region. It had become evident through informal contacts that across the academic disciplines – for example, from design to engineering, geography and environment to business, architecture to planning – people were working on sustainability, but as isolated pioneers. There was no coordinated forum for discussion, no unifying platform for external engagement to and from the university, and no top-level steer for these activities. As such, many opportunities for developing sustainability initiatives, research and consultancy, and new course developments were being missed. At that time, estates and facilities management took very traditional approaches – sometimes viewing sustainability as a 'green irritation' and failing to understand or be aware of forthcoming legislative drivers, sector- and in-house staff expectations or the potential benefits of sustainability management, for example financial savings, improved community relations, risk avoidance and reputational gain. Academic colleagues felt they were delivering education for sustainable development (ESD) in a vacuum of practical exemplars and leadership from the institution itself. Anecdotally, many support staff favoured recycling and adopted more sustainable procurement, waste management and energy-saving practices in faculty offices, but coordination through institutional governance and operational guidelines was lacking.

The bid success allowed appointment of a University Coordinator for Sustainability (0.75 full-time equivalent), initially for one year only. This post was subsequently extended for a second year, during which funds for a sustainability assistant, a graduate starter post, were also made available. This began a now-established tradition of sustainability staff giving temporary employment to a final-year student, or new graduate, to boost their initial career experience.

First steps were a university-wide curriculum audit and the invitation to like-minded colleagues from across the university to engage in a steering group to drive sustainability activity. From its inception, the Steering Group for Sustainability (SGS) comprised academics from every faculty, support staff and central administrators such as HR, technical services, IT, estates, procurement, facilities management, and, importantly, the Students' Union. It became a 'ginger group' of self-selected enthusiasts who worked together identifying priorities and guidelines, which still form the framework of sustainability work at Kingston. Its catholic composition ensured diverse lobbying opportunities (which, retrospectively, were vitally important) and brought together constituencies which otherwise rarely communicated. As the then head of Kingston's facilities management service company observed, '[T]he SGS is the one committee I take part in where I feel a valued equal, rather than obligatory appendage to (academic) discussion'.

This different ethos beneficially underpinned sustainability activity at Kingston; making the journey a shared enterprise removed barriers.

## Vision, baselines and progress monitoring

Establishing baselines is crucial to any project. In a bottom-up endeavour, it is particularly important to map progress and keep the 'big vision' in mind. It is also important to agree this vision. The incremental progress to be found in opportunist developments undertaken by people whose day job is different, for example to teach undergraduates or to run library services, can then be mapped to the main endeavour and remedial steps taken where progress is lacking. Regular discussions facilitated by the SGS and project team secured involvement and buy-in from people rarely consulted about operational concerns; it gave ownership across a wide constituency; it seemed empowering and it was fun. Never underestimate fun.

Making a difference, creating a Sustainable (Kingston) University – an institution fit, in every sense, to be a twenty-first century HEI – was the agreed big vision; incremental improvements the immediate aim. Five key areas were identified in which to progress and monitor developments:

1   curriculum and, thereby, the university's alumni
2   community and outreach
3   estates and facilities – all campus operations and developments
4   governance
5   research and consultancy.

Key objectives were to be:

- inclusive and accessible
- relevant, visible and responsive
- contributory in terms of the wider university.

This was to be an activity, a journey shared by the whole university community. Initially, progress was measured simply and mainly qualitatively through evidence of internal change (in practice and policy) and external engagement achieved at local, national and international levels, with the subsequent addition of external awards achieved. Winning external awards and generating income were unexpected morale-boosting bonuses.

These achievements raised the sustainability awareness of the university executive and governors, and played a significant part in the university executive's decisions in 2004–06 to core fund a Sustainability Facilitator post and two assistants on a permanent basis (by mid-2006, three people). This sustainability team undertook some teaching-linked activities, but their primary focus was to embed sustainability in campus operations – changes that the volunteer-based SGS found hard to achieve. The team also worked closely with the SGS to promote community engagement on all scales and to secure greater executive awareness of sustainability actions and governance changes needed within the university.

In 2009, the first Sustainability Hub was established, still funded as a project, through final-stage activity of the University's Centre of Excellence for Teaching and Learning, C-SCAIPE (see below), and based on a previously submitted development bid from the SGS. The sustainability team remained based within estates, though they worked closely with the Hub, which was charged with re-invigorating curriculum change for sustainability and coordinating sustainability research. Both groups reported to the executive, but through different pathways. As all the staff concerned recognised but the executive was initially reluctant to acknowledge (since it did not sit well with the university's reporting structures), for many activities, holistic, joined-up thinking, such as the SGS had long represented, was needed. In 2011, responding to these concerns and the growing reputation of Kingston's sustainability work, the university reorganised its sustainability structures to create one coordinated, fully integrated Sustainability Hub reporting directly to the deputy vice chancellor and based in the vice-chancellor's office (Figure 12.1).

All strands of sustainability were now formally embedded in the university's structure in one central department; sustainability was no longer a project, it was an acknowledged core activity. The Hub Director post was made a senior staff appointment, giving access to a fuller range of strategic decision-making committees and processes emphasising the university's commitment. The former SGS, retitled the Sustainability Forum, complements this structure; it remains a vital focus for sustainability discussion. Wider grassroots involvement is additionally secured through open Sustainability Exchange meetings (typically world-café style events[1]) around specific issues, while a network of Green Impact[2] teams secures day-to-day practical engagement with sustainability initiatives, such as energy saving, procurement and waste minimisation. While this recently achieved structure may look very professional and well thought out, along the route reporting arrangements and structures developed in an ad hoc fashion due to the 'temporary project' and 'volunteer-based' origins of sustainability activity at Kingston. Arriving at this new coordinated springboard for action shows what may be achieved when a grassroots initiative, seizing every opportunity, grasping every helping hand proffered along the way, takes action. As the anthropologist Margaret Mead is often quoted as saying, 'Never doubt that a small group of thoughtful, committed citizens can change the world. Indeed, it is the only thing that ever has.'

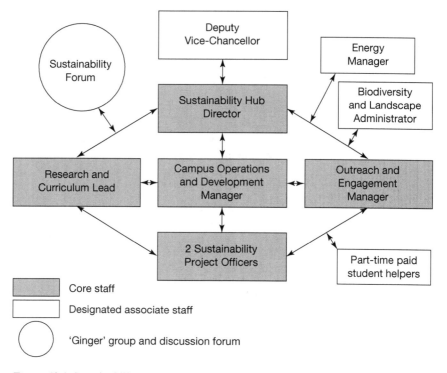

*Figure 12.1* Sustainability structure at Kingston University (August 2011)

The 'haphazard' developmental history typified by Kingston brings with it noteworthy grassroots benefits, which, providing they are not lost through future complacency as in, 'I'm too busy, the Hub can do that', bring significant future safeguards. The grassroots has custodianship since it 'built the boat'. It is important that, whatever governance structure emerges or is designed for an HEI (or any institution), the sustainability ethos so permeates the institutional culture, through staff and student awareness, that it is robust. In other words, it is not vulnerable to personality or institutional change, but is truly an embedded *développement durable*.

## Communicating sustainability

From the start, it was acknowledged that effective communication was essential for progress. The Sustainability Assistant created a website[3] within the university web framework. This is now in its fourth iteration and still uses the expertise of Sustainability Hub staff. This website has been useful for influencing and gently educating colleagues; sharing views; building a repository of, or easy pathway to, key materials; and for external visibility (visited from 178 countries, averaging around 60 hits per day). But a website does not do the total job. Busy academics often do not find time to look at it, especially if they are not already sustainability-aware; grounds staff and cleaners may have very limited time and facilities for computer access; students are focused on their next assignments, money-earning chances or relaxation opportunities. Supplementary means are needed and must suit the intended recipient. Sustainability communication at Kingston now includes a plethora of approaches: a monthly electronic newsletter, *Eco-bytes,* is delivered to in-house staff and students and local external subscribers; events and policy matters are advertised on the university's communally displayed electronic noticeboards and are visible to staff and students whenever they log in to the computing system; social media, such as Facebook, Twitter and blogs are increasingly used; and a new email circulation and exchange mechanism has been created for the Sustainability Forum. Sustainable behaviour information for students, such as the student eco-guide, is published electronically, supported by limited edition paper copies and an e-book edition is planned. Paper fliers and posters supplement electronic noticeboards in advertising immediately forthcoming events and specific campaigns, with targeted PowerPoint slides made available for lecturing staff. And, yes, still people report that, for example, they do not know about the Sustainability Hub, are unaware of sustainability events and support available, or have no idea that the latest new building uses a ground-source heat pump and sports a green roof. Communication at a busy multi-site university is not easy; but we are encouraged by the growing *Eco-bytes* subscriptions, the rapidity with which we can find volunteers to support events, and the consistent popularity of town–gown sustainability activities.

'The Hub is now very visible' (eco-beauty workshop participant, December 2011).

## Linking with partners: sharing and learning

Communicating externally has also been an important part of the Kingston sustainability journey. Workshops, seminars and academic publications; engaging with national debates; and making exchange visits locally, UK-wide and overseas have all played a part.

The benefits cannot be overstated. The SGS aimed at one major conference or workshop event annually – alternating between outward-focusing national and international events and in-house, local or specifically-themed participatory workshops and symposia (Table 12.1). The emphasis was on sharing experiences, learning from colleagues and building in-house sustainability awareness and professional capacity. These activities increasingly fed into external funding secured directly by SGS members or the Sustainability Team and, more recently, by Hub staff. This activity also encouraged new sustainability research and consultancy by showing the university had active engagement with sustainability and emerging methodologies. Between 2008–10, academic staff received over £1 million in grants for sustainability-related research from diverse organisations, including the Design Council, Leverhulme Trust, Investment Property Forum, UK research councils, government agencies and local authorities. Parallel developments in estates and facilities and governance, which endorsed the university's sustainability awareness and commitment, were important for this growing success and, during a similar timeframe, generated income approaching £500,000.

*Table 12.1* Kingston-hosted annual conferences, workshops and town–gown debates led by the SGS or Sustainability Hub – learning, sharing and developing sustainability expertise

| Date | Nature of event | Scope of event |
|---|---|---|
| 2002 | Joint sustainability workshop with Royal Borough of Kingston | In-house and local |
| 2003 | Sustainability in the Curriculum: Moving the benchmarks forward (a national conference) | National |
| 2004 | Workshops with local HEIs | In-house and local |
| 2005 | Joint sustainability workshop with Grand Valley State University | International |
| 2006 | Sustainability in Higher Education: Overcoming barriers (a national conference and workshop) | National |
| | Debating Climate Change: What can you do? | Town–gown |
| 2007 | London's first multicultural environmental awareness event for students | London Region |
| | Sustainability in Practice. From Local to Global: Making a difference (an international conference and workshops) | International |
| | Go Wild: Biodiversity in your back garden and beyond | Town–gown |
| | Sustainable Homes and Cities: Can you make a difference? | Town–gown |

| 2008 | Sustainable procurement strategy workshop with local government | Regional |
| --- | --- | --- |
| 2009 | Sustainability Hub launch exhibition | Regional |
| | Dinner's Dirty Secrets: Is it ethical to eat? | Town–gown |
| 2010 | Sustainability Strategy (conference and workshop): Sustainability in research and curricula at Kingston | In-house, but externally facilitated by HEA and BioRegional |
| | 21st Century – A Global Water Crisis | Town–gown |
| | Fairtrade in a Changing World | Town–gown |
| 2011 | Green Technologies in the Building Services Industry: Conference, workshop and exhibition | South London regional event with LSBU and South London Lifelong Learning Network, Moving Ambition, and industry representatives |
| | World Café event: Sustainable futures at Kingston University | In-house |
| | Can We Teach Happiness? | Town–gown |

An active policy of participation at external events was pursued. This has seen Kingston staff sharing and learning at events worldwide. Examples include work on ESD in Bilbao and in the Western Cape, South Africa; work with sustainability for built environment professions in Austria and Vancouver; fostering research and ESD links with the university's lead Erasmus partners (Dorich House Group); visits to Australia at Macquarie University and, in Sydney, sharing work on campus improvements and greening initiatives more generally; and facilities-led exchanges with Grand Valley State University, Michigan, also a Kingston partner university. These have been motivational tipping points generating significant feedback, new research links and exchanges, and prompting useful reflection in-house. Equally, visitors hosted at Kingston have enriched our understanding and experience, helping to progress sustainability initiatives.

## Gongs and awards

Winning prizes is clearly not the main point of a sustainability journey. Nevertheless, celebrating success and winning recognition from peers can be important for much more than the temporary glow of happiness generated. At Kingston, external success in sustainability activity helped to secure executive buy-in to continued funding and support. Noteworthy in this context was HEFCE's inclusion of Kingston's audit of sustainability in the curriculum (Dawe *et al.*, 2003; Sustainability and the curriculum, below in this chapter) and early SGS activities as a case study for its national debates and review of SD in higher education (HEFCE,

2005: 14–15). The associated invitation to the university's vice-chancellor to chair one of the linked discussion sessions substantially raised internal executive awareness of SGS activity and the importance with which it was viewed in the sector. Cogent argument and making the business case were, of course, also important. Writing any bid into funding or an external benchmark requires reflective thought on progress and achievement. It thus constantly reinforces in-house drive and direction and triggers progress reviews against stated objectives.

Even if unsuccessful in terms of 'gongs won', taking part generates useful reflective activity. For example, Kingston's strong grassroots action and initial pedagogic focus did it few favours in terms of the People and Planet Green League and the, initially, mainly estates-focused Green Gown Awards. On the other hand, the wish to participate in these and related schemes supported the early SGS's insistence on commissioning a whole-institution Environmental Management System (EMS) gap analysis, which demonstrated, through authoritative professional evaluation, the need for estates and facilities progress and the need for top-level commitment and a clear governance structure.

## C-SCAIPE: The Centre for Sustainable Communities Achieved through Integrated Professional Education

C-SCAIPE was established in 2006 as a Centre for Excellence in Teaching and Learning (a CETL) following a successful bid to HEFCE. It was one of four national sustainability-related CETLs. Though its primary focus was sustainability education for the professions, its presence is an important capacity-building block in the university's sustainability story. C-SCAIPE's learning space, designed to facilitate and promote discussion and project learning, gave a tangible and very visible sustainability space, becoming the 'venue of preference' for intermediate-size sustainability events, such as seminars, workshops and, especially, town–gown debates and exhibitions. In the latter phase of its formal operational period as a funded CETL, a strategic decision was taken to set aside some seed corn funds from C-SCAIPE to help establish the first Sustainability Hub in 2009 (as described above) and, thereby, fulfil wider C-SCAIPE ambitions. C-SCAIPE has been retained as a brand and focal space for sustainability events.

### Progressing sustainability

The SGS approach fostered strong links between activities in the five focal strands (see Vision, baselines and progress monitoring, above). This is very clearly seen in the real-world learning activities, which interweave all five in pedagogic developments. Interactions can be so close that it is difficult to label activity as specifically, for example, estates-led or research-led. There is constant exchange and team work, reflecting the 'bubble-up' grassroots origins of sustainability at Kingston. This approach is highly productive in generating ideas; it does not depend solely on the leadership of a small core group or an individual; and it gives ownership and developmental opportunities to a much wider community. The new Hub has retained this open participatory model; ideas sent into the Hub, from the wider community, are every bit as important as Hub-led invitations to engage with new developments.

## Sustainability and the curriculum

Under the terms of the 2002 successful bid for cross-university initiatives, an audit of sustainability in the Kingston curriculum was the starting point, alongside the emerging SGS discussions. The audit took the form of an indicative survey of staff drawn from all faculties and the executive to explore not only the courses and modules where sustainability was taught, but also whether staff knew and understood sustainability paradigms, as distinct from environmental concerns, and what, in their opinions, were the barriers (if any) to the delivery of sustainability in the curriculum. The structured interviews tested staff knowledge of a set of major topical environmental concerns and sustainability international treaties and explored perceived barriers to sustainability at Kingston (Dawe *et al.*, 2003).

Data showed that knowledge and understanding were patchy. Many staff felt there was no time to address such new material in the curriculum and that the space for this was a luxury compared with their subject heartland. Some considered students would find the material irrelevant to their studies. Others claimed that 'faculty fiefdoms' restricted new and, especially, interdisciplinary course development as did the rigid requirements of accrediting and professional bodies. That changing pedagogy could deliver sustainability literacy without necessitating major additional curriculum space was not widely understood. It was clear that if sustainable alumni were a Kingston goal (and they are), then engaging students through wider university activities outside the formal curriculum would be important.

This work gave impetus to pedagogic research and developments; real-world sustainability learning became a major feature of many courses (Box 12.2). Town–gown debates and guest lectures paved the way to community learning and engagement with international concerns and invited the question, 'What can you do?' on topics ranging from 'Climate Change' to 'Sustainable Food' to 'Happiness'. The audit also triggered the development of campus learning and volunteering activities and gave an additional impetus for campus sustainability improvements. Embedded sustainability learning, available to all, has become Kingston's main tool for progressing sustainability in the curriculum. Currently, around 25 per cent of courses and 33 per cent of modules across the university embrace sustainability concepts (Eames, 2012).

## Box 12.2 Pedagogic change for sustainability: two examples from Kingston

*Kingston's Lean Green Bike*

In 2008, final-year engineering students became engaged in a pilot project to design and build a zero emissions motorbike to compete in the Isle of Man TTX Grand Prix in June 2009. Students were introduced to the practicalities of industry-specific product Life Cycle Analysis, including issues of recycling and re-use for the bike chassis and for end-of-life batteries. Links with the Sustainability Hub introduced the students to wider social sustainability benefits beyond the main environmental and engineering arguments, enriching the activity. The engineering team decided on an electric solution and over the ensuing three years the design improved and new sponsorship was achieved. In 2011, major sponsorship to support further development was secured from the green energy company Ecotricity. Currently, Ecotricity has 54 per cent wind energy in its mix and the company specifically reserved wind-generated energy to charge the bike's batteries, effectively making the 2011 model a wind-powered machine. In 2011, Kingston's Lean Green Bike clocked 127 mph achieving a third-place podium finish, beating many commercial rivals. Enthusiasm generated for green technology has now fed back into wider developments in the faculty, including a new module for first-year students taught around the design of a more widely marketable, electric commuter bike (Brandon, 2010, 2011a, 2011b; King, 2007).

*Real-world, real-time learning for environmental/sustainability management*

Many examples linking learning in modules with campus needs and small and medium enterprises (SMEs) and charities in the local community have been developed at Kingston. Positive Environment Kingston (PEK) is one example (Corrigan, 2007). This scheme aimed to build the capacity of organisations in the Royal Borough of Kingston upon Thames (RBK) to improve their sustainability performance and through guidance on best practice help to reduce the borough's ecological footprint. The university's Sustainability Team worked with staff from the borough's environment and sustainability department and the University Volunteer Service to train volunteer students, from any academic discipline, as environmental mentors for local SMEs, small charitable organisations, schools, etc. to assist them in producing an Environmental Action Plan. In return for free training, volunteers mentored a partner organisation identified by the borough as needing help.

The scheme has since been adapted for use in the formal curriculum of undergraduate and postgraduate students working on environmental and sustainability auditing. Here, student teams undertake audits for local SMEs, charities, governmental organisations or in-house projects identified by estates and facilities managers and feedback recommendations, orally and by written report. Potentially, all participants 'win':

- employers win in terms of a free audit, exposure to sustainability ideas, and an opportunity to contribute to curriculum development
- students win in terms of professional experience to enhance CVs and employability and also through the contacts made to develop work-experience opportunities (and even immediate employment)
- university staff win in that employer links inform curriculum development and create contacts for research and consultancy opportunities

(Taylor *et al.*, 2011).

There are also wins in terms of town–gown goodwill, in-house capacity building and the 'feel good factor' of making a difference. Kemp (2008), trialling similar business-linked learning with Southampton students, also noted these wider benefits.

Formation of the SGS enabled staff to see more clearly the potential for cross-disciplinary sustainability learning and shared modules. A dedicated Bachelors' degree in Sustainable Development was launched in 2005, hosted by the School of Geography, Geology and the Environment but incorporating modules from Surveying and Planning, Life Sciences, Economics, and Business. Though restricted, in part, by practicalities such as timetable compatibility and mobility between campuses, this was a tipping point in terms of visibility and credibility for sustainability education at Kingston. Some new modules were designed reflecting the real-world learning ethos – for example, a work-shadowing module for final-year students linking with local business or in-house campus projects. The short, intense research phase within this module brings real benefits to host employers while students gain valuable professional skills. The module is highly rated by both parties and much more acceptable, in the current fee structures, than a full placement year.

Most advances in dedicated sustainability course development have occurred in the postgraduate curriculum, including a Sustainability Hub-coordinated suite of taught Masters degrees in which shared core sustainability modules are studied alongside discipline-specific modules. In this way, students learn about the sustainability perspectives of different disciplines and professional areas as an integral part of their programme. Potential barriers to communication through failure to understand and respect alternative professional practices are overcome. This part co-owned, part home-owned and Hub-supported approach has also overcome potential faculty objections, for example, around resourcing, timetabling, staff expertise, and the demands of professional organisations, as well as prompting debate and research developments among participating staff.

## Change for a sustainable campus

From the outset, creating a university where students would, every day, see best practice in sustainability management was a key aspiration for SGS colleagues. Academics wanted to deliver SD course materials and conduct sustainability research within the context of an institution that practised sustainability. Progress was driven by enthusiastic individuals and student demand – especially in terms of dissertations, personal staff research and consultancy, and estates- and facilities-linked real-world learning in taught modules (Box 12.2). Students and their tutors gathered the initial baseline information on campus biodiversity, waste, energy and travel. The need for a more comprehensive and coordinated approach mapped to an established EMS to help prioritise and monitor progress was clear, a point the 'gap analysis' (see Gongs and awards, above) reinforced. It also emphasised the need for top-level support and buy-in and, in 2005, following establishment of the Sustainability Facilitator post, the university secretary became the executive-level champion for sustainability. Previously, the Coordinator and, later, the Facilitator post, had reported to the SGS chair, as a project, but the SGS had no place in the university's formal line management system. Additional tipping points included the appointments of new facilities management and estates directors who gave higher priority to sustainability objectives. Campus developments are now routinely evaluated at the planning stage for their energy and carbon-saving potential and wider sustainability impacts.

Kingston's typically bottom-up, holistic engagement approach is seen in its end-of-year halls donation scheme (Smith, 2009). The scheme also charts transition from novice to best professional practice. The change reflects buy-in from hall managers, health and safety and facilities management staff, and the development of a protocol enabling the routine employment of student helpers to manage the process. Reusable items are taken by local charities or sold to incoming students at freshers' events the following autumn. Funds raised go towards the next year's scheme. Since 2005, when routine waste measurement began, about 29 tonnes of 'waste' have been reused in this way, reducing the university's environmental impact and waste disposal costs; benefiting local charities; helping new students set up with pots and pans, crockery, printers and stationery; and simultaneously raising student and staff awareness of waste management and sustainability.

## Towards zero waste at London School of Economics and Political Science (LSE)

LSE has adopted an innovative, yet simple, solution-based approach to waste management by incorporating reuse in operational activities such as: halls end-of-term reuse since 2005 (diverting an average of one tonne per hall of reuseable items); office moves and fit-outs since 2007 (keeping over 500 items in use, 20 tonnes of furniture out of landfill, saving 40 tonnes of $CO_2$ and at least £35,000); and formalising these requirements in the first 'zero waste' tender in 2008.

A three-step process was used to promote positive behaviour change for sustainability. This came to be known as RETHINK, which asks staff and students to REDUCE, REUSE and RECYCLE waste into resources.

At the operational level, pilot solutions were co-created with stakeholders. Results were carefully monitored and data collected to provide a business case with relevant indicators that gained the support of senior management. All actions were documented to produce clear instructions, training and engagement with all stakeholders in fun and interactive ways to make desired behaviour easy. Site visits, animation films and interactive games encourage desirable behaviour and bring about measurable continuous improvement

Zero waste implies questioning and adapting our current ways of thinking and doing; it generates behaviour change in order, not only to minimise negative impact, but to bring about positive impact. 'Out of the box' thinking and innovation is required. LSE has championed zero waste activities and demonstrated how they bring significant sustainability benefits for the environment, for the university's finances and to students and local community groups. A 'Reuse Guide and Film' has been made to share these activities with other universities. See: www2.lse.ac.uk/newsAndMedia/videoAndAudio/channels/sustainableLSE/Home.aspx

*Victoria Hands was Head of Environmental Sustainability at LSE and is now Director of the Sustainability Hub at Kingston University.*

External governmental and HEFCE demands for carbon reduction have served to further transform and invigorate sustainability in campus management. They have also established the engagement of student unions as a matter of routine good practice rather than individual enthusiastic endeavour. Most HEIs now have carbon management plans and have moved towards automated metering of energy and water use. Most have grappled with travel and waste plans; some have sustainable food policies, most embrace Fairtrade. Increasingly, universities are badged to formal EMS systems, such as ISO 14001, BS 8555 or the sector-specific EcoCampus. Kingston is working towards ISO 14001 through the BS

8555 route for its whole operation including its core business, its teaching, though others often focus on accreditation of operational concerns. Sustainable procurement and behavioural change to reduce Scope 3 emissions is the next big 'operations' challenge.

## The long and winding road of travel planning at Oxford Brookes

The environmental impact of the institution's travel has long been an issue grappled with at Oxford Brookes. A 'green commuter plan' was first adopted in 1999, proclaiming to the world that it would reduce the university's dependence on the private car. And, to a certain extent, it did. The university started charging for car parking in 2000 and then implemented a mixture of 'carrots and sticks' to change transport behaviour. Two subsequent plans included more exacting targets to reduce the number of people driving to campus.

In general, the 'sticks' have centred on the parking permit system – a charge linked to salary and an exclusion zone around the university's busiest site. The 'carrots' have been measures to promote the use of more sustainable transport modes. The BROOKESbus was introduced in 2003, which runs between all of the university's main sites. This has been a popular move; lauded for being the fastest-growing service in the southeast, with extra buses added to cope with demand. All students in halls of residence have a bus pass included in their accommodation package, encouraging good transport habits as soon as they come to Oxford.

Approximately 10 per cent of Oxford Brookes' students cycle to campus and 23 per cent of staff at our main campus; supporting cyclists is a key part of the strategy. Free bicycle maintenance sessions fix hundreds of bicycles each year and all campuses have showers for walkers, cyclists and joggers. Cyclists get free breakfasts during National Bike Week, have an online user group and are regularly asked how facilities could be improved.

The transport plan is monitored through a biennial survey of all staff and students. The results of the survey give stakeholder feedback to inform priorities for future work. Most importantly, progress is tracked. The latest measurement showed, for the first time, that the car is no longer the favourite choice for staff commuting to our Headington campus, a trend we hope will continue. The details of all three of the Oxford Brookes travel plans can be seen here: www.brookes.ac.uk/about/sustainability/about/resources

*Harriet Waters is Sustainability Manager at Oxford Brookes University.*

## Working for sustainable communities

Making a university more sustainable cannot be achieved through changes in the formal curriculum and campus management alone. The behaviour of staff and students, the in-house community, is crucial. Good communication, gentle education and engagement through fun activities are central to this. Building a communal sense of pride through awards achieved, charities supported or income generated by related research and consultancy is also pertinent.

At Kingston significant change came through linking with the Royal Borough of Kingston upon Thames' (RBK) annual Festival of Ideas by hosting sustainability-relevant town–gown events. The first, debating climate change and the concept of personal carbon allowances, gained a capacity mixed audience of 230 people. These events have proved sustainable in every sense. They continue despite cuts to the relevant RBK budget. Demonstrating how capacity-building secures embedded robustness, local citizens and Hub staff have liaised to maintain events; grassroots action has safeguarded the future. The borough and the university are both Fairtrade accredited and share some related activities and, as volunteers, university staff and students have worked with local residents to establish the RBK as a Transition Town. More formally, Sustainability Hub staff sit on RBK's committees for environment and sustainability and feed into the higher-level Strategic Partnership Group, thereby contributing to sustainability for London. Local partnership agreements have been made, for example with the South West London Environment Network and Thames Landscape Strategy. These links facilitate joint research and consultancy and develop new real-world learning opportunities for students. They enable the university to directly benefit its local hinterland, to be a good citizen. In this way, they support the SGS's aim to become a beacon in the region; they generate good will; facilitate a more collaborative approach to major planning developments; encourage local sponsorship; and engender public pride in the local university.

Beyond this local and immediate regional context, university staff who engage with sustainability contribute nationally and globally as external examiners, trailblazers for new professional practice, conference delegates, workshop leaders and speakers, and collectively influence progress for fairer sustainability-led societies.

### Conservation and development in sparsely populated areas: the CADISPA Trust

CADISPA was born in the late autumn of 1986 at Jordanhill College of Education in Glasgow as a partnership between WWF-UK and the college. Its task in the early days was to find a way to address the environmental education agenda in both the formal and non-formal sectors of education, developing materials for use in schools and informal settings that sought to bring to effect 'place-based' action learning. Following the Rio Summit in 1992, the EU approached the college to suggest that CADISPA should refocus around three key elements of the Rio Declaration: the non-formal learning sector, community development, and education for sustainable development (ESD) and action. In 1993, Jordanhill College of Education became the Faculty of Education at the University of Strathclyde.

All universities face a common problem in that their teaching can seem isolated from real life and, no matter how they seek to ground their teaching by the involvement of field practitioners as illustrations of practice, the criticism remains. Very few universities have brought central to their planning and activity the hands-on servicing of local communities by the same academics that teach the discipline of SD in the classroom. The CADISPA Project was such an illustration for over 21 years. With the support of the university managers, it showed how a university could service its wider role and responsibility to its host community while, at the same time, servicing the learning needs of its students. CADISPA supplied the University of Strathclyde with a test-bed for emergent ideas. It redefined the definitions of local SD, linking those principles more effectively to the lives of ordinary people. It harvested new knowledge, gathering it from the fields of activity across Scotland, and applying it directly, sometimes the very next day, to classroom teaching. Academic staff, from across the world, would come to sample a unique illustration of a late modernist, or post-modernist, interpretation of university teaching. That the university managers encouraged CADISPA to link the university so effectively into the hearts and minds of ordinary people in their host communities was an act of unparalleled professional faith.

CADISPA left the University of Strathclyde in 2009 and became the CADISPA Trust. In 2012, it continues to interrogate the linkages between sustainability and learning by finding routes into more sustainable living. CADISPA uses action-research and critical conversation as the process by which it supports social change in upwards of 50 communities across the CADISPA network. Its activity continues to be informed by the principles of both the UN Decade of ESD and the Millennium Development Goals.

For CADISPA, education is, and always has been, about useful learning and thoughtful action. Education as transformation, of both community living and personal behaviours, is always driven by those who truly know most about local sustainability – the people who live there. For 21 years, the University of Strathclyde (and Jordanhill College) shouldered the risk of allowing its academic staff to use their highly developed skills to the immediate benefit of ordinary people and to the lasting benefit of its student population. See: www.cadispa.org

*Dr Geoff Fagan is CEO of the Conservation and Development in Sparsely Populated Areas (CADISPA) Trust, which is an independent charity that helps people in rural Scotland to build sustainable communities 'for themselves – by themselves'.*

## Facilitating sustainability research and consultancy

It is not the remit of sustainability to determine the research ambitions of academics. Nevertheless, the growing sustainability expertise among staff has facilitated new collaborations and successful funding bids. Colleagues in the newly formed SGS developed a collaborative bid to the Higher Education Innovations Fund (HEIF) gaining income for work on sustainable design, planning, entrepreneurship and sustainability training for business. This supported an emerging lead-role in developing new professional standards for planners (Sayce and Lorenz, 2011), numerous design publications (*inter alia*, Chick and Micklethwaite, 2011) and follow-through funding for work with SMEs on sustainability audits (Baron, 2009). All these activities have fed back into sustainability course developments; they have been springboards for further research and consultancy, including the invitation to showcase work at the annual Ecobuild exhibition in London for the last two years.

Individual staff research bids are supported by the increasingly visible sustainability presence within the university. The university itself has invested by making sustainability research and consultancy one of three lead themes for its Dorich House Group of primary Erasmus partners. Participants work collaboratively on joint funding bids, including linking with industrial partners, and are developing a pan-European text on sustainability. Building on this increasing activity, Kingston's Sustainability Forum is now working to establish a distinctive research group for the forthcoming national research evaluation exercise.

## Building sustainability governance

Kingston's bottom-up sustainability activity initially had no governance structure. Much of the challenge, and the fun, lay in identifying a potential framework and effective lobbying strategies to achieve this. Explaining the business case was paramount. This was, and remains, not just a matter of finance and 'easy wins'. Big rewards also sit with reputational gain, potentially improved recruitment and retention of staff and students, as well as better town–gown relations, and the avoidance of expensive risks. In the project phase, each new bid for further university support funds also gave the opportunity to highlight forthcoming legislative burdens, essential risk avoidance strategies, and possible income generation or cost avoidance activities, for example, avoiding fines due to lack of compliance, or excess waste disposal bills. Since the SGS sat outside the university's formal reporting structure, this opened the door to informal direct communication with top-level executive members when urgent matters of changed governmental and/or HEFCE policy arose. Whenever sustainability visitors came, or conferences and workshops were held, executive colleagues were invited. Step-by-step awareness and understanding grew, and support followed. That the Royal Borough of Kingston was taking similar steps, and working with university colleagues to achieve sustainability, further influenced top-level engagement.

Key points along this route include the success of the SGS conferences, the CETL award, HEFCE's prioritisation of sustainability work, and local and national awards (such as the Green Gown Award) won. Each, in different ways, showed the value of sustainability awareness and behaviour. Sustainability policies are now in place at the highest level – though, typically for Kingston, they are embedded in other strategies such as equal opportunities, justice, citizen and environmental responsibility objectives, and the new 'led by learning' overall strategy. This approach is beneficial; it makes clear that sustainability is part of the fabric, in no sense a bolt-on extra. Feeding into the overarching strategy are, for example, the specific policies for education, environment, health and safety, wellbeing, Fairtrade and biodiversity.

## Reflections on the journey

What has enabled disparate bubble-up activities to generate change to a coordinated Sustainability Hub and an engaged top-level structure? Is this a repeatable story? There are many answers and some essentials are summarised in Box 12.3. Any transformation needs support and must reflect local circumstances and context. Showing how sustainability can link with institutional aims and objectives is a key step for generating support from top-level staff. 'Managing upwards', as essentially this process has been, needs a clear vision; an understanding of, and engagement with, the institution's routine processes; and, certainly initially, a touch of courage and a great deal of energy and persistence.

It is important to be clear what sustainability means; what concepts such as sustainable development embrace; and to listen, learn and explain generously. In the Kingston example, establishing a cross-university sustainability group, initially the SGS and now a focal discussion group, or Sustainability Forum, was an important first step. It built collegiality and support and by spanning academic, support staff and student communities, and embracing the whole university community, it gave authority to the requests for change. Building in-house expertise and understanding through workshops and seminars for staff, curriculum changes for students, and volunteering opportunities and informal learning exchanges with the local community, have all played a part and continue to be important. If all staff colleagues understand the aims, realising targets, for example, those around curriculum, waste management or sustainable travel, becomes much easier.

Linking to external partners and sharing the journey locally, nationally and internationally and across sectors has been significant and has facilitated research and consultancy developments, as well as informing best practice. Winning awards and getting noticed nationally as a leader for change undoubtedly helped to prompt top-level support and a realisation that failure to engage with sustainability might bring reputational embarrassment and resource loss. Local awards made a difference, too. That the university 'joined in' generated goodwill that is benefitting wider operations, from new building and

**Box 12.3: 'Bottom-up' pathways to a sustainable university: some key considerations**

- Have a clear vision – develop a route map.
- Make sure you have friends drawn from all sections of the university – form a support group.
- Ensure sustainability is understood – organise some gentle education from seminars to posters to guest speakers.
- Communicate effectively, in the way that people prefer to receive news. From websites to fun events, diverse approaches reach more people.
- Research your institution – know how it works and match your efforts to its structures.
- Research your baselines and regularly update on progress (or difficulties). Evaluate any disappointments and, if necessary, seek alternative approaches. Draw on the experience of external sector colleagues.
- Engage students and in-house colleagues to help.
- Link with local partners, community, business and other FE colleges and HEIs – generate wider support.
- Link to national and international debates, events and organisations, for example, EAUC, HEA, Fairtrade.
- Present the business case in terms that your executive expects, with some careful advance lobbying.
- Listen, engage and debate. Be responsive to institutional change and to people's personal circumstances – not everyone can cycle to work. Be flexible and encouraging to all.
- For sensitive campus operational changes, facilitate debate and discussion; always explain the reasons why.
- Address setbacks positively – for example, if photovoltaics are suddenly deemed uneconomic, find an alternative.
- Celebrate success – this will encourage new engagement and progress.
- Never preach – never.
- Make the journey fun.

refurbishments to grounds management and placements, and volunteering and research opportunities for students and staff.

Progress monitoring is essential; it helps to decide priorities and highlights successes. Establishing baselines needs to be a first step so that progress is demonstrable and the crucial areas for attention are evident and can be prioritised. Simple in-house evaluation systems work well initially, but subsequently entering external benchmarking exercises gives rigour and validates progress. Feedback from these exercises will support the business case for sustainability activity in economic and reputational terms. Making a robust business case is likely to be key to securing a small staff team whose priority is to focus on sustainability, as distinct from relying on the enthusiasm of staff whose professional remit lies in teaching, research, computing systems or elsewhere. For all that grassroots sustainability is important, the presence of dedicated staff brings extra safeguards and speedier action. In times of change, 'green issues' or 'sustainability objectives' may seem peripheral to colleagues otherwise focused on, for example, discipline survival, research rating or institutional financial viability. At these times, a dedicated sustainability team can beneficially maintain impetus and, where appropriate, redesign sustainability links to ensure their continued incorporation. Opportunism will be part of any institution's story; seizing chances presented by staff changes or structural reorganisations and proposed campus developments is essential. Building networks that encourage colleagues to bid into external funding opportunities for sustainability research and development, and making it easy to take part, will accelerate progress.

There will be setbacks but this is when internal and external support networks can make a difference and help to minimise disappointment, overcome nervousness or even suggest alternative strategies. There will be problems and challenges. For example, how do you quantify sustainability in the curriculum? Should you be counting numbers of courses, modules, assessments, research projects, and, if so, what are the keywords that denote sustainability is being taught or learnt? Or should you be examining change in student attitudes and how can this be measured? What to measure, and how, are current big questions. Furthermore, the answers may be sensitive to changes in academic structures, which may have little to do with sustainability concepts. Should you be aiming at a core sustainability module for all, irrespective of discipline, and would this work in your institution's academic structure? Do you have the staff expertise to facilitate this or is 'embedding' more appropriate? These are important on-going challenges. Many universities are currently revisiting modular structures. Recording sustainability presence in ten, fifteen, twenty or thirty credit units may give very different outcomes for essentially the same activity. Look beyond the surface of such metrics and accept that, in delivering sustainability in the curriculum, no 'one-size-fits-all' approach is likely to be effective. A plethora of good practice ideas should be embraced.

Making the sustainability journey fun will ensure wide participation and encourage support and debate. Sustainable fashion shows, campus farmers' markets, walks for health, community gardens and similar events all demonstrate the everyday relevance of sustainability and its fun 'pain-free' benefits. There are easy things to do, as well as more difficult lifestyle changes to make. Communicating effectively, internally and externally, builds support as well as enabling the learning and sharing of best practice; it is another key priority.

Engaging students is vital. They are the principal reason for this transformative action. Whether in the formal curriculum, through informal opportunities, or by the example of sustainability in institutional processes, it is their sustainability knowledge, understanding and skills that can achieve the ultimate goal of all this activity, a global sustainable future.

## Notes

1  A world café is a simple and effective format for hosting group dialogue. Using an informal café-style setting, it encourages relaxed group discussion around specific questions and taps into the collective intelligence of a group. Participants are often provided with pens to record their thoughts and ideas on paper tablecloths. At regular intervals participants change tables to cross-pollinate ideas. The events usually conclude with a plenary.
2  Green Impact Universities and Colleges is a national departmental accreditation scheme run by the NUS that encourages and rewards teams of staff and students who improve working practices and take practical action to 'green' their departments. See: www.nus.org.uk/en/campaigns/greener-projects/greener-unions/green-impact/
3  See: www.kingston.ac.uk/sustainability

## References

Annan, K (2001) *Secretary General Calls for Break in Political Stalemate over Environmental Issues.* Available at: www.unis.unvienna.org/unis/pressrels/2001/sgsm7739.html (accessed 21 May 2012)

Association of Commonwealth Universities (1993) *The Swansea Declaration.* Available at: www.iisd.org/educate/declarat/swansea.htm (accessed 15 March 2012)

Association of European Universities (1994) *Copernicus: The University Charter for Sustainable Development.* Available at: www.iisd.org/educate/declarat/coper.htm (accessed 15 March 2012)

Association of University Leaders for a Sustainable Future (1990) *The Talloires Declaration.* Available at: www.ulsf.org/programs_talloires_report.html (accessed 17 February 2012)

Baron, P (2009) *Surrey Sustainable Business Partnership: SSBP.* Poster presentation at the Sustainability Hub launch exhibition at Kingston University. Available at: www.kingston.ac.uk/sustainability/resources.html (accessed 21 June 2012)

Brandon, P (2010) *Development of an Electric Motorbike.* Available at: www.kingston.ac.uk/sustainability/resources.html (accessed 21 June 2012)

Brandon, P (2011a) *No Plan B for Brandon's Bike.* Available at: www.kingston.ac.uk/sustainability/resources.html (accessed 21 June 2012)

Brandon, P (2011b) *The Lean Green Racing Machine Takes Podium Place: Full story.* Available at: www.kingston.ac.uk/sustainability/resources.html (accessed 21 June 2012)

Chick, A and Micklethwaite, P (2011) *Design for Sustainable Change: How Design and Designers Can Drive the Sustainability Agenda*, Lausaunne: AVA Publishing

Clarke, Rt. Hon Charles (2003) 'Examination of Witness (Questions 200–219)', *UK Parliament Select Committee on Environmental Audit.* Available at: www.publications.parliament.uk/pa/cm200203/cmselect/cmenvaud/472/3032506.htm (accessed 21 May 2012)

Corrigan, N (2007) 'Positive environment Kingston', in *Green Gown Awards Winners Brochure*, 22. Available at: www.eauc.org.uk/2006-7_greeb_gown_awards (accessed 15 March 2012)

Corrigan, N, Sayce, S and Taylor, R (eds) (2009) *Sustainability in Practice. From Local to Global: Making a Difference*, Kingston: Kingston University Press

Dawe, G, Gant, R and Taylor, R (2003) *Sustainability in the Curriculum.* Available at: www.kingston.ac.uk/sustainability/resources.html (accessed 21 June 2012)

Eames, K (2012) *Kingston University: Sustainability in the Curriculum. Measuring and Monitoring Progress, a 2010–11 Snapshot. Internal Report,* Kingston: Sustainability Hub

EAUC (Environmental Association for Universities and Colleges) (2011) *Learning in Future Environments, LiFE.* Available at: www.eauc.org.uk/utc (accessed 3 March 2012)

Forum for the Future (2004) *Memorandum from Forum for the Future for Parliamentary Select Committee on Environmental Audit.* Available at: www.publications. parliament.uk/pa/cm200405/cmselect/cmenvaud/84/4121402 (accessed 17 February 2012)

HEFCE (Higher Education Funding Council for England) (2005) *Sustainable Development in Higher Education: Consultation on a Support Strategy and Action Plan,* Bristol: HEFCE

Higher Education Partnership for Sustainability (2004) *Learning and Skills for Sustainable Development: Developing a Sustainability Literate Society,* London: Forum for the Future

Kemp, S (2008) 'Embedding employability and employer engagement into postgraduate teaching: a case study from "environmental management systems"', *Planet,* 31: 47–52

King, J (2007) *The King Review of Low-Carbon Cars. Part 1: The Potential for CO$_2$ Reduction.* Available at: www.hm-treasury.gov.uk/d/pbr_csr07_king840.pdf (accessed 20 February 2012)

Sayce, S and Lorenz, D (2011) 'The value of sustainability: where are we now?' Presentation to the 40 percent Symposium, *Tackling the Impact of Sustainability on Commercial Property Value and Investment,* London: RIBA. Available at www.bre. co.uk/filelibrary/events/40%20Per%20Cent/Sarah_Sayce.pdf (accessed 21 March 2012)

Smith, H (2009) 'Kingston University end of term halls donations scheme', in *Green Gown Awards Winners Brochure,* 24. Available at: www.eauc.org.uk/2009_green_ gown_awards (accessed 15 March 2012)

Taylor, R, Smith, H, Ryall, C and Corrigan, N (2011) *Real-World Learning for Sustainable Environmental Management.* Available at: www.heacademy.ac.uk/assets/documents/ sustainability/kingston_sme_final.pdf (accessed 15 March 2012)

Toyne, P (1993) *Environmental Responsibility: An Agenda for Further and Higher Education,* London: HMSO

United Nations Educational, Scientific and Cultural Organisation (UNESCO) (2005) *United Nations Decade of Education for Sustainable Development 2005–14.* Available at: www.gdrc.org/sustdev/un-desd/implementation-scheme.pdf (accessed 4 March 2012)

Wright, T (2002) 'Definitions and frameworks for environmental sustainability in higher education', *Higher Education Policy,* 15: 105–20

# 13  Towards a Green Academy

*Heather Luna and Larch Maxey*

## Introduction

Some universities working on the sustainability agenda make headlines, and rightly so, like University of Bradford's Ecoversity (see Hopkinson and James, this volume) and Exeter University's One Planet MBA (although it is just one programme). The UK Higher Education Academy's pilot change programme, Green Academy,[1] has not gained significant media attention, but is leading to strategic collaborations, with initiatives feeding off each other across the participating universities, adding up to increased resources (both monetary and human) for sustainability thinking and action on campuses. In this chapter, we ask what truly *appropriate* responses to the full range of challenges facing people and the planet would look like for us individually, and for our institutions and organisations (see Sterling, this volume). In particular, could such appropriate responses include and take inspiration from Green Academy, or initiatives modelled upon it, to steer universities towards a 'strong' sustainability (Constanza and Daly, 1992) that really makes the most of the opportunities presented by these challenges?

Rather than replicate case study reports, which are publicly available (Kemp *et al.*, 2012; McCoshan and Martin, 2012), we investigate and learn from the personal values, goals, experiences and reflections of the people committed to this work. This chapter draws on Heather's participatory action research in the programme, interviews and email exchanges with the current team leader of seven higher education institution (HEI) teams. We sent them a list of questions, emphasising that their responses would be anonymous, thereby encouraging them to be frank. We were keenly aware that sustainability champions often self-censor certain feelings, experiences and phenomena when discussing the process by which they have progressed sustainability, for fear of 'rocking the boat'. Enabling change agents to speak freely is essential to maximising the learning applicable to HE in the UK and internationally. This can inform all change, including sustainability change and any future rounds of Green Academy or similar initiatives elsewhere.

## Engaging with universities

Sterling (Chapter 1) asks, 'What is a university for?' We might also wonder: Would university senior managers recognise the contribution of universities to the following goal?

> The goal is very simple and technical. And the goal is a delightfully diverse, safe, healthy and just world. With clean air, soil, water and power – economically, equitably, ecologically and elegantly enjoyed. Period.
>
> (McDonough, 2007)

Few people would dispute this is a fine goal, but would universities and the millions of staff and students within them worldwide claim it as theirs? We argue that neither academic freedom nor political-economic imperatives should prevent universities embracing this goal – and we are heartened by the many staff and students who would align themselves in this way. We know the goal is daunting. Many have cited the barriers to embedding sustainability in universities, such as lack of time, lack of resources, lack of knowledge, and the belief that sustainability is irrelevant to one's field (Dawe *et al.*, 2005). Further, university structures and wider societal structures often make change very hard to bring about (as reflected in a number of chapters in this volume). No blueprint exists for how a university can change so dramatically as to make an appropriate response to the scale of challenges we face, but there are pointers dispersed in best practice around the world. The following section considers the potential of Green Academy to offer several pointers within a model for effecting change.

## Green Academy: a change programme

### *History*

Green Academy is a programme that was launched in late 2010 to embed sustainability holistically within UK universities, 'to help institutions achieve sustainability in the curriculum goals' (HEA, 2012; see Kemp *et al.*, 2012; McCoshan and Martin, Embedding sustainability into the higher education curriculum, end of this chapter). We reflect here on its first cohort, which involved eight teams of five or six people. This Higher Education Academy (HEA) pilot initiative drew on the HEA's successful change management programme, Change Academy. Heather (in her former role with the HEA) conducted a scoping report and determined that no such change management programme yet existed for universities addressing this agenda. After much consultation with senior managers and academic champions, as well as Green Academy partners, the National Union of Students (NUS) and the Environmental Association for Universities and Colleges (EAUC), the format was developed by the main facilitator, Jimmy Brannigan, of ESD Consulting.

Each HEI team was required to include a senior manager, an academic champion and a student. The remaining two or three members were to be chosen from estates, finance, human resources or the careers office. Each team appointed one member as team leader.

The call to the sector consisted of an application, which included the following questions:

1   What is your institution's current approach to, and engagement with, sustainability and education for sustainable development?
2   What is your future vision of sustainability in the curriculum for the HEI?
3   How do you envisage participation in Green Academy will support your institution in achieving this vision? In other words, what does your institution hope to achieve through taking part?

Due to staff and monetary constraints, only eight of the twelve HEIs that applied could be chosen to participate. The selection was based on responses to the above questions, plus the strength of the team gathered. The chosen HEIs were: University of Bristol, Canterbury Christ Church University, Keele University, University of Nottingham, University of Southampton, Swansea University, University of Wales – Trinity St David, and University of Worcester. They represented different types of institutions, at different starting places.

The team leaders met in February 2011 at the HEA main office in York, mainly focusing on team dynamics. A month later, the full teams met in Leeds for a two-day residential. While some input was given by eight 'critical friends', (experienced practitioners in the field, including Shiel, Sterling, Taylor and Hopkinson, contributors to this volume), the majority of the time was allocated for the teams to work out their own strategy, in their own team room, and to consult with critical friends for advice. Two months later, the team leaders reconvened to check on progress.

To date, no evaluation or publication has included all eight HEIs due to the unavailability of a team leader at the eighth participating university. This highlights the importance of distributed leadership (Shiel, this volume), so that the task of carrying an initiative forward does not rely upon one individual. It also highlights the stresses sustainability advocates are frequently under as they juggle multiple responsibilities, often in addition to other workloads/'the day job' (see Sayce *et al.*, this volume). We approached the team leaders with the following questions, again, reassuring them that all comments would be anonymous. (A few would have preferred they were named, though, as they were quite pleased with their institution and what they were saying about it.)

The questions were:

1   What has been your own **personal goal** around the work? How does Green Academy (GA) fit within your own personal framework for reaching this goal? *Please note that we are not looking for the goals/visions/mission statements of your HEI. We are looking for your own vision/goals here. If*

*you want to mention how it is different from your HEI's own stated vision/ goal, that'd be great. But you might want to generalise it so we can't determine which HEI you are describing.*

2  How would you describe **your role** in this work? *Please reflect on a personal level here.*

3  What **personal lessons** have you learned (about working with people, about yourself, about the 'project')?

4  What would your **wish list** be to move the agenda forward in HEIs? What needs to happen next? What support would you need to make that happen? How might a future rollout of GA be improved to support this for you or for others coming up?

5  Apart from the support you had from the HEA, critical friends, and other GA teams, what were the **key resources** you used/are using during the process? What resources would you have like to have had, or would like to have?

We now pull together common and key themes that came out of the responses. In line with basic principles of interpretivism/qualitative research (Flick, 2006; Silverman, 2004), we are allowing the subjective meanings to be provided by the respondents themselves, emphasising their unique voices.

## *Starting out*

Green Academy institutions started from varying degrees of experience:

Green Academy saw us coming from nowhere to make this commitment.

[T]he team who had put in place significant ESD [Education for Sustainable Development] initiatives within the institution in the 1990s as young things, happened to have reached senior positions in [the] university just at the time of the HEA initiative. We had worked together for over a decade on sustainable development education and this process enabled us to think about how we could link together all the ESD networks within the university into an integrated network.

Less predictably, this variation extended to the starting points and background of the team leaders themselves. One had recently obtained a new senior management role, but had been working on sustainability at their university for nearly twelve years, another for fourteen years. They said:

There's this little sanctuary back where I come from that I used to go to escape to from stress, parents … It gave me a sense of peace and connection and I didn't even realize at the time that it was [a] connection with nature …. I would just walk around there and … that got me interested. But, actually, plate tectonics changed my life. As soon as I understood the concept of plate tectonics and the whole idea of the earth as an old entity and the way that even the plates are changing ... and the way the world has changed over time ... [Y]ou start putting a geologic time scale on the planet and then you look at what human beings are doing in that time and ... the horror of it all – that's what got me into the environment ... Always had that affinity with nature but it was putting a timescale on it all ... That's why now I'm into sustainability in the university.

Similarly, for another, their 'whole professional work [had been] shaped by the question of the carrying capacity of the planet'. For another, the field of sustainability was new, and an air of serendipity sent them into this position:

I was the member of the university's Teaching and Learning Board who expressed most interest when the new Director of Sustainability within Estates came to the Board to talk about shifting the sustainability agenda from the environment of the campus and procurement to the core activity of teaching and learning (I was sitting next to him).

Taking up the challenge of sustainability involved personal and professional re-assessment. For some participants this was challenging and unsettling:

I was worrying about the value and purpose of some of my teaching. First via the Director of Sustainability, and then via the Green Academy, I started to realise what [my teaching's] purpose was and to see how, looking beyond the core knowledge to think more deeply about application, pedagogy and process supported both myself and my students in changing our ways of thinking and acting. I suppose my personal goal was to understand sustainability beyond the environmental sphere. I do that now, but the downside is that my vision of sustainability is very big and perhaps too all encompassing. It has also led to me becoming more anxious about the future! I know that we need to change and adapt in many different ways but I don't see sufficient evidence of willingness for the West to do so.

I got into it because [I had been] reading student exams and essays [that said things], like, 'Humans are too insignificant to affect the climate.' I became deeply worried by this: that our students, who should be, at least, environmentally literate, are thinking like that.

More seasoned university change-makers were measured and strategic in their use of the programme:

[M]y personal agenda was about wanting a chance to work out how to move from all the great initiatives to something integrated.

*Defining sustainability*

A major challenge for integration across a university is language. What is meant by 'sustainability'? One team leader said:

> My personal goal has been to make sustainable development real.

But definitions in academia are notoriously contentious and participants were sensitive to this:

> The key thing is to find space for alternative visions, particularly for academics from different disciplines, but also for those who have a more environmentally focused view. I've struggled sometimes to sell Sustainability Literacy to those who see it as about carbon reduction – the environmental hold is still strong. It's explaining that sustainability isn't a subject or a topic, it is a way of thinking, acting, behaving – it is about empowering people to do things for themselves, not just about limiting impacts. So, I've learnt a lot about 'selling' new ideas – the need to avoid being polemical or preachy, the need to show positive benefits rather than negative problems.

Some recognised the limitations of sustainability's amorphousness and the value of clear targets and goals:

> [I]t is an area where the phrase is either neutral or benign and people are perfectly happy to use the word until they are asked to make a commitment – then they want a definition.

For senior managers and government bodies, sustainability offers challenges, risks … and opportunities:

> I have watched the journey from 'sustainable' being a shockingly radical word, where councils and developers would fight tooth and nail to avoid us getting it included in documents because it connoted change and discomfort, to its current vogue amongst those same people as a comforting pale green colour wash … [T]he language is now so universal, the currency so devalued that we are having to work to find alternative language that does deliver real change rather than the pale green colour wash of 'sustainable development'. I think I felt cynical about the way the language of Sustainable Development has moved from a focused radical language to being a green colour wash used in ways that are remote from the origins of the concept. But what I learnt [through the GA process at my HEI] was that, behind that fear, people had actually moved; that individuals across all disciplines were doing their own bit, and were open to being more explicit about it and being pushed to reflect a bit further.

### Defining the role

The team leaders unanimously were against being 'polemical' or 'preachy'. They saw their role as complex and astute, of one as listener and reflector. Their reflections indicate what it means to be a leader, in the broader sense of the word, echoing leadership themes Shiel raises in this volume. Some have specific sustainability roles; others are doing this work alongside their normal academic role.

Team leaders defined leadership in many ways, as:

- a catalyst and conscience;
- teacher ... mentor ... supporter;
- shock troops;
- key communicator;
- a practical role, not evangelical;
- to coordinate, integrate and celebrate;
- the invisible glue or string that ties things and people together.

Other comments indicate the depth and breadth of what is, and was, necessary:

> It is very much a partnership. I don't think it would have worked if it [were] just me or just him. [We] needed to be able to cover the academic community through me, and the more operations side of the community through him.

> My approach has been to build on the past and, to quote [Professor] Geoff Scott [PVC at University of Western Sydney, Australia], 'listen, link and lead'. My goal was and remains to uncover the leadership within that exists throughout our organisation.

> [Colleagues] welcome support, ideas, and gentle nudging just to ensure they do free up time to think about what they can do.

> I've had many, many hours of face-to-face conversations. I think I've had some one hundred or more face-to-face meetings, repeated meetings with senior management on responsibility: what they do, their aspirations, and then feed it back to them. [I] test if that's why they support it and how I can help support what they want to achieve on this agenda and how to put that into those headings ....You've got to invest in the time with the individual conversations in order to get that buy-in. But also, when people try to move away from the agenda, that's when it's critical that you've got the VC, or at least the PVC, on side so it's clear it is a strategic priority for the university. I see myself as the person who makes sure things happen, and these things are often the ideas and initiatives of others. And I'm the person who represents the ideas to senior management and ensures that sustainability feeds into the top-level strategies. I can provide the evidence base.

***Giving evidence***

As the last comment suggests, making the case for sustainability includes providing evidence:

> [D]emonstrating that value and good ethics make good financial, social, economic and environmental sense.

> This has been the biggest success for me since it has made the business case incontrovertible. It comes back to the selling argument. As a consequence, I've got surprising amounts of support from senior management.

> [I changed] my emphasis [by] demonstrating the business advantage to the university in terms of recruitment and global profile and also the benefit of linking to the local community .... My fundamental belief of where I want society to be and where I want the university to be and where I want the global community to be hasn't changed at all. I'm just looking at different mechanisms to get there. I don't care how we get there or how I help people get there. It's about getting there in the end. It's always been about that. It's all about strategy.

> [T]he issues around sustainability [are] just good business. Potentially it can reduce the costs of business operations in the longer term and therefore possibilities exist for making money in a manner that is more acceptable to the market that business serves.

***Measuring sustainability***

How should a university measure its success around sustainability? What evidence is 'valid'? Where do People and Planet's Green League and the EAUC's LiFE Index[2] fit into the question of appropriate metrics? Only two team leaders mentioned this contentious issue. One is in dialogue with a funding council 'as to what appropriate metrics might look like for HEIs on [this] agenda'. Another identified a fundamental contradiction between the emphasis on metric measurement of sustainability performance and the emergent, participatory nature of sustainability:

> There is a real need for us to rethink how we evaluate progress on ESD. At present it is being measured primarily in structural inputs – having policies, strategies, and measurement matrices – rather than change outcomes. And there is an emphasis upon the central top-down approaches, which make inputs measurable, rather than qualitative assessments of the strength of the grassroots network of initiatives, which change the culture of an institution from the bottom up. The [former] run counter to the decentralised, locally owned approaches to ESD, which give people the ownership of the change

they want to build. A question for us all is to reflect upon whether the measures of ESD success in HEIs are currently too driven by traditional managerialist hierarchic input measures such as central strategies ... Change and survival will come from the networking of people at the grassroots, creating dynamic, resilient networks of action and change – then we need to be mapping examples of that within HEIs and using those as exemplars and measures of success, not the more traditional management measures, which are indicators of static hierarchic organisations. Working out how to articulate ESD for HEIs that is networked, flexible and resilient, means letting go of some of the current measures of ESD success, and is challenging.

A different method was suggested:

We have adopted a social change governance approach, in which we seek to embed ESD within the community developing initiatives from the bottom up.

### *Working with others*

Counting on the 'bottom' tier to step up to the challenge of sustainability seems daunting, but team leaders such as the three quoted here offered this enthusiastic reassurance about engaging with everyone:

The most important thing I learnt, about the project, is that people are more ready for the change than I expected.

When people 'get sustainability' within the context of their curriculum, they are off and running before you know it.

There's always a great pleasure when you've managed to put your argument in a particular way that [pays off] – which could be called manipulating people. [smiles]

For sustainability advocates, the task thus becomes one of catalysing everyone's enthusiasm, creativity and potential to contribute.

### *Being surprised*

The generally positive, open reaction from colleagues was not the only thing that surprised the team leaders. In keeping with the original aims and approach devised by the HEA, the 'official' story of the Green Academy suggests that 'friendly competition' was a positive element of the programme and contributed to its success (Kemp *et al.*, 2012). However, not all participants experienced it in this way:

I was surprised by the competitive element that exists between universities and this is very evident now, which I think is a little unhelpful.

This concern over increasing competition within HE linked to concerns about its increasingly corporate nature (Blewitt, this volume). Another team leader identified the main learning from the programme as

[t]he morass of putridity that exists in the politics of management, across the whole HE sector. It is becoming very corporate in that regard, something I tried to leave behind in entering HE.

Some surprises were on a more personal level and reinforced the personal transformation and reassessment noted above:

I've … found a voice and a confidence that has been missing for a long time because I feel that I am now starting to be listened to and respected at a senior level.

Such transformations were not restricted to team leaders, as McCoshan and Martin's evaluation at the end of this chapter illustrates.

### *Learning*

The team leaders also offered tips and recommendations gleaned from their learning on the programme:

- Get senior management on board: 'If you are going to take an agenda as fundamental as this, you have to have a champion on the senior management team. The vice-chancellor has to be a champion because, in some areas, it might mean making difficult choices – particularly when they are financial choices.'
- 'Use the university calendar of meetings.'
- Move beyond the halls of academia: 'But not only do this internally, but also externally, because if you're making a big commitment like we are – in repositioning ourselves in the UK context – you've got to get out there and tell people what you're doing.'
- 'Don't be afraid to share ideas. I never wanted to share ideas externally because I've wanted to be protective. But what I've learned from GA is: Get out there, share ideas, and you can then create something better.'
- 'Take a framework that can be adapted; it can be very powerful.'
- 'Be careful about promising too much. That's my problem, I tend to say yes and have lots of ideas but [do] not necessarily manage to pull them all off because of time.'

And probably most importantly: 'There is no one answer, there is no right way.'

*Celebrating resources*

When asked about key resources, people were high on the list: 'interns, temporary staff, friends, networks within and without', but also students:

> The most important agent of change and support has been the student body, who are outstanding in their ideas and practical innovation – and interact seamlessly with the academics and institutional organisation. Students are the key resource without a doubt – for both ideas and action. They are unrestricted by the experience of knowing how difficult it can be to achieve things. I see my role as being to remove the obstacles for them, so they retain that freshness and innovation.

This is an important finding given the emphasis on 'student experience' and student as 'customer' (Blewitt, this volume). It demonstrates the potential to advance sustainability within HE's neoliberal turn (Maxey, 2009). While it has so far been very easy for permanent staff to ignore or dismiss the role of students, claiming they are naive, too transitory or unreliable, the above team leader's experience suggests that a more humble, empowering role opens up considerable scope for change.

In addition to human resources, two surprising material resources were identified in the form of policy statements that sit above the university, and which can be used to bring it to account:

> The government commitments. Every government will make commitments on this agenda. We need to hold government to them. We need reaffirmation for that commitment from the appropriate minister. I'm a great believer in winning arguments. Get full party support for initiatives. Get buy-in on the philosophy. Steam roller people! Take a radical position; this allows people to move more radically in that direction. It doesn't win any popularity contests but it does leave people thinking, 'Okay, well, maybe we do need to do a bit more'.

And:

> The UNESCO statement on ESD is particularly helpful to people in a diverse university because of its embracing scope, which enables all disciplines to see how they can play their part.

In addition, one team leader pointed out an important Green Academy resource:

> The critical friends helped a lot because I'd always been sort of aware of the work of, say, Stephen Sterling or Daniella Tilbury, but I'd never really appreciated how good it was and how useful it was. It's been good to get those resources and tap into those networks and, if I hadn't met them through the residential, then they wouldn't know who I am.

The book, *Sustainability Education: Perspectives and Practice Across Higher Education* (Jones *et al.*, 2010) also got a mention:

> The stuff from [the HEA's] ex-Subject Centres, particularly in terms of generating discipline-tailored work. But actually, the *Sustainability Education* book, with the disciplinary structure, is excellent and what is needed more of.

### Having time

Time, time, time. This echoed throughout the responses, and was identified as the key barrier and opportunity:

> The most important resource was space and time.

> There aren't enough hours in the day.

> [O]ur progress on GA has gone through the roof. I'm spending more time on it as a result – it's a bit of a double-edged sword.

> So then it just comes down to time.

> So the wish list is really about time and mechanisms to link sustainability more closely to teaching and research so that it doesn't require extra time, it is just a standard part of the job.

### Wishing

Next on the wish list, after time?

- Ask funding councils to institute requirements to implement sustainability.
- Harness student power: 'Support their interest and the rest of the institution has to follow.'
- Provide 'a [university] presence, a hub where we can start practising what we are preaching, acting as a venue for the student experience, an area that the community can use.'
- Provide more 'discipline-specific resources that are really easy to access .... There's a lot of supposed sharing of resources but the problem is that you start to share so many resources that it becomes impossible to find them'.
- Involve the public in open discussions about the role of the university in meeting sustainability challenges.
- Ask the HEA to implement a sector-wide senior management (VC) programme.
- Involve businesses, industries and local government in areas such as providing students with real-world learning opportunities and in informing

the curriculum to ensure it is relevant to local (and national) businesses engaging in the 'green economy'.
- Encourage local government to use universities as their 'port of call for project support'.
- Provide 'more empirical evidence of the benefits of sustainability'.

### Giving thanks

The HEA's Green Academy played a huge role in enabling the team leaders to carry forward sustainability initiatives at their universities:

> I have had tremendous support from the HEA. It has been really good how GA has meant that any ideas we have are given a wider hearing ... The HEA has actively looked at working with all kinds of different universities.

> Green Academy offers the opportunity for HEIs to look at whether they can do more. If GA has some members pushing, then others will too. It's a catalyst role.

> The project is a remarkable one in that there is little or no money or major investment of resources, but is powered by such good will and free individual energy.

> I needed the HEA to enable me to package my next steps work as part of my university management role, rather than additional to it, to get the space to do it.

> I had the remit and the desire to move us forward in curriculum development and, whilst I had a developing university strategy behind me, this was about me developing a new and creative approach, bringing people together and linking skills and expertise with university infrastructure and direction. Personally, the Green Academy has been a stepping stone to building the team that will help us build a curriculum for the twenty-first century.

> My university already talked about sustainability prior to Green Academy, but it is developing a clearer vision of sustainability within the curriculum now – although it is still a way off from ensuring this.

### Offering ideas

This raises the question of the future of Green Academy. The first cohort was in the academic year 2010–11. The HEA's formal support ended in May 2011, with the final team leaders' meeting. The following year (2011–12) was an 'evaluation year'. In October 2012, the HEA opened bids for a second cohort of Green Academy (GA2) for 2012–13. (The first cohort was heavily subsidised by the HEA; the second is less so.) The team leaders offer advice on how Green Academy could be improved and how it could help teams go further.

On networks and building a community:

> The next stage should have another eight HEIs, with each of the current eight acting as mentors ... We should become a network, and should talk to each other. We should formally be created as a network on the back of that. Therefore, through the HEA's wider work in all sorts of categories, people can then be referred to the GA network in terms of helping people make decisions in terms of xyz to move forward.

> Involving GA1 participants in GA2 would be good for both and would help to build a sense of community, which I have not seen much of.

> The next GA could do with more sustained contact between groups/team leaders. The residential was superb – everyone felt they achieved a huge amount, but although the team leaders met again, the whole teams haven't had the opportunity. Having said that, the pilot group have sustained contact and produced a conference paper [Kemp *et al.*, 2012] and one-day conference [Nottingham, May 2012], so they are doing what they should do, maintaining contact and using each other as a critical friend as needed.

On the process:

> Bring exemplar project practitioners together at regional events with potential future GA universities. This should happen before the bidding phase and should perhaps require teams to bring with them their university value statements, mission, strategic goals, etc., so that they can build a shared understanding of what their organisation is about and where it wants to go.

On the content:

> Via the HEA/NUS we had access to some research into student attitudes towards sustainability. However, there is a vast literature now on sustainability in the academic journals. The critical friends were authors of some of this, especially Stephen Sterling, but it would be good to have a facilitated discussion about more of this research literature as part of the formation of team strategies.

On stronger linking with the NUS:

> There should be closer alliance with the NUS. The GA HEIs should strongly support whatever the leading NUS initiative is on this agenda. The HEA should require that GA institutions commit their students' union going down the Green Impact route, looking at staff and students.

## Our recommendations

There have already been some spin-offs from the first round of Green Academy, albeit primarily limited to team leaders delivering joint workshops in international and national conferences (Kemp *et al.*, 2012; the national conference in Nottingham, May 2012). The fact that Green Academy teams, or at least their leaders, continued collaborating and organising in this way demonstrates the value and impact of the programme. It also indicates the significant benefit that could be unleashed with a little proactive support for the Green Academy's networking function, as some participants suggested.

Although it is by no means given as we write, perhaps another eight HEIs will join the GA programme (GA2) in late 2012, sending teams to start the process of integrating sustainability into the curriculum in earnest. Adding eight HEIs annually, even without the evaluation year taken between GA1 and GA2, would mean 40 of the 165 HEIs on board in five years. While that would be a lot better than nothing, it would not be an adequate response given the scale and significance of the challenges outlined earlier in this volume.

We recommend that Green Academy be sped up and expanded as a strategic and appropriate response:

- The main role of the Academic Lead for ESD at the HEA should be to focus on the Green Academy, rolling the programme out and supporting the creation of a vibrant network of cutting-edge ESD initiatives offering resources for HEIs throughout the UK and internationally.
- A minimum roll-out timetable would be to double GA teams annually, with each team mentoring a new team the following year and communicating the programme's value and importance to colleagues in HEIs not yet on board, so that, after five years, 128 HEIs would be involved. While this has the advantage of allowing activity to be progressively ramped up, it may not meet the scale and urgency of our global challenges. We, therefore, recommend a more rapid scale-up also be considered to ensure as many HEIs who want to participate, perhaps the full 165, are on board within three years.
- A growing pool of critical friends could operate across teams and work closely with both GA teams and the HEA ESD Academic Lead.
- Green Academy, itself, could be extended to include staff development on the ecological physical limits and boundaries we have breached and are breaching, and to critically appraise a goal such as McDonough's (see above).

- Action-plan timelines could be based on the real timelines of what is happening, particularly with regard to climate change, where the IEA (2011) concludes we have three to five years to prevent catastrophic climate change, beyond which the measures necessary become too expensive or unfeasible.

In line with this, the HEA could publicly recognise that its ESD theme is a necessary condition for the HEA and the other themes to exist.

Meanwhile, the senior managers involved in Green Academy (and those critical friends in the HEIs) could put together a statement not unlike that of the previous Blue Planet prize-winners (IIED, 2012). This would recognise the reality of the situation and call for structural change (most likely via HEFCE and the other national funding councils).

These recommendations are largely in line with those that came out of the Higher Education Academy's Policy Think Tank on universities, the twenty-first century, and the green economy (Luna *et al.*, 2012).

Despite the misgivings concerning sustainability metrics and rankings shared above, one driver for sustainability that some believe could help bring more HEIs into programmes such as Green Academy is sustainability rankings. All actors, including the HEA and GA HEIs, could support and lobby the International Ranked Expert Group (IREG) Observatory to rapidly agree and implement standards, auditing and certification for measuring sustainability performance in universities worldwide. They could also promote and participate in Universitas Indonesia Green Metric Ranking of World Universities. These steps would enable sustainability ranking to become a significant part of university rankings worldwide. Equally, existing national rankings could be supported, working with the organisations involved to promote them, including the UK's People and Planet Green League Table and EAUC's LiFE Index. However, it is worth noting that the pioneering US College Sustainability Report Card was suspended in March 2012 to focus resources on large-scale investment in energy efficiency through the Billion Dollar Green Challenge. This echoes the concerns of Green Academy team leaders that the emphasis should now be placed on action rather than monitoring, reporting and metrics.

Irrespective of progress on any of the above, any new participants in Green Academy, and indeed each of us, could take a radical position in our work with our university, and in outreach outside academia (as one team leader mentioned above).

## The future

Green Academy, and our attempts to morph it into a more strongly appropriate response, is just one possible response that we can make, once we gain the courage to 'stare hell in the face' (Abram, 2012). It is common for us to be scared or reluctant to look at the scale, speed and significance of what is happening: it can feel overwhelming and exacerbate feelings of powerlessness in the face of global challenges and seemingly immutable, unresponsive structures, such as national governments and hierarchical HEIs.

But these are just feelings. We can allow ourselves to feel and acknowledge them fully, knowing they need not control us. We each have the power to use our resources, voices, talents, ideas and spheres of influence. We can speak out clearly and thoughtfully and act in the most beneficial way in each situation, making an impact and bringing inspiration (Balanced View Team, 2011). We can be as bold and brave as if we are pulling our child out from the path of a speeding train.

Selby and Kagawa (2011) propose a 'return to first principles', wherein we remember the values behind our commitment to this agenda. More importantly, though, they ask:

> Is anything we are doing or saying – or anything we are not doing or saying – tantamount to trimming on our worldview for short-term influence? If so, what are the attendant dangers and likely consequences?

In other words, have sustainability champions at HEIs compromised their values, limited their vision and settled for small changes that fall short of the real changes that we know need to be made? At the May 2012 Green Academy conference, and in the evaluation at the end of this chapter, we hear about people 'already doing education for sustainable development'. Such phrases are echoed throughout universities as sustainability champions seek to be inclusive and bring colleagues into the agenda. But it is vital to balance such approaches with deep honesty regarding the scale of the challenges and the temptation to be complacent. If they are 'already doing it', then why do we need to do anything else? Why do we need interventions such as this book, or programmes such as Green Academy? Ultimately, as a sector, we are not sufficiently 'already doing it', as Sterling and Blewitt each demonstrate (this volume). Rather than telling academics they are already 'doing sustainability', or noting that it is happening in one corner of the university, we could say, 'All disciplines are *relevant to the agenda,* so please join us! Your expertise is crucial in providing students with these skills and attributes.'

Each of us, and our colleagues and students, are creative human beings. We can ensure that *everything* we create is not only good for us personally, but also good for everyone *and* the planet. This is possible and is demonstrated by 'Cradle to Cradle' theory and practice (McDonough and Braungart, 2002) and organisations such as the UK's Ellen MacArthur Foundation, through its activities and recent report (2012) on the circular economy. Such networks, organisations and initiatives are forming around the world: wherever you are, you can connect and work with them.

Language is important. Keep an eye on when you are not responding appropriately and understand why, put into place the elements you need to empower yourself, and then act. It's urgent, but keep it in perspective. We need lightness, too, so we leave you with a quote from one of the team leaders, on life lived under education for sustainability conditions:

It's been fun; it's been difficult; it's been rewarding. It always feels like I'm running down a hill with a big wheelbarrow full of frogs, trying to make sure I don't spill any of them.

## Embedding sustainability into the higher education curriculum: lessons from Green Academy

### *Introduction*

The Green Academy's first year of operation was independently evaluated in 2012. The evaluation methodology gave priority to exploring how Green Academy action plans, adopted by each university, had contributed to developing and enhancing approaches to the integration of sustainable development into the undergraduate and postgraduate curriculum and teaching and learning. The evaluation found that participants focused on five key areas, which are explained below.

### *Changing the institutional strategy*

All Green Academy teams addressed how to make sustainability part of institutional strategy and shared two features at the start of the process. First, the articulation and knowledge of sustainability curricula had previously been extremely scattered and rare beyond academic disciplines traditionally interested in the subject. Second, sustainability-related activity had often been estates-led and was based on an environmental definition.

The Green Academy teams, therefore, focused attention on pushing the wider definition of sustainability, which includes economic and social aspects, engaging with disciplines which, if not opposed to the idea, do not necessarily place it high on their agenda for a variety of reasons. They analysed current provision patterns and identified opportunities for working with staff in academic disciplines beyond the 'usual suspects'. Some conducted surveys, some audits – although systematic approaches were the exception rather than the rule.

In all cases, Green Academy had, in the words of one participant, 'given a boost to be more explicit about education for sustainable development in the strategic plan'. In one institution, the initiative developed around sustainability was seen by senior management as being 'incredibly valuable to delivering the strategic plan as a whole', including, for example, the development of local community partnerships. New institutions, in particular, tended to identify sustainability as a unique selling point. Where wider institutional changes were taking place, sustainability was used as a 'glue' to bind new structures. Some team leaders identified profound institutional changes: '[I]n mid-2010, I wouldn't have dreamt a chapter [on sustainability] in the strategic plan was possible.'

### Embedding sustainability in the curriculum

The teams adopted varied approaches, tailored to their particular contexts, with different degrees of compulsion and encouragement. Sustainability was regarded as so important to one institution's future that each faculty was required to offer one new undergraduate and one new postgraduate sustainability-related programme for the 2013 student intake, and embed sustainability into 15 per cent of all student experiences. Another was seeking to develop an elective in sustainability which would 'at a stroke ... put sustainability at the heart of the undergraduate curriculum', since it would be only one of a handful of electives available.

Other institutions rejected compulsion as it would not lead to genuine buy-in. One institution's earlier system of optional modules contained no incentive for departments/faculties to encourage students to take cross-disciplinary subjects like sustainability. A module might be compulsory to students, but would not be compulsory for staff and so would remain unembedded. Instead, the institutions are working with departments, demonstrating sustainability's relevance to each discipline and establishing its fit within departmental cycles of course (re)validation. Funds were made available within a planned initiative supporting staff time to develop new curricula, teaching and learning. This institution found sustainability was 'a wonderful catalyst for reframing the curriculum'.

Where the sustainability agenda had been strongly estates-led, Green Academy fostered connections between informal and formal curricula. This approach was advantageous where academic autonomy was particularly strong. Accreditation was important for stimulating take-up here, especially among mature students with domestic and employment commitments and limited time for informal curriculum activities.

Teams noted that, frequently, schools, departments or faculties had been doing education for sustainable development without realising it. All Green Academy participants worked closely with students and staff to link sustainability with their disciplines. Activity was brought to the surface, labelled and made more coherent by removing overlaps and filling gaps. For example, one business school agreed to move its business ethics module into a new sustainability elective.

### Developing the institutional narrative

All teams sought to raise the profile of sustainability. Some sustainability activity within institutions was already visible through recycling schemes and high-profile 'green' capital building programmes and so on. But the teams recognised the need to develop an institutional narrative around the wider definition of sustainability. Green Academy was widely publicised within every institution as part of the profile-raising process. In the words of one participant, Green Academy had been used to 'move ESD above the radar'. The university's sustainability profile was a key reason for doing Green Academy; sometimes, while the team knew that their institution was doing well, the institution itself did not.

Developing the institutional narrative involved tackling stereotypes. Sustainability was often identified with particular subjects rather than being seen as relevant to all disciplines. Some also attached negative perceptions to the word 'green', adopting other descriptors including 'sustainability' and/or 'education for sustainable development' in communications. Others offered presentations/seminars to faculty boards or groups throughout the university to explore the language of sustainability and offer ways forward. However, these were not evaluated.

Teams discovered that how the message is communicated is as important as the message itself. Methods included: discussions with key committees, profile-raising events, presentations and discussions with faculties, and the generation of case studies. Often, these were with subject areas not normally associated with sustainability. Teams reported the need for the institution to demonstrate its commitment to sustainability visibly on campus, as well as in curricula.

### Engaging management

Presentation by Green Academy teams of a systematic and formally organised message to faculty and departmental heads was vital to strategy implementation, narrative development and curricular integration. The key challenge was to convince those not traditionally focused on sustainability that it is as important for them and their students as anyone else. Some teams launched their own programmes, others charged faculty heads with ensuring that each department identifies the role it can play with respect to sustainability. Good practice examples helped both approaches. As one informant commented, 'Once people saw they could do it within the discipline, it took off.'

### Enabling students

Including students on Green Academy teams emphasised their importance in developing sustainability activities and supporting change. Participation enabled students to develop their own perspectives and empowered them to play significant roles. In one institution, the fact that students themselves articulated demand for a sustainability curriculum enhanced the impact on their academic audience.

The important role of students in stimulating demand for sustainability was harnessed by some institutions. In one institution, the student union gathered views on ten possible elected pathways: three were popular, including sustainability.

Students have sometimes been key to developing the informal curriculum on campus, often led or facilitated by university sustainability teams based in estates departments. This demand is now extending beyond seeing sustainability as a subject matter to seeing sustainability as an opening up of new teaching and learning methods. Students were reported as wanting sustainability to be discovery- rather than fact-based, with learning through activity. This resonates with higher education institutions increasingly recognising students as dynamic 'co-producers' of curriculum and learning, rather than as 'consumers'.

## Conclusions

Thanks to Green Academy, institutions did things more quickly, in different ways, across a broader front, and/or on a bigger scale than would otherwise have been possible. It inspired participants to engage strategically with their institutions, embedding sustainability within strategic planning where before it had either been absent or confined to a narrower environmental definition. Individuals gained confidence to engage with senior managers, and to implement action plans to stimulate curriculum developments. Green Academy heightened awareness of the student role and raised awareness of sustainability, in some cases supporting the development of a comprehensive sustainability narrative within institutions.

Only so much can be achieved in a year, however. Familiar obstacles remain: lack of time and resources, discipline silos, etc. Action in some areas has yet to take place in most institutions, notably, in the areas of leadership, mapping, target setting and monitoring. And yet foundations have been built, and Green Academy teams have shown what can be achieved with small-scale resources in a context of financial constraint. Solid bases have been established on which further progress can be made.

*Dr Andrew McCoshan is Associate Fellow, Centre for Education and Industry, University of Warwick. Professor Stephen Martin is Visiting Professor, Pedagogic Research and Scholarship Institute, University of Gloucestershire, and Chair, Higher Education Academy Education for Sustainable Development Advisory Group.*

## Notes

1  For more information on Green Academy, please see Kemp *et al.*, 2012 and McCoshan and Martin, 2012. Also see: www.heacademy.ac.uk/projects/detail/esd/esd_green_academy
2  For People and Planet's Green League, see: http://peopleandplanet.org/greenleague and for EAUC's LiFE Index, see: www.thelifeindex.org.uk/

## References

Abram, D (2012) (panellist) 'Paul Kingsnorth and friends discuss "Confessions of a Recovering Environmentalist"', *Orion Magazine* (podcast). Available at: www.orionmagazine.org/index.php/audio-video/item/paul_kingsnorth_friends_discuss_confessions_of_a_recovering_environmentalis/ (accessed 3 June 2012)

Balanced View Team (2011) *Open Intelligence: Changing the Definition of Human Identity*, Mill Valley: Balanced View Media. Available at: www.balancedview.org/images/books/open_intelligence_human_identity.pdf (accessed 29 May 2012)

Constanza, R and Daly, H (1992) 'Natural capital and sustainable development', *Conservation Biology*, 6 (1): 37–46

Dawe, G, Jucker, R and Martin, S (2005) *Sustainable Development in Higher Education: Current Practice and Future Developments*, York: Higher Education Academy. Available at: www.heacademy.ac.uk/resources/detail/sustainability/dawe_report_2005 (accessed 3 June 2012)

Ellen MacArthur Foundation (2012) *Towards the Circular Economy*. Available at: www.thecirculareconomy.org/ (accessed 3 June 2012)

Flick, U (2006) *An Introduction to Qualitative Research*, London: Sage

HEA (Higher Education Academy) (2012) *Green Academy: Curricula for Tomorrow*. Available at: www.heacademy.ac.uk/projects/detail/esd/esd_green_academy (accessed 15 June 2012)

IEA (International Energy Agency) (2011) *World Energy Outlook 2011*, Paris: OECD/IEA

IIED (International Institute for Environment and Development) (2012) *Blue Planet Prize Winners Call for Transformational Change to Achieve Sustainable Development*. Available at: www.iied.org/blue-planet-prize-winners-call-for-transformational-change-achieve-sustainable-development (accessed 3 June 2012)

Jones, P, Selby, D and Sterling, S (2010) *Sustainability Education: Perspectives and Practice Across Higher Education*, London: Earthscan

Kemp, S, Scoffham, S, Rands, P, Robertson, A, Robinson, Z, Speight, S, Raghubansie, A and Luna, H (2012) 'A national programme to support education for sustainable development in the United Kingdom: university experiences of the HEA Green Academy programme', in Leal, W (ed.) *Sustainable Development at Universities: New Horizons*, Frankfurt: Peter Lang Scientific Publishers

Luna, H, Martin, S, Scott, W, Kemp, S and Robertson, A (2012) *Universities and the Green Economy: Graduates for the Future*, York: Higher Education Academy

Maxey, L (2009) 'Dancing on a double-edged sword: sustainability within university corp', *ACME: An International E-journal for Critical Geographies,* 8 (3): 440–53. Available at: www.acme-journal.org/vol8/Maxey09.pdf (accessed 1 June 2012)

McCoshan, A and Martin, S (2012) *Evaluation of the Impact of the Green Academy Programme and Case Studies*, York: Higher Education Academy

McDonough, W (2007) (interviewee) *Waste = Food* (video). Available at: http://vimeo.com/15266520 (accessed 19 April 2012)

McDonough, W and Braungart, M (2002) *Cradle to Cradle: Remaking the Way We Make Things*, New York: North Point Press

Selby, D and Kagawa, F (2011) 'Development education and education for sustainable development: are they striking a Faustian bargain?' *Policy and Practice: A Development Education Review*, 12 (Spring): 15–31. Available at: www.developmenteducationreview.com/issue12-focus3 (accessed 3 June 2012)

Silverman, D (2004) *Doing Qualitative Research*. London: Sage

# 14 The sustainable university

## Taking it forward

*Stephen Sterling and Larch Maxey*

The focus of this book is not higher education. It is *higher education-in-context*, that is, the nature of the relationship between higher education and the wider world. Taking a whole systems view, we argue that this relationship needs to change for the benefit of both; that there needs to be much better congruence between these two realities – on one hand higher education systems, on the other the socio-economic and ecological supra-systems in which HE is embedded. At stake is the health, or viability, of both higher education and the global context. A viable system is one which is 'able to survive, be healthy, and develop in its particular environment' (Bossel, 1998: 75), and therefore is able to anticipate and adapt to significant change. Seen from a systems perspective, a higher education system oriented towards sustainability would benefit both HEIs and society, by increasing health, security and well-being at organisational and individual levels right across the piece.

We are not yet there. The backdrop to this book is the kind of future that this generation and its successors will experience or, through our collectively changing course, *could* experience. The difference between these two pathways is passive and unthinking entry into a world full of uncertainty and threat, *or* consciously and determinedly shaping a world that is more sustainable and resilient, ecologically, socially and economically.

We have the choice: to continue 'business as usual' based on individualism, high levels of consumerism and competition, and continuing depletion of natural resources, or to take a co-evolutionary pathway based on diversity, a circular economy, collaboration, sufficiency and well-being within planetary limits, as outlined in Great Transition scenarios (Raskin *et al.*, 2002). Only one of these pathways is viable in the long term.

The high-level reports outlined in the Introduction to this book make it clear what needs to be done, but also that time is short. They underline the need for urgent policy change and political leadership, but the United Nations Conference on Sustainable Development (UNCSD) Rio+20 Summit of 2012's final document, *The Future We Want,* was widely criticised for acknowledging the severity of the issues while falling short of the kinds of commitments that the expert reports state are necessary (Watts and Ford, 2012). Watts and Ford report Greenpeace International's Executive Director Kumi Naidoo as saying, 'We didn't get the Future We Want in Rio, because we do not have the leaders we need'. Some years ago, the young policy-makers-to-be were perhaps poorly educated for the decisions they later faced at Rio.

The rhetoric is fine however, as shown by the first paragraph of *The Future We Want*, under the subheading of 'Our Common Vision':

> We, the heads of State and Government and high level representatives, having met at Rio de Janeiro, Brazil, from 20–22 June 2012, with full participation of civil society, renew our commitment to sustainable development, and to ensure the promotion of an economically, socially and environmentally sustainable future for our planet and for present and future generations.
>
> (UNCSD, 2012: 1)

Education is endorsed and recognised in the document as critical to achieving this vision, echoing UNESCO's vision…

> Sustainable development cannot be achieved by technological solutions, political regulation or financial instruments alone. Achieving sustainable development requires a change in the way we think and act, and consequently a transition to sustainable lifestyles, consumption and production patterns. Only education and learning at all levels and in all social contexts can bring about this critical change.
>
> (UNESCO, 2012: 13)

… and indeed, echoing many such high-level documents of the last three decades. What is less common is a *critique* of why education seems so slow to take up this challenge. This quotation from the United Nations Economic Commission for Europe is therefore unusual:

> Education should play an important role in enabling people to live together in ways that contribute to sustainable development. However, *at present, education often contributes to unsustainable living.* This can happen through a *lack of opportunity for learners to question their own lifestyles and the systems and structures that promote those lifestyles.* It also happens through *reproducing unsustainable models and practices.* The *recasting of development, therefore, calls for the reorientation of education towards sustainable development.*
>
> (UNECE, 2012; our emphases)

A crucial question then is how can education generally, and higher education in particular, reorient and renew its purposes, policies and practices in order to make a positive rather than a negative difference to securing a safer, more sustainable future; to assist the transition towards such a future? Clearly, developing sufficient commitment is critical.

## Response and commitment

We suggest the contextual challenges that face society require the higher education community – policy-makers, staff, students and stakeholders – to move towards a *culture of critical commitment* in their thinking and practice. This requires a reimagining of higher education until it is engaged enough to make a real difference to socio-ecological resilience and sustainability, and reflexively critical enough to learn constantly from experience and to keep options open in working for sustainability transformations. But achieving this culture of critical commitment itself requires a transition. In the somewhat young history of the response of HE to sustainability, four positions have emerged that can be characterised as follows (see Table 14.1):

(A) is the dominant *business-as-usual* position, which involves little or no engagement with sustainability and involves little or no critique of dominant socio-economic and environmental values. At the same time, there is a growing awareness that sustainable development needs to be 'covered' or 'delivered' by educational policy and practice, and some progress towards this might be in evidence.

(B) is the *advocacy position*, championed by enthusiasts and NGOs energised by the urgency of sustainability issues, and frustrated by the slow response of education. There is an explicit critique of (A), but there tends to be an instrumental emphasis on universal 'sustainability literacy' rather than on deeper implications for change in educational thinking, learning and practice.

(C) is the *liberal position*. This is critical of (A) for not taking sufficient notice of sustainable development, and of (B), suggesting that a rush towards developing sustainability ignores different interpretations of what this might mean, and if done badly, comes at the cost of academic freedom and quality pedagogy and learning. This position holds that sustainability is best advanced through critical appraisal of all views and alternatives.

*Table 14.1* Four positions in HE's transition to a culture of critical commitment

|  | *Uncritical* | *Critical* |
|---|---|---|
| *Non-committal* | A – *Business as usual position* Mainstream, little or no critique of dominant assumptions or evidence of sustainability, although growing awareness of the need for some response | C – *Liberal position* Embraces the need for sustainability, but adopts a critical line. Favours pluralism and rationalist, liberal approach, but puts prime value on learning and educational process |
| *Committed* | B – *Advocacy position* Stresses urgency and need for universal sustainability literacy as self-evident, and this is the prime focus, rather than the educational process by which it might be achieved | D – *Transformative position* Sustainability is seen as implying and requiring change in cultural paradigm, interpreted from a committed but critically reflexive stance |

Based on Sterling and Gray-Donald (2007: 242)

(D) is the *transformative position,* which sees unsustainability as arising from outdated, deep-seated cultural assumptions and norms, and articulates the need for urgent cultural change based on systemic, ecological or relational thinking, which is also self-critical, necessarily exploratory and capable of multiple interpretation within different contexts. It recognises the starting points of many (A), the urgency of the situation (B), the need for quality learning and education and to respect multiple voices (C), and embraces these in its articulation of paradigm change.

This is, of course, an over-simplification: the four positions are not exclusive or watertight, and there is a degree of interaction and overlap between them, but the model perhaps clarifies some of the tensions, both in the debate and as experienced within any one institution. There is also some validity in each position. Therefore, the model might help map a journey of any individual, group or HEI towards whole institutional change. If starting at position (A), a group might begin to take heed of the sustainability context as pressed by (B), consider critically the arguments and discourses involved as advocated by (C), and begin to shift towards a culture of critical commitment as supported by (D). This last transformative position is, of course, the intentioned paradigmatic change that many authors in this volume and that we as editors believe is necessary if HE is to play a full part in the sustainability transition. It has to start with individuals (see Box 14.1):

> It is evident that transformation will only come about when a large number of people set up different priorities in both the large and the small arenas of the university, establishing new routines and structures despite local conflicts and set-backs.
>
> (Sharp, 2002: 133)

---

**Box 14.1 Commitment case study: international travel**

The internationalisation agenda can both support sustainability, through global citizenship, diversity and inclusion, and challenge it, through increasing social dislocation, transportation and emissions, particularly long-haul flights (Barr *et al.*, 2011). University staff are increasingly engaging with these issues personally and professionally, yet this has remained largely outside university policy debate to date. Interviews with activists and academics carried out for this book indicate that their critical commitment to sustainable travel has supported individuals' movement from (A) to (D) in the above typology. This reader/associate professor, might be seen as somewhere between (C) and (D), having been a regular flyer previously:

> My position is no short-haul flights, I insist invitees allow time and resources for the train … long-haul flights are accepted once every three to four years, and then I'm selective and choose where I will make an impact. I don't fly to preach to the converted, but to do face-to-face talks to social movements or community organisers who will actually benefit from my presence.

This professor's commitment had moved him to (D) and was continuing to transform his thinking and practice:

> Over the last decade I've become an expert in train travel … Most colleagues have looked at me in slight bewilderment and amazement … but when they give me more time to explain, then they usually get it. It's actually been a very personal thing about wanting to calm down, do less, and not jet set around without connecting to any local space … I've become very involved with local politics … I don't think that decision [not to fly] has had a negative impact on my career; almost the opposite. I've had more time to do things, rather than flying around sitting in airport lounges. But yes there is of course massive pressure to fly … The more senior you become the less time you have to spend travelling on slow trains. But it can be done, and I'm determined to continue to do it.

These stories demonstrate the ways in which we are all capable of beginning to create aspects of the sustainable university in each moment, building congruence between our ideas, values and actions (Sharp, 2002). In doing this we form new skills, ideas and connections. All we need is the commitment and to begin.

## Critique, vision and design

In addition to commitment, purposeful change in any system requires three elements: an awareness and informed *critique* of a problematic situation, a *vision* of how it might be changed for the better, and a *design* or strategy for moving towards the vision. This applies at any system level, and successful change – inevitably – involves a learning process, at both individual, group and organisational levels. Each element – critique, vision and design – has been elaborated and described by our authors in different ways throughout the book, and we hope that their lessons are helpful for those wishing to advance towards the sustainable university. The critique of HE, recognising its shortfalls but also positive progress, has been evident in the book. We hope the possibility and potential of visioning for change is also apparent. The vision picture in Box 14.2 represents an interpretation of the student view, by the UK National Union of Students.

## Box 14.2: A sustainable university

A sustainable university needs to be doing much more than reducing the negative impacts of campus estates. Institutions need to realise that they can play a crucial leadership role in greening society by producing graduates equipped with the knowledge and skills to make the world a more sustainable place. In the UK alone, there are seven million learners in further and higher education, and this population refreshes itself every two to three years, providing our sector with a unique chance to install pro-environmental attitudes, behaviours and habits in cohort after cohort of future decision-makers. But how can universities do this?

Institutions need to engage their students and staff in sustainability, creating bottom-up peer-to-peer behavioural initiatives, which will help to create a social norm of sustainability for students at a key moment of change in their lives. However, to fulfil the social norm of sustainability, institutions must demonstrate leadership on embedding sustainability through the curriculum regardless of discipline. Only when we have visible sustainability in dormitories, departments and lecture theatres will we start to produce graduates *en masse* who are enthused and equipped to help make society more sustainable.

*Jamie Agombar, Ethical and Environmental Manager, National Union of Students UK*

We have said less about the value base for change other than indicating that the paradigm change increasingly called for is rooted in a more relational, systemic and ecological worldview. Repeatedly, in the sustainability discourse, stress is put on developing more systemic, integrative, participative approaches to address the conditions of complexity, uncertainty and unsustainability that now characterise our times. This is accompanied by calls for a more holistic ethic, such as is exemplified in depth by the Earth Charter (www.earthcharterinaction. org). This provides a powerful ethical framework, which can help a purposive re-orientation of higher education in terms of values and vision.

The need for vision and action to go together is a recurring theme in this book; inevitably, as vision alone is never sufficient:

> Vision without action is useless. But action without vision is directionless .... Vision is absolutely necessary to guide and motivate. More than that, vision, when widely shared and firmly kept in sight, does *bring into being new systems*.
>
> (Meadows *et al.*, 2005: 272; authors' italics)

Action is also guided by design or strategy, the third element of the change approach. It is not appropriate to go into strategy theory here, but by way of summary, some common elements of developing systemic change that can inform strategy, are shown in Box 14.3. These, again, are 'whole systems' ideas that are reflected across the chapters.

---

**Box 14.3: Elements of developing systemic change**

- Building on what already exists.
- Ensuring feedback.
- Understanding systemic change.
- Harnessing opportunities .
- Being opportunistic – but also pragmatic and tactical.
- Avoiding working in silos – forging new connections and partnerships.
- Encouraging reflection.
- Developing high levels of connectivity and communication.
- Being inclusive, participative and respectful of diverse views.
- Developing a shared ethos.
- Identifying, linking and working with the energy of existing enthusiasts – students and staff.
- Winning external recognition.
- Getting senior management buy-in.
- Creating alliances outside the system.
- Learning from others' and creating your own exemplars.
- Providing leadership and creativity.
- Using and developing channels to spread innovation.
- Practising an appreciative culture.
- Being persistent and patient – it takes time.
- Continually learning how the organisation works.
- Providing resources and support.
- Maintaining a 'critical intelligence' of where things stand and where needs are.
- Celebrating success.

---

Bartlett and Chase (2004: 17–21), commenting on the experience of change towards sustainable universities in the United States, conclude that:

- personal relationships are critical;
- trust, which emerges from strong relationships, drives the change we seek;
- success is not always related to the numbers of people involved;
- different paths and starting points are fine;
- leadership emerges from many sources;

- support from above is critical;
- spontaneity (taking advantage of opportunity) and persistence are really necessary.

The response to a poll on barriers and pathways conducted by the GUNI network and involving the views of some 200 experts in sustainability in HE worldwide, led to the identification of a number of priority areas, summarised here as:

- developing an institutional understanding, vision and mission on sustainable development;
- changing the incentive system and quality indicators for encouraging interdisciplinary work;
- building a culture of sustainability by involving all stakeholders including the local community;
- engaging internal stakeholders to encourage ownership, empowerment and participation;
- monitoring the design and implementation of sustainable development related curricula and training for all academic and administrative staff.

(based on Granados-Sánchez *et al.*, 2012: 202)

The process can start here and start now. Let's personalise this for a moment before concluding the book. Box 14.4 presents a brief example of a visioning exercise, which can be used and adapted for many contexts.

---

**Box 14.4: Visioning the sustainable university**

Take a moment to completely relax body and mind. Close your eyes if you wish.

Now it is five years from today and you and your colleagues have created the sustainable university! Allow yourself to see it, hear it, smell it, feel it and interact with it all around you. Follow your progress through a typical day in this sustainable university.

Now describe and report on your day and the sustainable university.

It is helpful to communicate, share and update your visions regularly and use visual methods, including drawings, photos, computer graphics/ animation as well as words.

(Adapted from Meadows (undated) and Balanced View Team, 2011)

---

## Going forward

The challenge is whole-system transformation, which needs to take place in a co-evolutionary way in tandem with changes in wider society. The simplistic question 'of how can education change people's behaviour' that has informed much education for change in the past, is superseded by a systemic view, which generates a very different question: 'how can education and society change together in a *mutually affirming* way, towards more *sustainable patterns* for both?' (Banathy, 1991: 129). This signals a change from education focusing on maintaining the existing state and operating as a rather closed system, towards helping shape society 'through co-evolutionary interactions, as a future-creating, innovative and open system' (Banathy, *ibid*).

How likely is this? According to the most recent UNESCO survey of sustainability education internationally, carried out as part of the UN Decade of Education for Sustainable Development (DESD) monitoring process, '[w]hole-institution approaches – which require the active engagement of multiple actors in the joint redesign of basic operations, processes and relationships – are increasingly put forward as a mechanism for making meaningful progress towards sustainability' (Wals, 2012: 5). Secondly, the author notes that:

> The boundaries between schools, universities, communities and the private sector are blurring as a result of a number of trends, including the call for lifelong learning; globalization; information and communication technology (ICT)-mediated (social) networking education; (and) the call for relevance in higher education and education in general.
>
> (Wals, 2012: 5)

Shifts are occurring, offering opportunities for sustainability to occupy a central place in thinking and practice. Not least, the whole green skills/low carbon economy agenda offers significant openings and reasons for higher education to embrace the implications of sustainability, particularly in relation to employability (BITC, 2010; BIS, 2011). The blurring Wals refers to offers scope for more accessible, permeable universities, which not only open their campuses, libraries and resources to local residents, but fully and directly serve their communities. This links with trends towards empowering, participatory, collaborative research and teaching noted throughout the book. Also important are social movements beyond the academy mentioned by Blewitt (this volume) including the free education, Occupy and critical pedagogy movements in creating new learning spaces. However, university-led free online platforms offer transformative potential, too, such as Harvard and the Massachusetts Institute of Technology's $60m (US) edX, and Stanford and Princeton's Coursera, which predicts that over one million students, from over 190 countries, will complete courses in autumn 2012 (EAUC, 2012).

On the margins of mainstream HE policy and practice, 'transition universities', off-shoots of the global Transition Movement, are gaining ground. This illustrates a synergy between social movements and universities that helps to build more sustainable universities. Such initiatives have the potential to unite students, staff and wider communities (see Higgins *et al.*, this volume; Hazan, this volume), creating innumerable opportunities to enhance learning, student experiences, research and every facet of the university. For example, the Transition Movement has pioneered skillshares (Hopkins, 2011), pop-up, self-organised learning and teaching events which would translate well to our campuses and indicate what else may be possible.

Similarly, research within the sustainable university will be transformative on all levels: personally, professionally, institutionally and societally. This significant shift from current research paradigms (White, this volume; Maxey, 2009) requires new ways of proposing, developing, conducting, reporting, sharing and evaluating research. Emerging trends such as governments and funders emphasising research impact, relevance and collaborative work with stakeholders make this task more achievable. As with free education, emerging organisations such as the Transition Research Network (Maxey, this volume) and the Occupy Research Collective (http://occupyresearchcollective.wordpress. com/) bridge social movements/communities and academia, offering innumerable opportunities.

The *Higher Education Sustainability Initiative for Rio+20* (see shaded Box on page 315) has been signed by many institutions from nearly 50 countries. With sound (if not particularly radical) steps towards the sustainable university, the initiative itself heralds shifting ground in this direction. A stronger set of aspirations, following international best practice, might be outlined as in Box 14.5. In the spirit of the sustainable university, these are offered for discussion and reflection, not prescription.

---

**Box 14.5: A sustainable university identification guide**

Working towards a stage where:

1　The whole institution is a learning organisation committed to sustainability *congruence*.
2　It is open to change and is *responsive, creative*, solution-focused and dynamic in every sphere.
3　The *curriculum*, both formal and 'hidden', equips tomorrow's leaders fully with the confidence, humility, commitment, and systemic and critical thinking skills required to contribute to sustainability transitions.
4　*Pedagogies* foster deep learning and enquiry.
5　The *informal curriculum*, architecture, and healthy learning and working environments foster sustainability.
6　*Alumni* are beacons of and leaders in society's sustainability transitions.
7　All *research* is sustainable research contributing to and/or taking account of society's sustainability transitions.
8　*Estates* model sustainability, for example through circular economy operations, minimal or zero carbon footprint, biodiversity and green space, intelligent and human scale buildings.

---

9  *Financial management* maximises contributions to sustainability transition, through purchasing, investment, pensions, etc.
10 *Partnerships* and community links striate all spheres, with open, free learning and teaching opportunities, library and campus access, etc.

… and all successes are *celebrated.*

Sustainability is about making educational systems and the education that they offer fit for purpose and fit for the future: more relevant, more responsible and better in quality; about encouraging critical reflection on the values, dispositions, understanding and competencies that both staff and students need to deal with significant and mounting challenges in uncertain times; and about the transformation of educational systems so that they might become transformative and contribute to a safer and better world. Given the real-world context reviewed in this volume (see Introduction and Chapter 1) this is a necessary but also a positive shift, beneficial to institutions, their staff and students, and thereby to society:

> The necessity of taking the industrial world to its next stage of evolution is not a disaster – it is an amazing opportunity. How to seize the opportunity, how to bring into being a world that is not only sustainable, functional, and equitable but also deeply desirable is a question of leadership and ethics and vision and courage, properties … of the human heart and soul.
>
> (Meadows et al., 2005: 263)

The sustainable university is an idea whose time has come. And it has come just in time – but only if it can be realised throughout the sector and with sufficient speed and effect that it can make a positive difference to the sustainability transition. Some people are optimistic about the future and the human condition, while others see things very differently. Alan AtKisson puts it this way:

> the good news is that this … [sustainability] transformation [is] already underway. The bad news comes in the form of a challenge: *How fast can we make … beneficial changes happen?*
>
> (AtKisson, 2011: 21; author's emphasis)

This book was conceived and planned at Jurys Inn, Plymouth. At the time the group wondered how far higher education could, and would, respond to the sustainability challenge and opportunity. With a slight smile, we said 'the jury's out'.

## DECLARATION

As Chancellors, Presidents, Rectors, Deans and Leaders of Higher Education Institutions and related organizations, we acknowledge the responsibility that we bear in the international pursuit of sustainable development. On the occasion of the United Nations Conference on Sustainable Development, held in Rio de Janeiro from 20–22 June 2012, we agree to support the following actions:

- *Teach sustainable development concepts*, ensuring that they form a part of the core curriculum across all disciplines so that future higher education graduates develop skills necessary to enter sustainable development workforces and have an explicit understanding of how to achieve a society that values people, the planet and profits in a manner that respects the finite resource boundaries of the earth. Higher Education Institutions are also encouraged to provide sustainability training to professionals and practitioners.
- *Encourage research on sustainable development issues*, to improve scientific understanding through exchanges of scientific and technological knowledge, enhancing the development, adaptation, diffusion and transfer of knowledge, including new and innovative technologies.
- *Green our campuses* by: i) reducing the environmental footprint through energy, water and material resource efficiencies in our buildings and facilities; ii) adopting sustainable procurement practices in our supply chains and catering services; iii) providing sustainable mobility options for students and faculty; iv) adopting effective programmes for waste minimization, recycling and reuse, and v) encouraging more sustainable lifestyles.
- *Support sustainability efforts* in the communities in which we reside, working with local authorities and civil society to foster more liveable, resource-efficient communities that are socially inclusive and have small environmental footprints.
- *Engage with and share results through international frameworks*, such as the UN Decade of Education for Sustainable Development, led by UNESCO, the UN University system, the UN Academic Impact, the Global Compact, the UN-supported Principles for Responsible Management Education initiative and the UN Environment Programme's Environmental Education and Training initiatives, in order to exchange knowledge and experiences and to report regularly on progress and challenges.

The Rio+20 Directory of Committed Deans and Chancellors, 2012.

## References

AtKisson, A (2011) *The Sustainability Transformation: How to Accelerate Positive Change in Challenging Times*, London: Earthscan

Balanced View Team (2011) *Open Intelligence: Changing the Definition of Human Identity*, Mill Valley: Balanced View Media. Available at: http://www.balancedview.org/images/books/open_intelligence_human_identity.pdf (accessed 29 May 2012)

Banathy, B (1991) *Systems Design of Education*, NJ: Educational Technology Publications

Barr, S, Gilg, A and Shaw, G (2011) '"Helping people make better choices": exploring the behaviour change agenda for environmental sustainability', *Applied Geography*, 31: 712–20

Bartlett, P and Chase, G (2004) *Sustainability on Campus: Stories and Strategies for Change*, Cambridge, MA: MIT Press

BIS (2011) *Skills for a Green Economy: A Report on the Evidence*, London: Department for Business, Innovation and Skills. Available at: www.bis.gov.uk/assets/biscore/further-education-skills/docs/s/11-1315-skills-for-a-green-economy (accessed 1 July 2012)

BITC (Business in the Community) (2010) *Leadership Skills for a Sustainable Economy*, London: BITC. Available at: www.bitc.org.uk/resources/publications/leadership_skills.html (accessed 3 June 2012)

Bossel, H (1998) *Earth at a Crossroads: Paths to a Sustainable Future*, Cambridge: Cambridge University Press

EAUC (2012) 'Top Universities Put Their Reputations Online'. Available at: www.eauc.org.uk/top_us_universities_put_their_reputations_online (accessed 30 June 12)

Granados-Sánchez, J, Wals, A, Ferrer-Balas, D, Waas, T, Imaz, M, Nortier, S, Svanstrom, M, Van't Land, H and Arriaga, G (2012) 'Sustainability in higher education: moving from understanding to action, breaking barriers for transformation', in Global Universities Network for Innovation (GUNI) (2012) *Higher Education's Commitment to Sustainability: From Understanding to Action*, World in Higher Education Series No. 4, Barcelona, Spain: GUNI

Hopkins, R (2011) *The Transition Companion: Making Your Community More Resilient in Uncertain Times*, Totnes: Green Books

Maxey, L (2009) 'Dancing on a double edged sword: sustainability within university corp', *ACME: An International E-Journal for Critical Geographies*, 8 (3): 440–53

Meadows, D (undated) *Visioning Exercise Template*. Available at: http://leadershiplearning.org/system/files/VisioningTemplate_DonellaMeadows.pdf (accessed 4 July 2012)

Meadows, D H, Meadows, D L and Randers, J (2005) *Limits to Growth: The 30-year Update*, London: Earthscan

Raskin, P, Banuri, T, Gallopin, G, Gutman, P, Hammond, A, Kates, R and Swart, R (2002) *Great Transition: The Promise and Lure of the Times Ahead*, Boston: Stockholm Environment Institute/Tellus Institute

Rio+20 Directory of Committed Deans and Chancellors (2012) *Higher Education Sustainability initiative for Rio+20*. Available at: http://rio20.eurome d-management.com/wp-content/uploads/2012/03/Higher-Education-Sustainability-Initative-for-Rio-The-directory-of-Deans-Chancellors-committed.pdf (accessed 16 June 2012)

Sharp, L (2002) 'Green campuses: the road from little victories to systemic transformation', *International Journal of Sustainability in Higher Education*, 3 (2): 128–45

Sterling, S and Gray-Donald, J (2007) 'Editorial', *International Journal of Innovation and Sustainable Development*, 2 (3/4), 2007: 241–8

UNCSD (2012) *The Future We Want*, Final Document of the Rio+20 Conference, United Nations Conference on Sustainable Development. Available at: www.uncsd2012.org/content/documents/727THE%20FUTURE%20WE%20WANT%20-%20FINAL%20DOCUMENT.pdf (accessed 29 June 2012)

UNECE (2012) *Education for Sustainable Development: Introduction,* United Nations Economic Commission for Europe, Education for Sustainable Development. Available at: www.unece.org/?id=24234 (accessed 1 July 2012)

UNESCO (2012) *From Green Economies to Green Societies: UNESCO's Commitment to Sustainable Development*, Paris: UNESCO. Available at: http://unesdoc.unesco.org/images/0021/002133/213311e.pdf (accessed 3 July 2012)

Wals, A (2012) *Shaping the Education of Tomorrow: 2012 Full-length Report on the UN Decade of Education for Sustainable Development*, Paris: UNESCO, DESD Monitoring and Evaluation

Watts, J and Ford, L (2012) 'Rio+20 Earth Summit: campaigners decry final document', *Guardian*, 23 June 2012. Available at: www.guardian.co.uk/environment/2012/jun/23/rio-20-earth-summit-document (accessed 25 June 2012)

# Selected websites

## United Kingdom

http://www.eauc.org.uk/
Environmental Association for Universities and Colleges (EAUC)

http://www.heacademy.ac.uk/ourwork/teachingandlearning/sustainability
The Higher Education Academy

http://www.thelifeindex.org.uk/about-life/
Learning in Future Environments

http://www.goodcampus.org/index.php
Higher Education Environmental Performance Improvement

http://www.sustainableuni.kk5.org/
A UK-based one-stop shop initiative by Asitha Jayawardena

http://efsandquality.glos.ac.uk
Guide to Quality and Education for Sustainability in Higher Education

## US and Canada

http://www.secondnature.org/
Second Nature

http://www.aashe.org/
Association for the Advancement of Sustainability in Higher Education

http://www.iisd.org/educate/
International Institute for Sustainable Development

http://www.esdtoolkit.org
Education for Sustainable Development Toolkit

## *International*

http://www.copernicus-alliance.org/
The Copernicus Alliance is the European Network on Higher Education for Sustainable Development

http://www.uncsd2012.org/index.php?page=view&type=510&nr=147&menu=20
Global Universities Partnership for Environment and Sustainability (GUPES)

http://talloiresnetwork.tufts.edu/
The Talloires Network

http://www.iau-aiu.net/content/sustainable-development
International Association of Universities

http://www.guninetwork.org/
Global University Network for Innovation (GUNI)

http://www.ias.unu.edu/sub_page.aspx?catID=108&ddlID=183
Regional Centres of Expertise on Education for Sustainable Development

http://www.international-sustainable-campus-network.org/
International Sustainable Campus Network (ISCN)

http://www.unep.org/training/publications/Rio+20/Greening_unis_toolkit%20
120326.pdf
UNEP Greening Universities Toolkit

http://ugaf.eu/page/ugaf-pledge
UNICA Green Academic Footprint

http://www.sustainability.edu.au/
Australian HE site

http://www.acts.asn.au/
Australasian Campuses Towards Sustainability

http://www.ariusa.net/
Latin American Network Alliance of Universities for Sustainability and the Environment

# Index